APPLIED STATISTICS FOR ENVIRONMENTAL SCIENCE WITH R

APPLIED STATISTICS FOR ENVIRONMENTAL SCIENCE WITH R

Abbas F. M. Alkarkhi
Wasin A. A. Alqaraghuli

Elsevier
Radarweg 29, PO Box 211, 1000 AE Amsterdam, Netherlands
The Boulevard, Langford Lane, Kidlington, Oxford OX5 1GB, United Kingdom
50 Hampshire Street, 5th Floor, Cambridge, MA 02139, United States

Library of Congress Cataloging-in-Publication Data
A catalog record for this book is available from the Library of Congress

British Library Cataloguing-in-Publication Data
A catalogue record for this book is available from the British Library

ISBN 978-0-12-818622-0

For information on all Elsevier publications visit our website at https://www.elsevier.com/books-and-journals

Publisher: Joe Hayton
Acquisition Editor: Marisa LaFleur
Editorial Project Manager: Redding Morse
Production Project Manager: Omer Mukthar
Cover Designer: Greg Harris

Typeset by SPi Global, India

Dedication

Abbas
To the memory of my parents (deceased)
To my children Atheer, Hibah, and Farah

Wasin
To the memory of my father (deceased)
To my mother

Contents

Preface

Applied statistics for environmental science with R was written in an easy style to introduce some statistical techniques that are useful to students and researchers who work in environmental science and environmental engineering to choose the appropriate statistical technique for analyzing their data and drawing smart conclusions. The explanation of the R output is carried out in a step-by-step manner and in an easy and clear style to enable non-statisticians to understand and use it in their research.

A step-by-step procedure is employed to perform the analysis and the interpretation of results by matching the results to the field of study where the data were obtained. The book focuses on the applications of univariate and multivariate statistical techniques in the field of environmental science. Furthermore, real data obtained from research over more than fifteen years of work in environmental science were employed to illustrate the concepts and analysis.

The book uses R statistical software to analyze the data and generate the required results. R is open source and provides facilities to provide feedback and produce a high-resolution plot. Furthermore, it is easy to get online assistance provided by various communities. R is available over the internet under the General Public License (GPL) for the Windows, Macintosh, and Linux operating systems.

Finally, we wish to thank our families, friends, and colleagues for their continuous support. We would like to extend our thanks to the R software community and R family (R users and contributors to R) providing the software for free. This book would not have been possible without the information provided online, which is easy to obtain. We thank the University of Kuala Lumpur (Unikl MICET) for its support.

Abbas
Wasin

CHAPTER

1

Multivariate Data

LEARNING OBJECTIVES

After careful consideration of this chapter, you should be able:

- *To describe the concept of environmental statistics.*
- *To understand the importance of environmental statistics for making intelligent conclusions.*
- *To describe the concept of multivariate analysis.*
- *To know the advantages of employing multivariate methods for analyzing environmental data.*
- *To organize the outcomes of multivariate data to be prepared for analysis.*
- *To explain the distinction between univariate and multivariate notions.*
- *To explain how and where to employ multivariate data.*
- *To understand the univariate normal distribution.*
- *To understand the multivariate normal distribution.*

1.1 THE CONCEPT OF ENVIRONMENTAL STATISTICS

Environmental statistics is the application of various statistical methods, including procedures and techniques in the field of environmental science and environmental engineering, such as weather, air, water quality, climate, soil, fisheries and other environmental activities. Statistical methods are used in designing environmental projects and for the analysis and interpretation of environmental data to help and guide scientists draw useful and meaningful conclusions in various aspects of the environment. Environmental statistics can help in describing environmental problems in terms of mathematical modeling to understand the impact of the chosen variables under study and show the direction of change—increase or decrease—or the nature of the relationship—positive or negative. Furthermore, statistical techniques can identify the general trend and entangle the hidden relationships as well, which would help scientists understand the process and have a clear picture regarding all relationships to avoid risk and guide management to properly plan environmental projects.

Environmental statistics can help in understanding the importance of variability and oscillation in the data, employing various measures and methods to show the influence of variability, and lead scientists to search for scientific explanations. Thus, scientists should at least learn and understand basic statistics to help understand the importance of the results and the analysis that guides informative conclusions.

1.2 THE CONCEPT OF MULTIVARIATE ANALYSIS

The concept of univariate statistical analysis covers statistical techniques for testing a data set with one variable. However, most research projects need to measure several variables for each research unit or individual (sampling units or experimental units) in one or more samples.

For example, consider assessing the water quality of a river based on monitoring certain parameters such as pH, dissolved oxygen (DO), electrical conductivity (EC), turbidity, biological oxygen demand (BOD), chemical oxygen

Applied Statistics for Environmental Science with R
https://doi.org/10.1016/B978-0-12-818622-0.00001-0

demand (COD), and total suspended solids (TSS), to make a decision about the pollution status of the river. In this case, there are seven variables to be measured for each sample, which are generally highly correlated. If one variable at a time is considered to analyze the results of multivariate data, the relationships between the variables would be ignored and a different picture would be reflected regarding the true behavior of the chosen parameters (variables) in the presence of other parameters (variables). Thus, we should use a method that takes into account the correlation between chosen variables to untie the overlapping input (information) by the correlated variables to understand the behavior of the chosen variables properly.

Therefore, data sets with several variables can be analyzed, employing multivariate methods that consider the relationship between the chosen variables. Multivariate methods are a collection of techniques that can serve several purposes in the field of environmental science and engineering, which include cluster analysis for recognizing groups of similar observation (e.g., individuals, objects); principal components analysis and factor analysis as data reduction methods to reduce the number of variables to a smaller number of dimensions, called components (factors), which are uncorrelated without losing valuable information; discriminant analysis, which is applied to separating the data into various groups based on the measured variables; multivariate analysis of variance (MANOVA), employed to perform statistical hypothesis testing based on multivariate data (several variables); and multivariate multiple regression analysis, which is employed for making predictions based on the relation among the variables.

1.3 CONFIGURATION OF MULTIVARIATE DATA

We can organize multivariate data for various variables measured from a number of samples (items) in a table. The number of samples (items) specify the number of rows and the number of variables specify the number of columns of the table.

In general, Table 1.1 shows the configuration of n samples (representing the rows) and k variables (representing the columns) measured for each sample.

Note: The total number of variables measured for each sample (k) is usually smaller than the total number of samples, n.

1.4 EXAMPLES OF MULTIVARIATE DATA

We can illustrate the concept of multivariate data by providing some real examples of multivariate data with regard to environmental science that may be useful in more easily explaining the associated computations. Furthermore, the interpretation of these examples will be given, including the scenario of each case.

TABLE 1.1 The Configuration of n Samples and k Variables in Multivariate Form

	Variable					
Sample	Y_1	Y_2	...	Y_j	...	Y_k
1	Y_{11}	Y_{12}	...	Y_{1j}	...	Y_{1k}
2	Y_{21}	Y_{22}	...	Y_{2j}	...	Y_{2k}
.	.	.	...	.	...	.
.	.	.	...	.	...	.
.	.	.	...	.	...	.
i	Y_{i1}	Y_{i2}	...	Y_{ij}	...	Y_{ik}
.	.	.	...	.	...	.
.	.	.	...	.	...	.
.	.	.	...	.	...	.
n	Y_{n1}	Y_{n2}	...	Y_{nj}	...	Y_{nk}

EXAMPLE 1.1 ASSESSMENT OF SURFACE WATER

An assessment of water quality in the Juru and Jejawi Rivers in the Penang State of Malaysia was conducted by monitoring 10 physiochemical parameters; namely, the electrical conductivity (EC) and temperature were measured employing an HACH portable pH meter, and dissolved oxygen (DO) was measured with a YSI 1000 DO meter. The chemical oxygen demand (COD), biochemical oxygen demand (BOD), total phosphate concentrations, and total nitrate were analyzed employing a spectrophotometer (HACH/2010). The turbidity was measured employing a nephelometer. The total suspended solids (TSS) were analyzed gravimetrically in the laboratory. The APHA Standard Methods for the Examination of Water and Wastewater were applied to analyze the concentration of the abovementioned parameters. The data obtained from 10 different sites are each given in Table 1.2.

TABLE 1.2 Physiochemical Parameters for the Juru and Jejawi Rivers

Location	Temperature	PH	DO	BOD	COD	TSS	EC	Turbidity	Total phosphate	Total nitrate
Juru	28.15	7.88	6.73	10.56	1248.00	473.33	42.45	13.05	0.88	12.45
Juru	28.20	7.92	6.64	10.06	992.50	461.67	42.75	14.11	0.55	15.05
Juru	28.35	7.41	5.93	6.01	1265.00	393.34	29.47	25.95	0.88	15.80
Juru	28.40	7.84	6.23	14.57	1124.00	473.34	42.35	22.00	1.26	12.45
Juru	28.30	7.86	6.29	7.36	1029.50	528.34	40.70	12.36	0.61	12.90
Juru	28.30	7.89	6.36	5.27	775.00	458.33	42.50	17.90	1.01	19.25
Juru	28.55	7.40	5.67	4.81	551.00	356.67	29.51	27.70	1.02	25.55
Juru	28.75	7.41	5.26	4.96	606.00	603.33	28.45	29.35	0.84	17.90
Juru	29.30	7.25	4.18	6.61	730.00	430.00	26.25	35.35	0.73	12.85
Juru	29.55	7.18	5.14	5.11	417.00	445.00	25.79	27.75	0.72	17.15
Jejawi	27.80	7.40	4.96	36.65	82.35	815.00	29.55	1.01	1074.00	10.51
Jejawi	27.70	7.53	5.86	37.20	122.90	858.34	21.80	0.77	947.50	6.31
Jejawi	27.85	7.45	5.52	41.50	46.35	823.34	27.55	1.12	1194.50	7.36
Jejawi	27.85	7.41	5.27	36.75	52.65	821.67	16.45	0.46	988.50	5.87
Jejawi	27.65	7.27	5.71	37.95	99.95	736.67	21.10	1.01	653.00	16.36
Jejawi	27.85	7.36	5.16	34.85	36.30	373.34	11.50	0.73	911.00	11.26
Jejawi	27.85	7.37	5.08	37.05	33.50	385.00	13.35	0.71	632.00	3.75
Jejawi	27.75	7.40	5.23	37.10	36.35	403.34	10.85	0.57	787.00	3.31
Jejawi	27.70	7.38	5.30	38.30	30.00	550.00	13.90	0.75	879.50	3.15
Jejawi	27.65	7.36	5.17	38.40	26.15	363.34	12.65	0.62	1339.00	8.56

In this example, 20 samples (rows) are collected from the Juru and Jejawi Rivers (10 different sites each), and 10 physiochemical parameters (columns) are measured from each sample. It is very difficult for the researchers to make a decision about the pollution status of the rivers or about the behavior of the different parameters based on the chosen parameters under study because there are many different values, and the differences (fluctuations) between the values of the parameters from sample to sample will mislead the researcher, making it difficult to make a correct decision. Thus, these data should be analyzed employing an appropriate multivariate method that meets the objective of the study to untangle the information and to understand the body of the phenomenon. The objectives of this research were to determine the range of similarity among the sampling sites, to recognize the variables responsible for spatial differences in river water quality, to determine the unobserved factors while demonstrating the framework of the database, and to quantify the effect of possible natural and anthropogenic sources of the chosen water parameters of the rivers.

EXAMPLE 1.2 ASSESSMENT OF LANDFILL LEACHATE TREATMENT

A researcher wishes to investigate the concentrations of 12 heavy metals (magnesium (Mg), calcium (Ca), sodium (Na), iron (Fe), zinc (Zn), copper (Cu), chromium (Cr), cadmium (Cd), lead (Pb), arsenic (As), cobalt (Co), and manganese (Mn)) and eight physiochemical parameters (chemical oxygen demand (COD), biochemical oxygen demand (BOD), total dissolved solids (TDS), total suspended solids (TSS), electrical conductivity (EC), pH, ammoniacal-N (NH3N), and dissolved oxygen (DO)) in three ponds (collection, aeration, and stabilization) of a landfill leachate. The data for the three ponds are given in Table 1.3.

TABLE 1.3 The Results of Physiochemical and Heavy Metals Obtained From Three Ponds of Landfill Leachate

	Parameter									
Pond	**pH**	**EC**	**TDS**	**TSS**	**COD**	**BOD**	**NH3H**	**DO**	**Mg**	**Ca**
Collection	7.19	6242.00	4530.00	43.26	944.67	9.57	81.66	6.90	12.58	14.28
Collection	7.36	6062.00	4392.00	29.45	845.33	8.77	61.33	6.69	17.91	13.17
Collection	7.41	6164.00	4476.00	31.52	897.00	9.21	71.00	6.48	14.44	15.54
Collection	7.92	6710.33	4710.00	39.36	810.00	8.66	52.43	6.85	19.99	88.25
Collection	7.84	6690.00	4709.83	36.46	686.67	9.66	55.80	6.85	20.50	81.78
Collection	7.83	6600.33	4654.83	37.00	1365.00	10.33	52.80	7.23	21.18	85.18
Collection	8.46	6893.00	4849.00	47.33	692.00	8.47	110.67	2.34	27.23	24.94
Collection	8.47	6850.00	4860.00	50.67	967.00	9.27	93.00	2.82	19.01	14.38
Collection	8.48	6873.00	4859.00	49.00	833.00	8.87	103.00	2.67	24.66	19.99
Aeration	8.98	2501.00	1792.00	101.67	502.67	76.33	3.27	7.88	10.98	17.04
Aeration	8.97	2438.00	1750.00	89.00	543.67	93.67	8.90	7.62	10.99	9.84
Aeration	9.07	2422.00	1762.00	92.00	526.67	71.00	10.80	8.31	10.98	15.00
Aeration	8.21	2255.00	1594.00	148.83	566.67	4.66	4.30	7.23	18.51	54.30
Aeration	8.38	2092.83	1577.83	151.00	618.33	15.33	3.80	7.37	20.14	54.08
Aeration	8.32	2240.67	1578.83	162.67	573.33	10.00	4.40	7.55	21.77	55.30
Aeration	9.97	1513.00	1065.00	51.00	339.00	9.10	20.08	8.08	7.96	5.25
Aeration	10.18	1497.00	1049.00	42.00	322.00	9.32	23.27	8.01	6.85	4.44
Aeration	10.21	1532.00	1084.00	38.00	324.00	8.95	23.26	8.12	8.53	6.78
Stabilization	8.67	1491.00	1087.00	115.00	521.33	17.33	1.37	6.95	20.82	19.99
Stabilization	8.60	1521.00	1104.00	105.33	504.00	29.67	7.82	6.42	20.50	23.10
Stabilization	8.58	1545.00	1123.00	108.33	532.00	47.33	6.00	5.71	20.48	27.13
Stabilization	8.82	1321.83	932.67	98.00	253.67	13.66	6.60	7.55	11.92	66.96
Stabilization	8.92	1352.88	965.00	89.39	520.00	15.00	6.20	7.23	12.46	65.28
Stabilization	8.93	1314.83	942.83	101.61	395.00	15.00	6.20	7.23	12.34	62.13
Stabilization	9.09	652.00	462.00	57.00	248.00	4.98	2.60	6.74	4.47	1.96
Stabilization	9.19	669.00	476.00	51.00	257.00	5.61	2.59	6.71	4.11	2.31
Stabilization	9.43	680.00	480.00	47.00	273.00	4.92	1.93	6.66	3.87	2.30

TABLE 1.3 The Results of Physiochemical and Heavy Metals Obtained From Three Ponds of Landfill Leachate—cont'd

	Parameter									
Pond	**Na**	**Fe**	**Zn**	**Cu**	**Cr**	**Cd**	**Pb**	**As**	**Co**	**Mn**
Collection	978.37	0.35	14.46	9.80	31.50	0.31	2.71	32.28	6.34	5.44
Collection	978.37	0.23	17.89	9.03	22.37	0.32	3.35	20.55	5.14	9.55
Collection	978.37	0.62	11.50	9.59	10.70	0.01	2.16	9.75	3.95	16.60
Collection	284.44	0.56	23.80	57.10	18.12	0.85	8.03	11.52	6.26	52.72
Collection	273.47	0.56	24.00	57.40	17.04	1.73	7.60	14.41	7.43	51.82
Collection	283.80	0.56	23.10	56.80	17.58	1.71	7.04	10.09	6.61	50.92
Collection	673.45	1.34	61.01	64.92	42.36	0.04	0.70	25.21	21.30	109.34
Collection	633.35	1.98	49.15	53.45	59.06	0.56	3.91	14.81	18.78	133.59
Collection	665.61	0.79	27.48	58.75	52.27	0.06	2.08	17.07	23.67	122.59
Aeration	980.00	0.25	151.10	0.43	47.82	0.01	2.82	10.72	6.74	22.94
Aeration	975.55	0.73	259.08	2.83	3.67	0.39	2.72	11.12	2.61	32.44
Aeration	970.55	1.21	203.88	5.23	91.97	0.08	2.62	10.92	10.81	41.94
Aeration	525.57	0.87	425.46	69.42	86.18	0.56	7.17	21.66	16.18	4.97
Aeration	516.21	0.93	424.43	66.68	89.81	0.35	7.57	21.36	15.99	5.32
Aeration	517.13	0.85	423.45	63.63	85.56	0.16	7.37	20.68	15.78	5.96
Aeration	676.98	0.77	28.05	13.66	27.88	0.32	2.58	6.73	5.73	8.52
Aeration	630.31	0.94	18.69	15.97	37.11	0.08	4.90	10.12	5.61	10.34
Aeration	665.12	0.57	34.44	16.75	30.65	0.11	3.86	7.66	6.75	9.52
Stabilization	975.37	1.33	117.47	3.90	55.15	0.30	3.80	22.23	15.01	11.50
Stabilization	978.37	0.81	278.47	5.21	70.29	0.01	4.12	18.10	20.84	21.61
Stabilization	973.37	0.29	175.47	6.51	54.57	0.00	2.90	12.15	17.93	30.51
Stabilization	411.60	1.37	229.90	23.93	48.47	0.28	6.45	16.55	5.19	6.44
Stabilization	439.04	1.43	227.39	25.23	49.26	0.61	5.25	15.75	4.75	6.02
Stabilization	494.26	1.23	221.89	25.03	48.83	0.01	5.89	17.35	4.62	6.82
Stabilization	141.00	0.33	149.11	6.42	12.88	0.02	2.48	5.71	1.28	1.63
Stabilization	103.81	0.87	240.77	7.91	8.76	0.14	4.63	3.32	1.54	1.93
Stabilization	81.20	0.51	190.68	6.64	11.00	0.04	3.78	2.37	1.62	1.66

The leachate samples were collected from collection, aeration, and stabilized ponds in the ATLS leachate collection system. The leachate samples were collected three times during the period between August 2017 and January 2018, with three sampling points at each pond. The samples were manually gathered and placed in 500ml polyethene containers. The samples were immediately transported to the laboratory and cooled to 4°C to reduce biological and chemical reactions (Japan International Cooperation Agency (JICA)).

In this example, 27 samples (rows) were collected from the three ponds (collection, aeration, and stabilization), and 20 parameters (columns) were measured from each sample. It is not easy for the scientist to make a decision about the treatment process of the landfill or about the behavior of the different parameters under study because there are many different values (27 samples × 20 parameters = 540 values). Thus, the relationship among the different chosen parameters should be investigated and studied properly to understand the differences (fluctuations) in the parameters from one pond to another. These data should be analyzed employing an appropriate technique to achieve the objective of the project. The first objective was to assess whether the treatment process of the landfill leachate worked properly, and the relationship among the chosen parameters should be investigated for more information on the behavior of each variable in the presence of other chosen variables, which would help to identify the source of the variation. The second objective was to assess the effect of the landfill on the groundwater and surface water in the chosen area (the data for groundwater and surface water are not presented to save space). Furthermore, the contribution of each chosen variable (parameter) in illustrating the total variation in the collected data was identified employing multivariate methods. This research may help in estimating the impact of the landfill on groundwater and surface water in the chosen area.

EXAMPLE 1.3 INORGANIC ELEMENTS IN THE PARTICULATE MATTER IN THE AIR

A researcher wishes to investigate the concentrations of nine inorganic elements in the particulate matter (PM_{10}) in the air of an equatorial urban coastal location. In 2009, air pollution levels were studied during the summer and winter monsoon seasons employing high-volume sampling techniques. Atomic absorption spectrophotometry was employed to collect PM_{10} samples, with an average time of 24h. The parameters were the particulate matter PM_{10}, aluminum (Al), zinc (Zn), iron (Fe), copper (Cu), calcium (Ca), sodium (Na), manganese (Mn), nickel (Ni), and cadmium (Cd). The data are given in Table 1.4.

TABLE 1.4 The Results of Inorganic Elements in Particulate Matter in the Air (μg/m^3)

Season	PM_{10}	Al	Zn	Fe	Cu	Ca	Na	Mn	Ni	Cd
Summer	37.71	0.01	2.32	0.97	0.00	1.37	7.94	0.05	0.01	0.01
Summer	48.03	0.01	2.05	0.75	0.00	1.56	8.46	0.05	0.01	0.04
Summer	67.87	0.01	3.43	0.44	0.00	1.41	9.98	0.06	0.10	0.05
Summer	39.01	0.01	4.24	0.45	0.00	1.12	12.93	0.01	0.07	0.00
Summer	38.33	0.01	3.51	0.50	0.02	1.60	9.92	0.02	0.00	0.05
Summer	29.70	0.01	2.78	0.58	0.02	1.38	12.13	0.04	0.06	0.02
Summer	53.66	0.01	2.40	0.73	0.02	1.59	7.81	0.23	0.11	0.02
Summer	132.28	0.01	2.34	0.55	0.02	1.59	9.01	0.03	0.00	0.01
Summer	66.31	0.01	2.13	0.45	0.02	1.23	8.76	0.02	0.00	0.05
Summer	69.20	0.01	2.13	0.50	0.03	1.95	8.66	0.03	0.08	0.05
Summer	78.17	0.01	1.96	0.56	0.04	1.35	8.88	0.03	0.02	0.07
Summer	31.63	0.01	2.21	0.42	0.04	1.24	9.02	0.02	0.08	0.05
Summer	66.73	0.01	1.46	0.51	0.03	1.33	8.53	0.02	0.13	0.03
Summer	113.56	0.01	2.07	0.60	0.03	1.51	7.65	0.02	0.06	0.05
Summer	123.40	0.01	1.75	0.57	0.03	1.75	7.45	0.02	0.12	0.04
Summer	72.39	0.01	1.61	0.57	0.03	1.54	8.40	0.02	0.07	0.02
Summer	51.85	0.01	1.16	0.51	0.18	1.72	7.44	0.01	0.09	0.00
Summer	77.59	0.01	1.66	0.81	0.04	2.04	7.89	0.03	0.08	0.03
Summer	30.30	0.01	1.09	0.61	0.03	1.80	6.28	0.02	0.05	0.03
Summer	100.40	0.01	1.82	0.47	0.04	1.70	7.05	0.02	0.05	0.02
Summer	132.98	0.01	1.86	0.52	0.03	1.74	7.69	0.01	0.10	0.03
Summer	126.38	0.01	0.34	0.02	0.00	0.01	1.36	0.01	0.18	0.00

TABLE 1.4 The Results of Inorganic Elements in Particulate Matter in the Air (μg/m^3)—cont'd

Season	PM_{10}	Al	Zn	Fe	Cu	Ca	Na	Mn	Ni	Cd
Summer	31.82	0.01	1.41	0.52	0.03	1.93	7.04	0.02	0.04	0.00
Summer	110.53	0.01	1.25	0.67	0.04	2.29	7.39	0.02	0.03	0.00
Summer	38.41	0.01	1.55	0.79	0.12	2.21	7.74	0.02	0.06	0.03
Summer	124.26	0.01	2.24	0.00	0.00	1.56	9.65	0.07	0.06	0.01
Summer	53.62	0.01	1.97	0.01	0.00	1.39	7.07	0.03	0.04	0.03
Summer	23.30	0.01	1.28	0.02	0.00	1.80	6.39	0.03	0.00	0.00
Summer	67.34	0.01	0.91	0.02	0.00	0.80	7.14	0.00	0.08	0.00
Summer	22.58	0.03	1.29	0.02	0.01	0.92	7.21	0.02	0.05	0.02
Summer	54.52	0.03	1.37	0.04	0.01	1.48	7.98	0.01	0.03	0.03
Summer	112.83	0.01	1.88	0.03	0.00	1.46	8.44	0.01	0.02	0.00
Summer	175.28	0.01	1.68	0.03	0.00	1.38	8.13	0.00	0.03	0.00
Summer	47.30	0.01	2.20	0.03	0.00	1.13	11.39	0.01	0.03	0.01
Summer	57.08	0.01	1.69	0.04	0.00	1.04	8.34	0.02	0.07	0.00
Summer	15.16	0.01	1.58	0.04	0.01	1.77	6.69	0.05	0.08	0.02
Summer	272.98	0.01	1.88	0.04	0.01	1.60	7.05	0.04	0.08	0.00
Summer	101.82	0.01	1.86	0.02	0.01	1.79	6.02	0.05	0.13	0.00
Summer	59.90	0.01	1.73	0.02	0.01	1.70	7.25	0.05	0.00	0.02
Summer	31.07	0.01	1.92	0.02	0.02	1.40	4.92	0.05	0.03	0.04
Summer	107.43	0.01	2.03	0.02	0.02	1.48	5.19	0.04	0.06	0.00
Summer	30.51	0.01	2.42	0.03	0.03	2.08	5.70	0.07	0.00	0.04
Summer	84.02	0.01	2.10	0.22	0.01	1.61	5.85	0.03	0.00	0.00
Summer	7.37	0.01	2.63	0.35	0.02	1.24	3.90	0.03	0.00	0.00
Summer	7.65	0.01	1.38	0.34	0.01	1.11	4.20	0.04	0.00	0.00
Summer	15.05	0.01	1.56	0.21	0.02	0.35	2.47	0.03	0.00	0.00
Summer	84.13	0.01	1.89	0.04	0.02	0.41	4.44	0.02	0.00	0.02
Winter	82.87	0.01	1.66	0.15	0.02	0.64	4.79	0.02	0.06	0.00
Winter	138.25	0.01	2.53	0.30	0.03	0.42	3.02	0.02	0.00	0.03
Winter	82.67	0.01	2.66	0.31	0.04	0.03	3.16	0.03	0.00	0.00
Winter	46.08	0.01	3.39	0.52	0.02	0.21	2.34	0.03	0.00	0.00
Winter	15.33	0.01	2.63	0.23	0.02	0.06	4.38	0.03	0.05	0.00
Winter	30.78	0.01	1.98	0.17	0.01	0.40	3.68	0.01	0.00	0.00
Winter	155.00	0.01	3.72	0.15	0.02	1.21	4.24	0.02	0.00	0.00
Winter	61.75	0.01	2.93	0.00	0.03	0.92	4.29	0.01	0.00	0.00
Winter	88.90	0.01	4.91	0.11	0.02	0.93	6.65	0.02	0.00	0.00
Winter	38.47	0.01	3.46	0.15	0.02	0.73	4.25	0.01	0.01	0.00
Winter	24.46	0.01	1.97	0.19	0.01	0.75	1.96	0.01	0.00	0.00
Winter	14.29	0.01	2.76	0.11	0.01	0.67	9.55	0.03	0.00	0.00
Winter	65.52	0.01	2.56	0.32	0.01	0.54	2.70	0.04	0.00	0.00
Winter	63.33	0.01	1.21	0.16	0.02	0.53	3.40	0.04	0.03	0.00
Winter	62.28	0.01	3.52	0.43	0.01	0.80	7.78	0.02	0.11	0.00

Continued

TABLE 1.4 The Results of Inorganic Elements in Particulate Matter in the Air (μg/m^3)—cont'd

Season	PM_{10}	Al	Zn	Fe	Cu	Ca	Na	Mn	Ni	Cd
Winter	65.87	0.01	3.48	0.25	0.02	0.99	11.06	0.01	0.03	0.00
Winter	122.83	0.01	3.24	0.41	0.03	1.09	11.47	0.01	0.06	0.00
Winter	50.18	0.00	2.40	0.49	0.02	1.16	7.90	0.02	0.11	0.00
Winter	60.19	0.01	3.02	0.38	0.02	0.84	9.51	0.02	0.06	0.01
Winter	28.28	0.01	3.68	0.48	0.06	1.18	12.21	0.01	0.09	0.00
Winter	42.50	0.01	1.86	0.46	0.06	1.34	10.51	0.03	0.07	0.00
Winter	49.68	0.01	2.29	0.32	0.04	1.57	11.02	0.02	0.00	0.04
Winter	39.91	0.01	3.30	0.45	0.05	1.46	8.61	0.03	0.08	0.02
Winter	37.94	0.01	3.57	0.31	0.08	1.59	9.47	0.04	0.00	0.05
Winter	27.97	0.01	3.14	0.09	0.06	1.46	12.08	0.03	0.00	0.05
Winter	45.34	0.01	2.59	0.26	0.06	1.59	11.43	0.03	0.06	0.00
Winter	58.45	0.01	3.46	0.12	0.06	1.66	8.98	0.04	0.10	0.00
Winter	43.30	0.01	1.58	0.09	0.05	1.46	7.59	0.03	0.00	0.00

Table 1.4 provides a huge and complex data set. It can easily be observed that the table does not provide helpful data (information) for making a decision. The objective of the study was to assess the air quality of Penang, Malaysia, in terms of PM_{10} and inorganic elements, and to recognize the main sources of PM_{10} and inorganic elements, whether crustal or noncrustal. The goals included investigating the relation between the different chosen inorganic elements and PM_{10} during the summer and winter monsoons and determining the similarities between the chosen parameters.

EXAMPLE 1.4 HEAVY METALS IN SEDIMENT (mg/L)

A researcher wishes to investigate the concentrations of eight heavy metals (cadmium (Cd), iron (Fe), copper (Cu), zinc (Zn), chromium (Cr), mercury (Hg), manganese (Mn), and lead (Pb)) in sediments obtained from two sites. The two sites, with 10 sampling points at each site, were Kuala Juru (the Juru River) and Bukit Tambun (the Jejawi River) in the Penang State of Malaysia. The sediment samples were gathered at low tide with an Eijkelkamp gouge auger from each of the 20 chosen sampling points (10 sampling points from each river estuary). A flame atomic absorption spectrometer (FAAS; Perkin Elmer HGA-600) was employed for the analysis of Cu, Pb, Zn, Cd, Mn, Fe, and Cr, and a cold vapor atomic absorption spectrometer (CV-AAS) method was used for Hg analysis after sample digestion in an acid solution. The data are given in Table 1.5.

TABLE 1.5 The Concentration of Heavy Metals in Sediment (mg/L) for Juru and Jejawi Rivers

	Cu	Zn	Cd	Cr	Fe	Pb	Hg	Mn
Juru	0.63	1.89	1.95	0.15	26.17	0.48	0.10	0.01
Juru	0.73	1.79	1.99	0.17	22.95	0.35	0.12	0.02
Juru	0.35	0.90	1.94	0.09	25.41	0.38	0.12	0.03
Juru	0.76	2.19	1.98	0.20	24.13	0.17	5.19	0.01
Juru	0.60	1.74	1.94	0.14	23.16	0.36	0.49	0.02
Juru	0.36	3.38	1.95	0.14	27.20	0.17	0.14	0.02
Juru	0.63	3.12	1.98	0.12	26.84	0.61	0.12	0.03
Juru	0.52	3.42	1.93	0.14	25.98	0.32	0.12	0.03

TABLE 1.5 The Concentration of Heavy Metals in Sediment (mg/L) for Juru and Jejawi Rivers—cont'd

	Cu	Zn	Cd	Cr	Fe	Pb	Hg	Mn
Juru	0.55	2.86	1.97	0.19	25.60	0.30	0.12	0.02
Juru	0.47	3.71	1.92	0.13	26.26	0.80	0.11	0.03
Jejawi	0.38	3.27	1.60	0.13	38.72	0.16	0.17	1.63
Jejawi	0.43	2.47	1.69	0.12	38.66	0.18	0.23	1.68
Jejawi	0.39	1.59	1.62	0.17	38.94	0.15	0.15	1.75
Jejawi	0.30	2.12	1.64	0.20	37.74	0.18	0.15	1.75
Jejawi	0.49	2.28	1.66	0.28	39.52	0.28	0.56	1.90
Jejawi	0.45	2.24	1.68	0.11	38.49	0.28	0.24	1.67
Jejawi	0.48	1.76	1.76	0.16	38.22	0.15	0.17	1.71
Jejawi	0.46	2.51	1.75	0.23	39.01	0.42	0.22	1.77
Jejawi	0.23	1.80	1.71	0.21	39.61	0.41	0.24	1.85
Jejawi	0.20	1.45	1.73	0.17	38.80	0.15	0.18	1.70

In this study, the researcher wanted to investigate and understand the interrelationship between the chosen parameters and to extract information about the resemblance or differences between the different sampling sites, identification of the variables (heavy metals) accountable for the spatial differences in river estuaries, and the effect of the possible sources (natural and anthropogenic) on the chosen heavy metals of the two river estuaries.

1.5 MULTIVARIATE NORMAL DISTRIBUTION

The normal distribution is the most important distribution in statistics. The normal distribution is very important because most of the tests used in statistics require that the assumption of normality be met; the data are gathered from a normally distributed population.

A brief explanation of the univariate and multivariate normal distributions is given below.

1.5.1 Univariate Normal Distribution

Suppose Y is a random variable that follows the normal distribution. Then, the probability distribution function of the univariate normal distribution is given in Eq. (1.1).

$$f(Y) = \frac{1}{\sqrt{2\pi\sigma^2}} e^{-(Y-\mu)^2}/2\sigma^2, \ -\infty < Y < \infty \tag{1.1}$$

where

μ is the mean; and
σ^2 is the variance.

The univariate normal distribution can be written as $Y \sim N(\mu, \sigma^2)$.

The normal distribution (bell-shaped) curve is presented in Fig. 1.1. R statistical software was used to generate the curve. The commands and built-in functions for creating a normal distribution curve are presented in the Appendix.

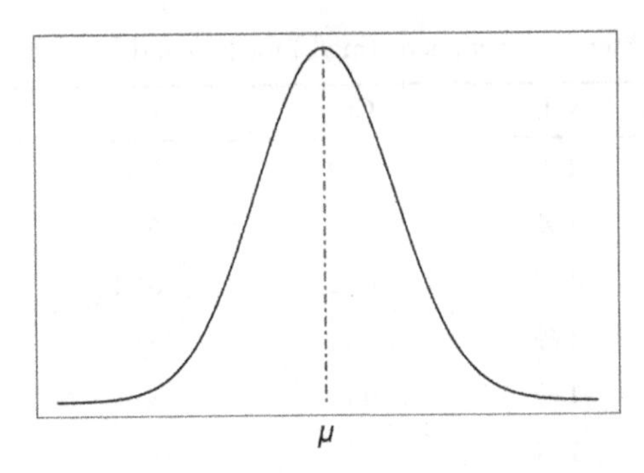

FIG. 1.1 Normal distribution plot.

1.5.2 Multivariate Normal Distribution

Suppose $\boldsymbol{Y}$ is a random vector of k random variables that follows the multivariate normal distribution. Then, the multivariate normal distribution is given in Eq. (1.2).

$$g(Y) = \frac{1}{(2\pi)^{k/2}\left|\sum\right|^{1/2}} e^{-(\boldsymbol{Y}-\boldsymbol{\mu})'\sum^{-1}(\boldsymbol{Y}-\boldsymbol{\mu})/2} \tag{1.2}$$

where

k is the number of variables;
$\boldsymbol{\mu}$ is the mean vector;
$\sum$ is the covariance matrix; and
$(\boldsymbol{Y}-\boldsymbol{\mu})'\sum^{-1}(\boldsymbol{Y}-\boldsymbol{\mu})$ is the Mahalanobis distance (statistical distance).

The multivariate normal distribution can be denoted as $\boldsymbol{Y} \sim N_k(\boldsymbol{\mu}, \sum)$.

Note:

- A suitable transformation of the data should be used if the normality assumption is violated by one or more variables under investigation, as when the data are highly skewed with several outlier (extreme) values (high or low) or repeated values.
- If all the individual variables follow a normal distribution, then it is supposed that the combined (joint) distribution is a multivariate normal distribution.
- In practice, the real data never follow a multivariate normal distribution completely; however, the normal density can be employed as an approximation of the true population distribution.

Further Reading

Alkarkhi, A.F.M., Alqaraghuli, W.A.A., 2019. Easy Statistics for Food Science with R, first ed. Academic Press.
Alkarkhi, A.F.M., Ismail, N., Ahmed, A., Easa, A.M., 2009. Analysis of heavy metal concentrations in sediments of selected estuaries of Malaysia—a statistical assessment. Environ Monit Assess 153, 179–185.
Banch, T.J.H., Hanafiah, M.M., Alkarkhi, A.F.M., Amr, S.S.A., 2018. Statistical evaluation of landfill leachate system and its impact on groundwater and surface water in Malaysia.
Blogger, 2011. R graph gallery: A collection [Online]. Available, http://rgraphgallery.blogspot.my/2013/04/shaded-normal-curve.html.
Bryan, F.J.M., 1991. Multivariate Statistical Methods: A Primer. Chapman & Hall, Great Britain.
Daniel & Hocking, 2013. Blog Archives, High Resolution Figures in R [Online]. R-bloggers. Available, https://www.r-bloggers.com/author/daniel-hocking/.
Johnson, R.A.W., W, D., 2002. Applied Multivariate Statistical Analysis. Prentice Hall, New Jersey.
Rencher, A.C., 2002. Methods of Multivariate Analysis. J. Wiley, New York.
Yusup, Y., Alkarkhi, A.F.M., 2011. Cluster analysis of inorganic elements in particulate matter in the air environment of an equatorial urban coastal location. Chemistry and Ecology 27, 273–286.

CHAPTER

2

R Statistical Software

LEARNING OBJECTIVES

After careful consideration of this chapter, you should be able:

- *To know R software.*
- *To know RStudio.*
- *To describe how to set up R packages.*
- *To know how to write variables, vectors, sequences, and matrixes in R language.*
- *To use R commands and built-in functions for environmental data.*
- *To understand how to call stored data files into the R Environment.*
- *To know how to write down simple scripts.*
- *To choose suitable R commands to write the scripts.*
- *To know how to generate high resolution plots in R.*
- *To understand the concept of working directory and how to set a new working directory.*
- *To describe R output and extract a useful report.*

2.1 INTRODUCTION

R is a software environment for statistical computing and graphical programming languages. The R software has been employed by professionals at various organizations, colleges, survey institutions, and others. Considerable statistical packages are provided for various statistical analyses; however, researchers, scientists, and others who are concerned in data analysis, designing, modeling, and producing beautiful and high-resolution plots prioritize employing R. R is offered for free as an open-source software; under the terms of the GNU General Public License, "R is an official part of the Free Software Foundation's GNU project, and the R Foundation has similar objectives to other open-source software foundations like the Apache Foundation or the GNOME Foundation." R language is similar to the S language and environment that was developed by Bell Laboratories. R is employed by millions of researchers around the world, and the number of R users continues to increase. R has become a substantial, engaging, singular, and new statistical software for the following purposes:

- R is free (open-source) software including many packages, and many sources around the world permit downloading and installation of the software, regardless of the position and the institution you work with, or whether you are affiliated with a public or private organization.
- R offers many built-in functions to help make the steps of the analysis simple and easy. We can carry out data analysis in R by providing scripts to know the required variables and asking built-in functions in R to carry out the required process, such as computing the correlation, average, variance, or other statistical values.
- Clear, high-resolution, and unique graphs can be produced by R that meet specific standards or reflect a particular opinion of the task (convey thoughts to the plot).
- R can easily be used by researchers without programming proficiency.
- People can download and install R for various operating systems such as Windows, Linux, and MacOS.

Applied Statistics for Environmental Science with R
https://doi.org/10.1016/B978-0-12-818622-0.00002-2

- Many statistical and graphical packages are provided by R library for data analysis, and different computations and graphical applications generate high-quality plots.
- It is easy to interact with online community around the world, interchange thoughts, and receive assistance.
- R codes, commands, and functions are available online for free; moreover, considrable sources offer demonstrations regarding R for free, containing courses, material, and responses to inquiries, which other software packages do not offer.
- More facilities are offered by RStudio to operate R, and it is simpler and friendlier to employ than R.

The concepts and related terms to the R statistical package are addressed to provide a starting point for readers who are novice to the R language and environment. Beginners will be guided on how to download and install the software (R and RStudio), comprehend some ideas and related concepts employed in R, and write simple and easy scripts in R. We attempted as much as we could to make the procedure and directives simple and comprehensible to everybody. Considerable examples are provided to lead the researchers step-by-step and make the procedure interesting.

We can download and install R statistical software and its packages simply in a few steps, and the required packages related with R software are then installed. R provides considerable packages to carry out various statistical methods. After installing R software, we can download and install RStudio that can be employed to operate R effectively and in a more friendly way than R.

2.2 INSTALLING R

Consider we have not installed R statistical software yet. The software can be installed and downloaded for free by employing the six steps below.

1. The reader can type https://cran.r-project.org/ and then click "Enter"; "the Comprehensive R Archive Network" will appear as shown in Fig. 2.1.
2. The Screen for "The Comprehensive R Archive Network" offers three choices to download R software based on the operating system of the computer, as presented below:
   (1) Download R for Linux
   (2) Download R for (Mac) OSX
   (3) Download R for Windows
   If we choose R for Windows, click install R for the first time (or base) as presented in Fig. 2.2.

CRAN
Mirrors
What's new?
Task Views
Search
About R
R Homepage
The R Journal
Software
R Sources
R Binaries
Packages
Other
Documentation
Manuals
FAQs
Contributed
The Comprehensive R Archive Network
Download and Install R
Precompiled binary distributions of the base system and contributed packages, Windows and Mac users most likely want one of these versions of R:
Download R for Linux
Download R for (Mac) OS X
Download R for Windows
R is part of many Linux distributions, you should check with your Linux package management system in addition to the link above.
Source Code for all Platforms
Windows and Mac users most likely want to download the precompiled binaries listed in the upper box, not the source code. The sources have to be compiled before you can use them. If you do not know what this means, you probably do not want to do it!
The latest release (2018-07-02, Feather Spray) R-3.5.1.tar.gz, read what's new in the latest version.
Sources of R alpha and beta releases (daily snapshots, created only in time periods before a planned release).
Daily snapshots of current patched and development versions are available here. Please read about new features and bug fixes before filing corresponding feature requests or bug reports.
Source code of older versions of R is available here.
Contributed extension packages
Questions About R
If you have questions about R like how to download and install the software, or what the license terms are, please read our answers to frequently asked questions before you send an email.

FIG. 2.1 Comprehensive R Archive Network.

CRAN
Mirrors
What's new?
Task Views
Search

About R
R Homepage
The R Journal

Software
R Sources
R Binaries
Packages
Other

Documentation
Manuals
FAQs
Contributed

R for Windows

Subdirectories:

base	Binaries for base distribution. This is what you want to install R for the first time.
contrib	Binaries of contributed CRAN packages (for R >= 2.13.x; managed by Uwe Ligges). There is also information on third party software available for CRAN Windows services and corresponding environment and make variables.
old contrib	Binaries of contributed CRAN packages for outdated versions of R (for R < 2.13.x; managed by Uwe Ligges).
Rtools	Tools to build R and R packages. This is what you want to build your own packages on Windows, or to build R itself.

Please do not submit binaries to CRAN. Package developers might want to contact Uwe Ligges directly in case of questions / suggestions related to Windows binaries.

You may also want to read the R FAQ and R for Windows FAQ.

Note: CRAN does some checks on these binaries for viruses, but cannot give guarantees. Use the normal precautions with downloaded executables.

FIG. 2.2 Showing the instructions for installing R for Windows.

CRAN
Mirrors
What's new?
Task Views
Search

About R
R Homepage
The R Journal

Software
R Sources
R Binaries
Packages
Other

Documentation
Manuals
FAQs
Contributed

R-3.5.1 for Windows (32/64 bit)

Download R 3.5.1 for Windows (62 megabytes, 32/64 bit)
Installation and other instructions
New features in this version

If you want to double-check that the package you have downloaded matches the package distributed by CRAN, you can compare the md5sum of the .exe to the fingerprint on the master server. You will need a version of md5sum for windows: both graphical and command line versions are available.

Frequently asked questions

- Does R run under my version of Windows?
- How do I update packages in my previous version of R?
- Should I run 32-bit or 64-bit R?

Please see the R FAQ for general information about R and the R Windows FAQ for Windows-specific information.

Other builds

- Patches to this release are incorporated in the r-patched snapshot build.
- A build of the development version (which will eventually become the next major release of R) is available in the r-devel snapshot build.
- Previous releases

Note to webmasters: A stable link which will redirect to the current Windows binary release is
<CRAN MIRROR>/bin/windows/base/release.htm.

Last change: 2018-07-02

FIG. 2.3 The Screen to download R-3.5.1 for Windows (32/64 bit).

3. We can click on the available version. The newest available version for R software is R-3.5.1, as shows on the Screen, or there may be other versions. Click on the "Download R-3.5.1 for Windows (62 megabytes, 32/64 bit)" as shown in Fig. 2.3.
4. Click on the "Download R-3.5.1 for Windows (62 megabytes, 32/64 bit)"; in the lower bottom-left corner, there is a message "R-3.5.1-win.exe" (Fig. 2.4), this indicates that the file starts downloading to the computer.
5. Click on download file once the download of R is finished to open another screen, and then click on "run" as shown in Fig. 2.5. The next step is to follow the guidance given to finish the installation.
6. R is installed on the computer and ready to be used. We can start using R by double clicking on the R icon.

2.2.1 R Material

It is highly useful to have a handbook or manuals to direct readers employing the new software, particularly for novices who are new to the R software. This service is provided for free; we can use on-line websites to download manuals and notes such as https://cran.r-project.org/, or other authorized R sources. The handbooks, manuals and notes can be obtained offline by employing the help button in the upper row of the R environment, as presented in Fig. 2.6; the help button provides many options, one of which is Manuals (PDF).

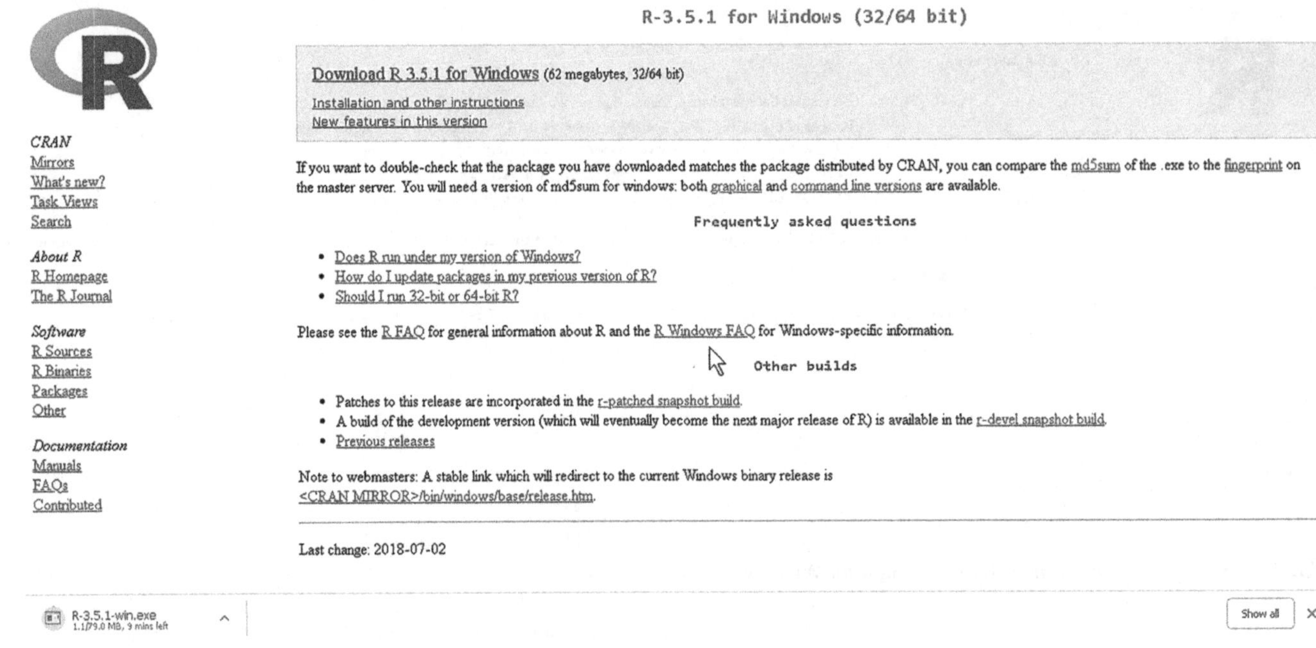

FIG. 2.4 Showing the place of downloaded file.

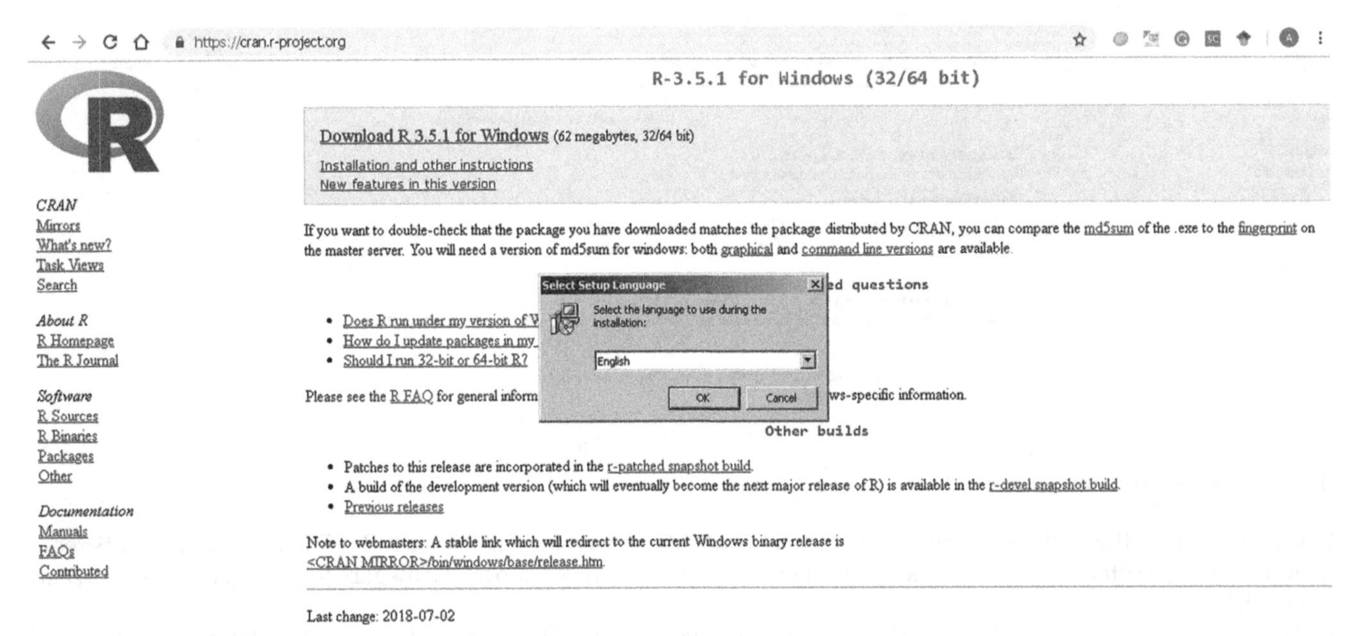

FIG. 2.5 Showing the directives to download the software.

We have found that the handbooks and associated notes are helpful and offer clear direction, particullary for novice readers. Some documents have been produced in various languages, such as Russian, Chinese, and German.

2.2.2 R Packages

R offers various statistical packages, and some packages are built-in packages (standard/base packages, loaded packages once R installation is finished). Users can download other packages from the upper row of the R Console ("Packages"). The packages can be downloaded by clicking "Install package(s)," and then select the site you wish to download. A list of Packages are provided in R software, thus you can select the package you require to install. The search `()` function is employed to display some of the loaded packages in your computer when R starts.

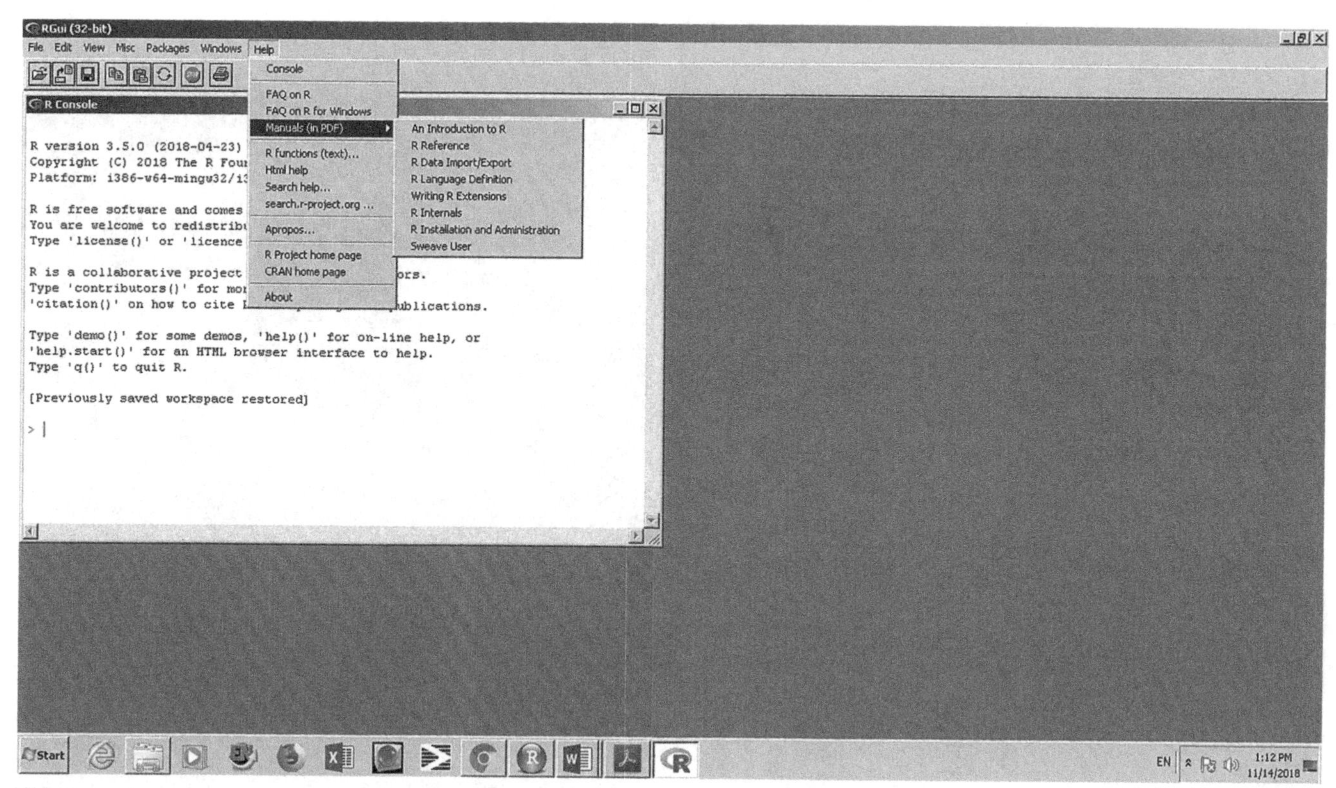

FIG. 2.6 Showing the steps to download notes and manuals.

`Search ()`

The function `search ()` is used to show the loaded packages.

```
> search()
[1] ".GlobalEnv"          "package:stats"  "package:graphics"
[4] "package:grDevices" "package:utils"  "package:datasets"
[7] "package:methods"   "Autoloads"      "package:base"
```

2.3 THE R CONSOLE

The Screen where we place the commands; then run the scripts to perform the analysis or any wanted computations is called R Console. There is a symbol called the "command prompt" by default >, which is placed at the end of the Console. The R scripts should be written after the command prompt. For example, 8 + 3 is defined and then followed by "Enter" to obtain the output (11) as presented in Fig. 2.7.

The buttons for File, Edit, View, MISC, Packages, Windows and Help are shown in the upper row of the Screen; each button has choices to carry out a particular job.

Note:

- The output lines for running any script are preceded by [1].
- Built-in functions are called in to perform the required job, as we have seen in the preceding example for 8 + 3; pressing "Enter" will invite the built-in function for addition to perform the proposed task. Names are used to call the built-in functions in R followed by the argument in parentheses; then press "Enter," which is an instruction to perform the requested job. Usually, the built-in functions are available in the memory of the computer; for instance, to quit R, we should write the built-in function `q ()`.

```
> q ()
```

- Asking for assistance is achieved by employing the function `help ()` to turn on a new Screen concerning the requested subject. For example, the `help (quit)` command will offer the ready input concerning to the quit (terminate an R session) in R library containing a characterization of the function, application, arguments, references, and models.

```
> help (quit)
```

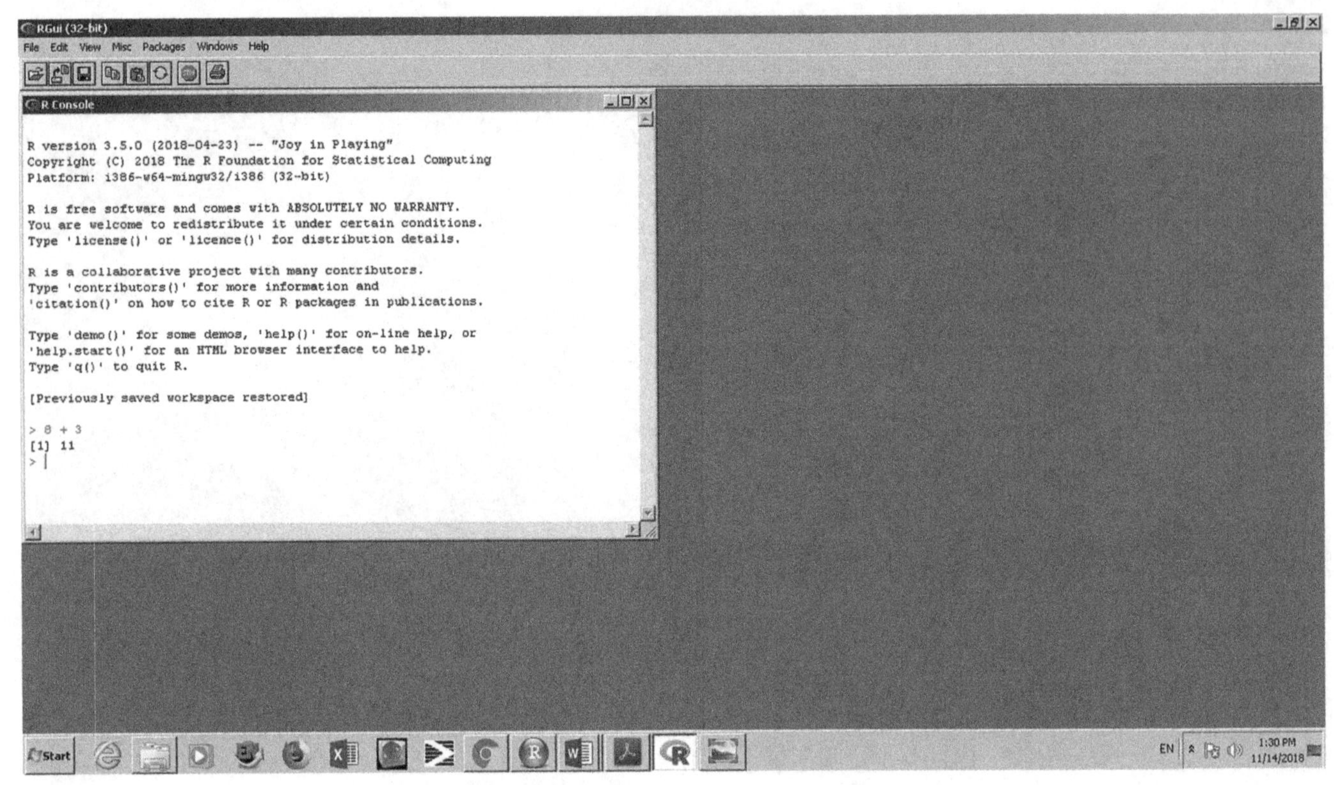

FIG. 2.7 Showing R commands placed after the command prompt.

The question mark (?) is another built-in function employed for requesting assistance, which may be utilized as a shortcut for asking assistance, as in ?quit. The two functions help(quit) and ?quit are valent. The screen for calling assistance employing both functions is presented in Fig. 2.8.

- Writing the name of a variable (parameter) as a command will work similarly to the built-in function print (); for example, the outcome of B = 6 + 4 can be printed (displayed on screen) either by writing B or writing print (B) as presented below.

```
> B = 6 + 4
> B           #Print the value of B
[1] 10
> print (B) #print the value of B
[1] 10
```

- We can transfer the output to a file by employing R built-in functions such as write.table(), or the simplest method is to copy and paste from the R Console to a word document.

2.4 EXPRESSION AND ASSIGNMENT IN R

This section covers expression and assignment including arithmetic operators, mathematical functions, and relational operators.

1. Arithmetic operators refers to the standard arithmetic operators, which are ^, +, -, *, and /, and every operator performs a specific (particular) action. The job of each operator is shown below:

   - ^ or **: exponentiation operation
   - + : addition operation
   - - : subtraction operation
   - * : multiplication operation
   - / : division operation

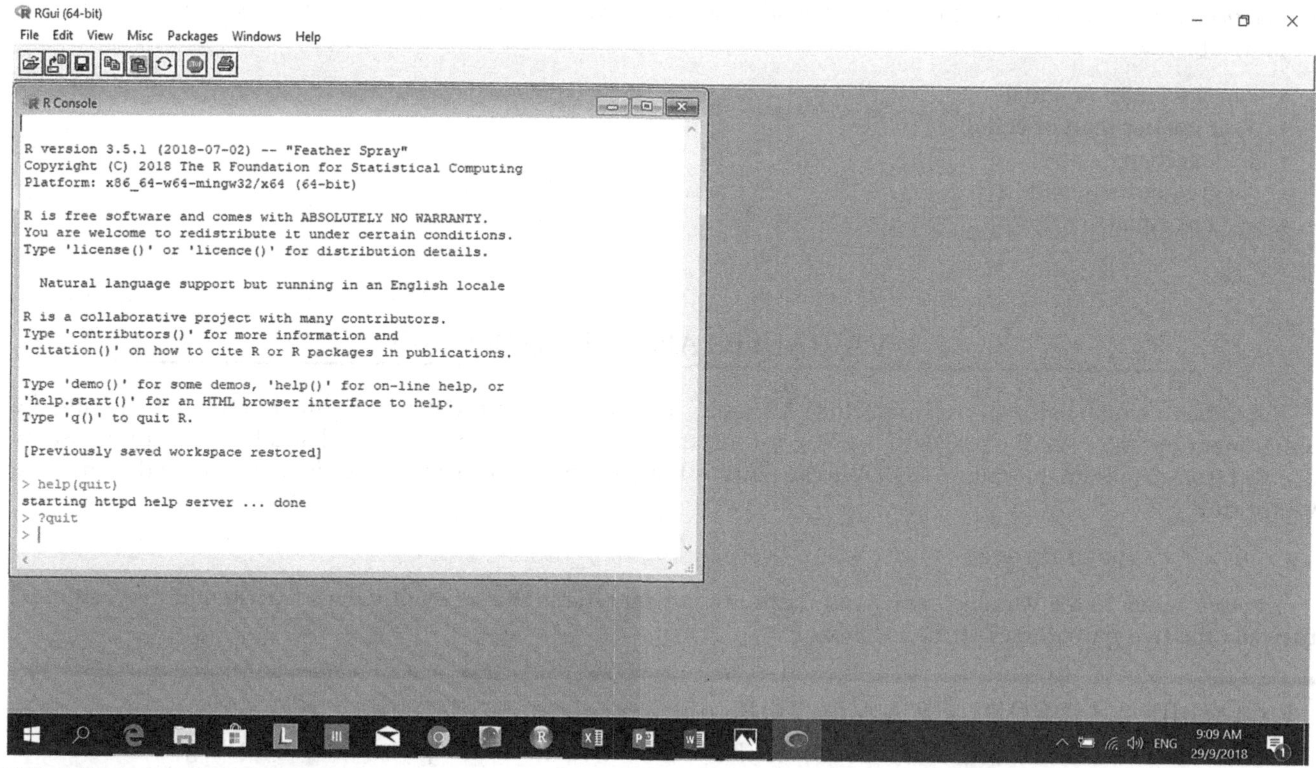

FIG. 2.8 Showing help command.

The priority of arithmetic operators in R follows the standard precedence; i.e., ^ is the highest, and addition and subtraction are the lowest. Parentheses are employed to control the order of the arithmetic operators.

EXAMPLE 2.1 EMPLOYING ÷, ^ AND + FUNCTIONS

Compute 9 ÷ 3, and 3^2, 5+3 employing R.

R operates as a calculator with arithmetic operators to perform the requested jobs. The output of employing R built-in functions for performing division, exponentiation, and addition are shown below:

```
> 9/3
[1] 3
> 3^2
[1] 9
> 5+3
[1] 8
```

2. R also deals with mathematical functions such as exp (exponential), sqrt (square root), and log (logarithm); more built-in functions are offered by R in various packages.

EXAMPLE 2.2 EMPLOY LOG FUNCTION

Compute log (3 ÷ (1 + 0.5)) The function `log ()` is employed to compute the value of log.

```
Log (3 / (1+.5))
[1] 0.6931472
```

3. Relational operators are also offered by R statistical software like >=, <=, <, >, and !=. The task of every operator is shown below:

   - <= : less than or equal,
   - >= : greater than or equal,
   - < : less than,
   - > : greater than, and
   - != : not equal.

2.5 VARIABLES AND VECTORS IN R

R handles a single value as well as vectors. Variables and vectors (matrix) can be defined in R by employing the assignment operator <- or equality sign (=); for instance, $Y < -6$ and $Y = 6$ are equal and bear exactly the same concept (sense) (assign the value 6 to Y). We can employ the function `c ()` to define a series of values in the form of a vector in R.

```
> y <- c (data frame)
```

where y refers to the variable name, and `data frame` represents the series of values that should be positioned between the two parentheses of the function `c ()`.

EXAMPLE 2.3 PRODUCE A VECTOR

Place the values 10, 7, 14, 8, 11, 12 in a vector form called Y. The function `c ()` is employed to represent the given values in a vector form.

```
> Y <- c (10, 7, 14, 8, 11, 12)
```

Note:

- The assignment operator <- takes place side by side or performs as a "variable-defining" operator, which is the valent of the operator "=".
- A comma is usually employed to separate the data values in the vector (?, ?, ...).
- Any command can be performed by pressing "Enter."
- X and x represent two different variables in R.
- Consider X and Y are two vectors of the same length, then a new vector will be produced by $X+Y$ ($X-Y$) with values representing the sum (difference) of the corresponding values of X and Y.

EXAMPLE 2.4 TWO VECTORS OF THE SAME LENGTH

Consider $X = 5, 3, 2$ and $Y = 4, 2, 6$, then $X+Y$ equals to 9, 5, 8 and $X-Y$ equals to 1, 1, -4 of the same length. The R commands are employed to compute $X+Y$ and $X-Y$.

```
> X <- c (5, 3, 2)
> X
[1] 5 3 2
> Y <- c (4, 2, 6)
> Y
[1] 4 2 6
> X + Y
[1] 9 5 8
> X - Y
[1] 1 1 -4
```

Consider X and Y are two vectors of different length, then a new vector will be produced by $X+Y$ ($X-Y$), repeating the shorter vector as needed. The number of values (observations) in the $X+Y$ ($X-Y$) generated vectors is equal to the extended vector.

EXAMPLE 2.5 TWO VECTORS OF DIFFERENT LENGTH

Consider $X = 5, 3$ and $Y = 4, 2, 6$, then $X-Y$ equals to 1, 1, -1. The R commands and built-in functions are employed to compute $X-Y$.

```
> X <- c (5, 3)
> Y <- c (4, 2, 6)
> X - Y
[1] 1 1 -1
Warning message:
In X - Y : longer object length is not a multiple of shorter object length
```

One can observe that the value 5 of the vector X is employed two times (5 - 4) and (5 - 6), while the value 3 is employed only one time to make the length of X the same as Y.

EXAMPLE 2.6 CREATE A VECTOR WITH A ZERO IN THE CENTER

Form a vector with $2n+1$ values representing two copies of X with a zero in the center. Four values 3, 6, 4 and 8 represents the variable X. The commands employed to create a vector with a zero in the middle are:

```
> X <- c (3, 6, 4, 8)
> z <- c (X, 0, X)
> z
[1] 3 6 4 8 0 3 6 4 8
```

Note: Calling a single value can be achieved by using the function `vector name []` to locate the position.

```
> vector name [ position of the element]
```

EXAMPLE 2.7 EXTRACT A VALUE

Extract the second value, and then extract the fourth value of a vector $y = 10, 5, 8, 4$. The function `y [ ]` in R is employed to extract the requested values.

```
> y <- c (10, 5, 8, 4)
> y [ 2 ]
[1] 5
> y [ 4 ]
[1] 4
```

R offers specific functions to extract successive values of a vector. Successive values of the vector can be extracted by using colon (:) operator.

```
Vector name [A : B]
```

where A refers to the starting value and B refers to the ending value.

EXAMPLE 2.8 EXTRACT SUCCESSIVE VALUES

Extract the last two values of a vector y presented in Example 2.7. The command for extracting the two successive values is `y [3:4]`.

```
> y [3:4]
[1] 8 4
```

Excluding a value from a vector can be achieved by adding a negative subscript associated with the built-in function as given below.

```
Vector name [- position of the value]
```

the value at that position will be excluded from the vector, and the call for excluding sequential values of the vector requires to locate two values in the vector and put them between two brackets, as shown below:

```
vector name [- (A : B)]
```

where, A refers to the starting value and B refers to the ending value.

EXAMPLE 2.9 ELIMINATE A VALUE OR VALUES

Employ the data set presented in Example 2.7 to eliminate the third data value and then to eliminate the first and second data values. The function `y [-3]` is employed to eliminate the third value, and the function `y [ - (1:2)]` is employed to eliminate the first and second values.

```
> y <- c (10, 5, 8, 4)
> y [ -3 ]
[1] 10 5 4
> y [ - (1 :2) ]
[1] 8 4
```

The function `y [y < or y >]` can be employed to call values that are more or less a given element.

```
y [ y < Specific value] or y [ y > Specific value]
```

EXAMPLE 2.10 LESS THAN AND MORE THAN

Locate the values of y that are less than 7, less than 10, and more than 15 respectively, where $y = 8, 14, 13, 7, 10, 15, 18, 20$. The function `y [y < or y >]` was employed to locate the values that are less than or more than a value, as shown below.

```
> y <- c (8, 14, 13, 7, 10, 15, 18, 20)
> y [ y < 7 ]
numeric(0)
> y [ y < 10 ]
[1] 8 7
> y [ y > 15 ]
[1] 18 20
```

2.5.1 Matrix in R

Multivariate data for various variables can be presented in a table including rows and columns, called a matrix. So far, we have experimented with scalars and how to form a vector employing built-in functions in R. The next section shows how to employ R commands to generate a matrix from available data.

The built-in function `matrix ()` is usually employed to produce a matrix in R. R offers space to produce a matrix by rows or by columns.

```
M = matrix (data, nrow = Number, ncol = Number, byrow = TRUE)
M <- matrix (data, No.rows, No.columns, byrow = TRUE)
```

The command `byrow =?` Has two options either `TRUE` or `FALSE`.

EXAMPLE 2.11 PRODUCE A MATRIX

Generate a 3-by-3 (three rows and three columns) matrix for the variable *X*. The variable *X* consists of nine values 5, 7, 6, 4, 10, 7, 8, 9, and 7. The commands and built-in functions for generating the 3-by-3 matrix are presented below:

```
X <- c (5, 7, 6, 4, 10, 7, 8, 9, 7)
M = matrix (X, 3, 3, byrow = TRUE)
M
```

Moreover, the output of employing the built-in function `matrix ()` is:

```
> X <- c (5, 7, 6, 4, 10, 7, 8, 9, 7 )
> M = matrix (X, 3, 3, byrow = TRUE)
> M
     [,1] [,2] [,3]
[1,]    5    7    6
[2,]    4   10    7
[3,]    8    9    7
```

This situation produces the matrix by rows, which is similar to the order `M = matrix (X, 3, 3)`.

The two charts displayed with the result of R `[, number]` and `[number, ]` refers to the number of columns and the number of rows of the generated matrix.

Note: extracting a value of a matrix can be achieved by locating the row and column employing the built-in function `matrix name [row, column]` in R.

```
Matrix name [row, column]
```

EXAMPLE 2.12 EXTRACT A VALUE OF A MATRIX

Use the matrix M generated in Example 2.11 to extract the value in row 3 and column 2. The function `M [3, 2]` can be employed to extract this value of the matrix M.

```
> M [3, 2]
[1] 9
```

The entire column or row of a matrix can be called by employing the function `matrix name []` in R

```
matrix name [row number,]
```

and the call for extracting a column of a matrix is:

```
matrix name [, column number]
```

EXAMPLE 2.13 EXTRACT A COLUMN OR A ROW

Extract the second column and then the second row of the matrix M produced in Example 2.11. The output of employing the function M [,] is given below.

```
> M [,2 ] #Extract the second column of the matrix
[1] 7 10 9
> M [2, ] #Extract the second row of the matrix
[1] 4 10 7
```

The matrix multiplication operator value by value is %*%.

```
%*%
```

Matrix inversion can be computed by employing the `solve ()` function.

```
> solve (matrix name)
```

The transpose of a matrix can be found by using the `t ()` function.

```
t (matrix name)
```

2.6 BASIC DEFINITIONS

1. The function `seq ()` in R can be employed to generate a sequence of values pointing out the starting point, ending point and step size.

   ```
   seq (from = A, to = B, by = C)
   ```

   where, A refers to the starting value, B refers to the ending value, and C refers to the step size.

EXAMPLE 2.14 GENERATE A SEQUENCE

Generate a sequence starting from 1 to 10 using the value 1 as the step size, and then the value 2 as the step size. The result of employing the function `seq ()` is:

```
> seq (1, 10, 1)
[1] 1 2 3 4 5 6 7 8 9 10
> seq (1, 10, 2)
[1] 1 3 5 7 9
```

2. Repeating a variable (vector) or a specific value in R requires employing the function `rep ()`.

   ```
   > rep (variable name, times)
   ```

   For example, `rep (3, 4)` refers to recurring the number 3 four times.

EXAMPLE 2.15 RECURRING A VECTOR

Consider that the values 5, 4, 3, 2 represent a variable X. Generate a vector by recurring the data set two times. The result of using the function `rep ()` in R to repeat the vector X two times is shown below.

```
> x <- c (5, 4, 3, 2)
> rep (x, 2)
[1] 5 4 3 2 5 4 3 2
```

EXAMPLE 2.16 RECURRING VALUES

Recur each value of 2:6 three times. The function `rep ()` in R is employed to repeat each value three times.

```
> rep (2 : 6, c(3, 3, 3, 3, 3)) #The first option to repeat the values
  [1] 2 2 2 3 3 3 4 4 4 5 5 5 6 6 6
> rep (2 : 6, each = 3) #The second option to repeat the values
  [1] 2 2 2 3 3 3 4 4 4 5 5 5 6 6 6
```

2.7 PLOTS IN R

R offers unique graphic functions to create shapely, high-resolution diagrams readily. Plots are produced by using built-in functions; each built-in function will perform a specific job correlated with a sketching graph. We can use the function `plot ()` to produce a graph in R as given below.

```
> plot ()
```

EXAMPLE 2.17 SIMPLE GRAPH

Graph the values 28, 31, 25, 27, 33, 23. First, we should know the variable, say temperature (T), and then employ the built-in function `plot ()`.

```
T <- c (28, 31, 25, 27, 33, 23)
Plot (T)
```

The plain graph for temperature data set is given in Fig. 2.9.

Fig. 2.9 shows a very simple graph, we can improve this graph by adding color and lines to link various points and characters.

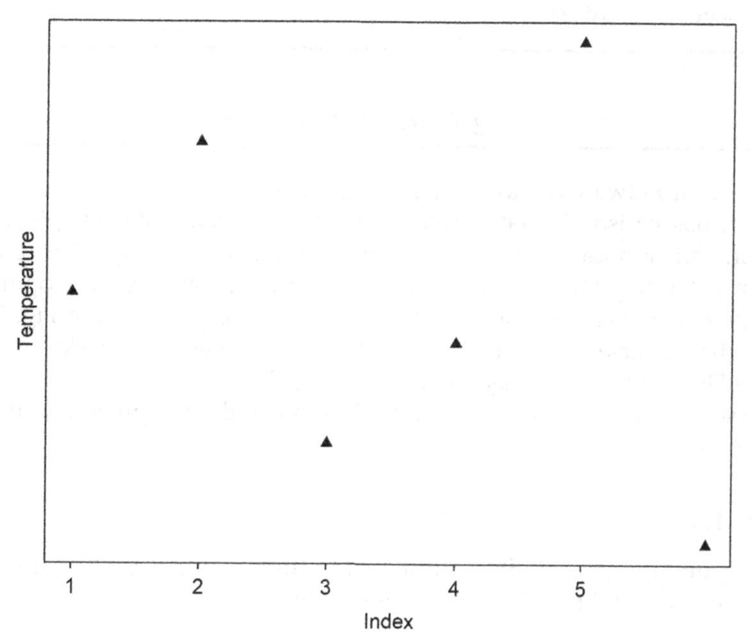

FIG. 2.9 The plot for temperature (T).

EXAMPLE 2.18 COLORED AND CONNECTED GRAPH

Connect different values and give color to the graph generated in Example 2.17. A plain plot was generated to represent temperature values (Fig. 2.9). The graph can be improved by adding two functions, the first function is `type = ""`, which is employed to link between various points, while color can be added by employing the second built-in function `col = ""`.

```
> T <- c (28, 31, 25, 27, 33, 23)
> plot (T, type = "o", col = "blue")
```

"o" is employed to represent overplotted points and lines (R offers other symbols for various characterizations). The graph, with linked points and blue color, is shown in Fig. 2.10.

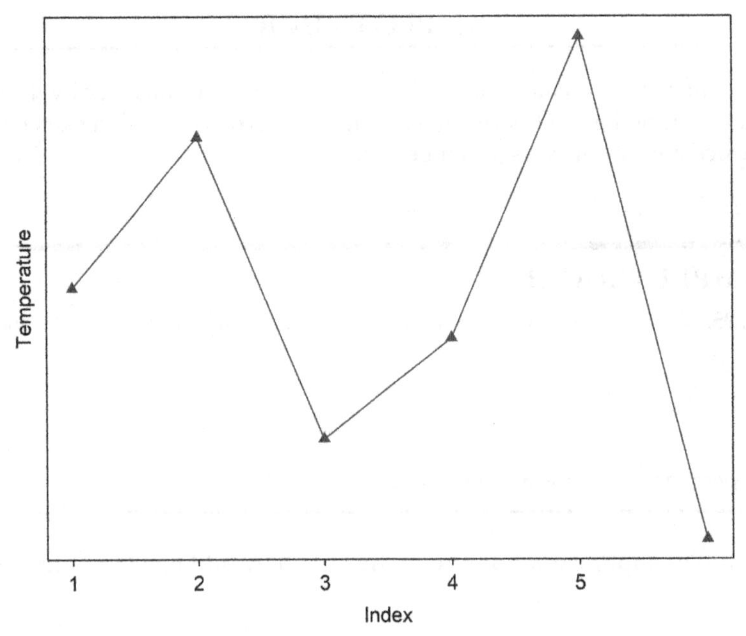

FIG. 2.10 The plot for temperature with linked points.

2.8 RSTUDIO

We have installed R statistical software version R-3.5.1 and prepared for employing the built-in functions and other R commands to perform various statistical data analyses, as well as producing shapely plots and other required computations. RStudio is a software application that offers more features and is simpler to run jobs and tasks with for operating R. RStudio is an IDE (Integrated Development Environment) software development application. RStudio is available for free at https://www.rstudio.com/products/RStudio/ as presented in Fig. 2.11.

We can download Rstudio by clicking on the "Desktop Run RStudio on your desktop" in Fig. 2.11; move to the "DOWNLOAD RSTUIDO DESKTOP" button as presented in Fig. 2.12.

After downloading the software, we can install RStudio Desktop on the computer, and then keep track of the next steps to finish the installation.

2.8.1 Navigating RStudio

RStudio is a user-friendly application that offers additional features in operating R scripts. The name and the job of the four parts of the RStudio screen (Fig. 2.13) are presented below:

1. The first part is called R Script, which is represented by the upper-left part of the RStudio screen, where scripts are placed; we can amend errors and perform the codes easily by clicking on the icon run in the upper-right corner of the top row. We can generate a new R Script either from ***File*** > ***New File*** > ***R Script*** or by clicking on the button "+" sign on the upper-left corner of the second row.

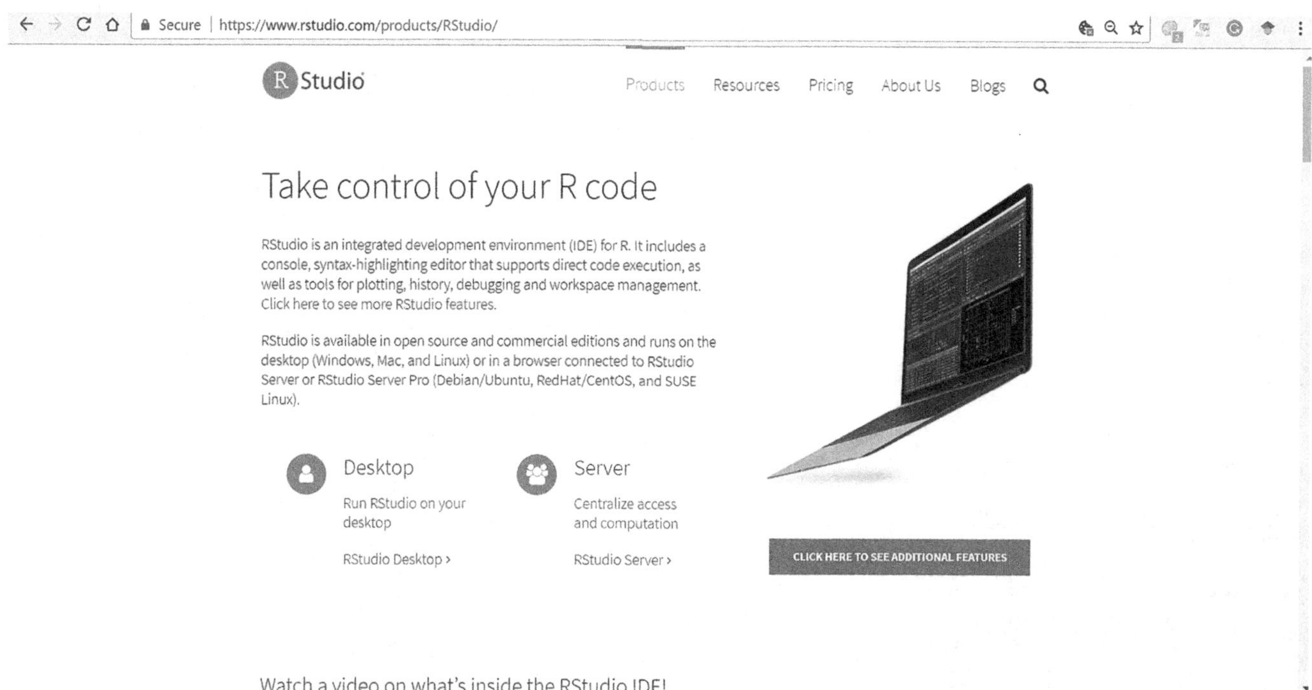

FIG. 2.11 Showing the screen for installing RStudio.

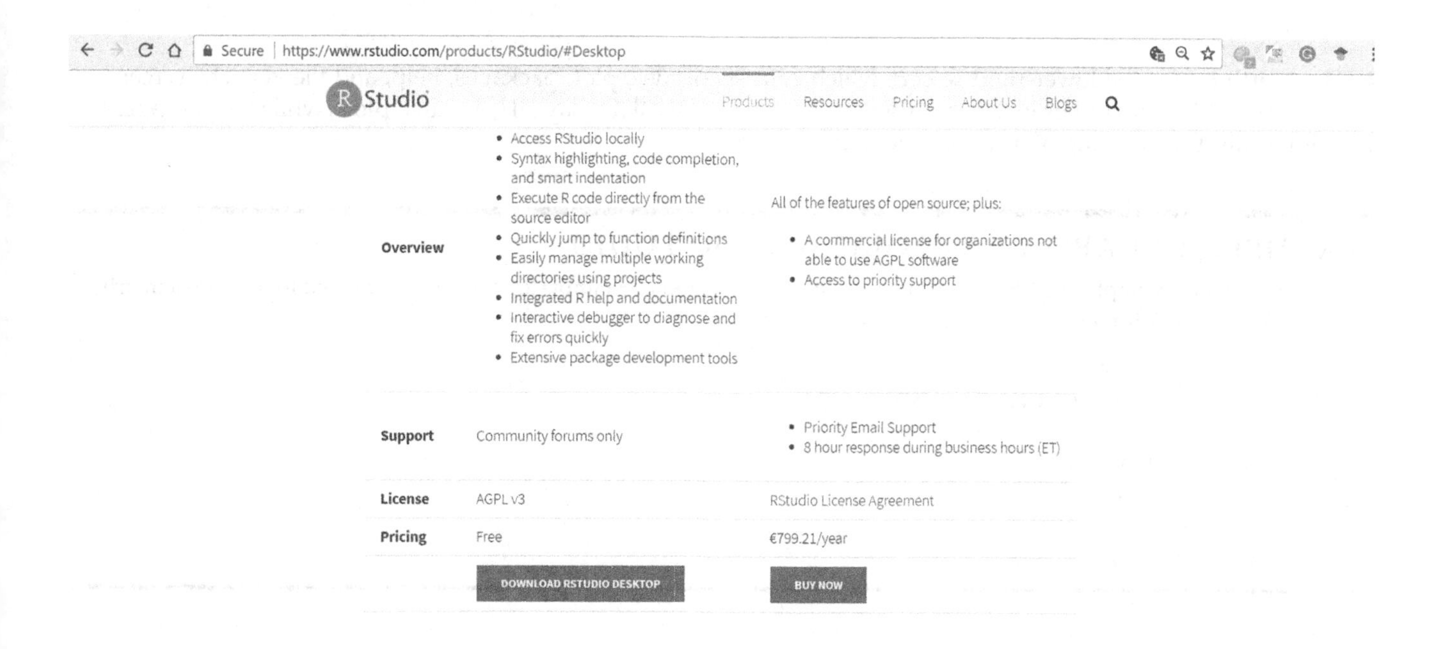

FIG. 2.12 Showing RStudio server.

2. The second part includes environment (workspace), history, and connections, which is located at the upper-right of the screen. This part is usually employed to save any results, values, and functions, or anything that is generated during the R session.
3. The third part is called the R Console, which is represented by the lower-left part of the screen where the outcomes of performing functions and commands are displayed. Moreover, functions and scripts can be written, and then click "Enter" to perform the intended job.

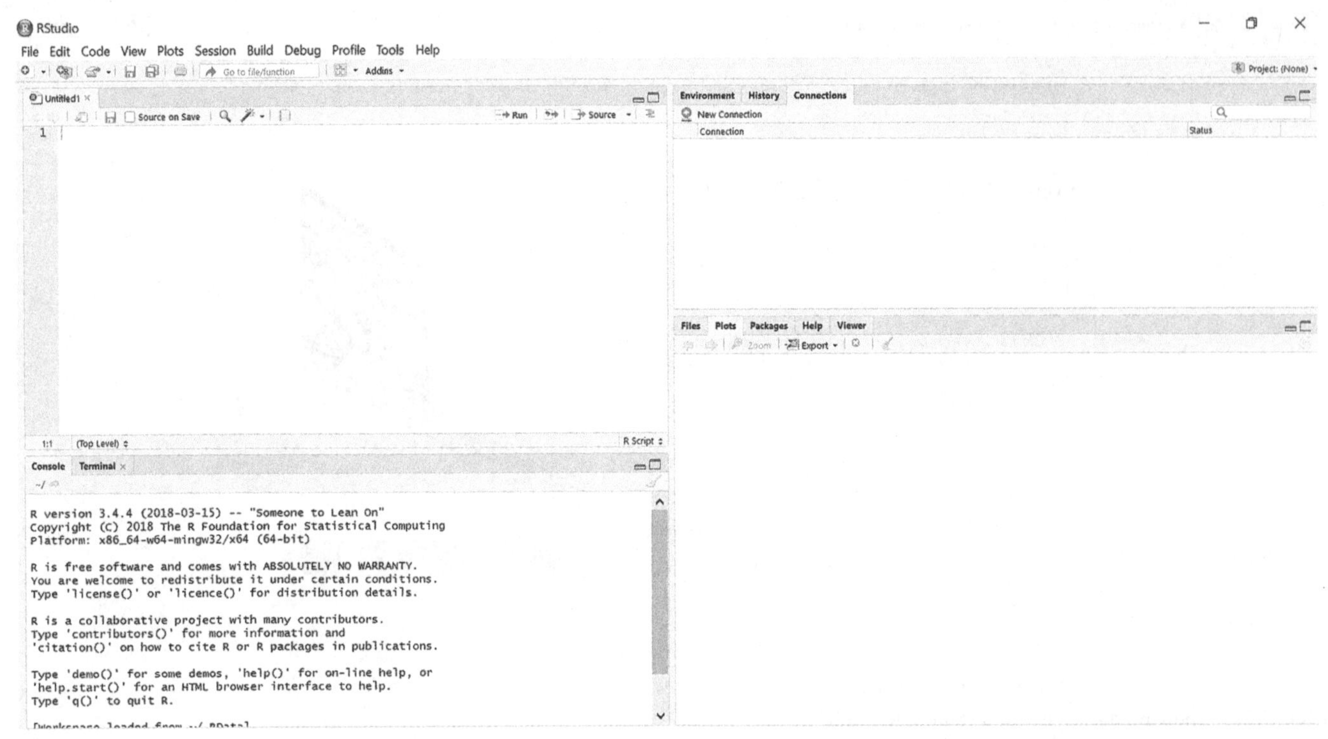

FIG. 2.13 The four sections of the RStudio screen.

4. The fourth part is the lower-right screen, which consists of files, plot, packages, help, and viewer. The screen displays all files and folders in the environment; the plots tab will display all generated plots, while the packages tab will permit downloading of any package that is requested.

EXAMPLE 2.19 APPLY R COMMANDS IN RSTUDIO

Produce a matrix employing RStudio, using the data set presented in Example 2.11. RStudio was used to produce a matrix, the produced matrix is the same as was created in Example 2.11.

```
> x<- c (5, 7, 6, 4, 10, 7, 8, 9, 7 )
> M = matrix (x, 3, 3, byrow = TRUE)
> M
     [,1] [,2] [,3]
[1,]    5    7    6
[2,]    4   10    7
[3,]    8    9    7
```

2.9 IMPORTING DATA

Data can be keyed in directly in R Console and then the order performed, or we can store the data set in a file. We can call the stored data into the R environment by informing the computer of the place of the saved data; importing a file should be achieved if we know the working directory. The working directory of the computer can be identified by employing the function `getwd ()`. The function `setwd ()` can be employed to set a new working directory where the data is stored.

EXAMPLE 2.20 WORKING DIRECTORY

Consider that there are files need to be imported from the Desktop where the files have been stored. As a first step, we need to know the current working directory employing the function `getwd ()`. The second step if to employ the function `setwd ()` to set a new working directory to the place where the data have been stored.

```
> getwd ()
[1] "C:/Users/MICET/Documents"
> setwd ("C:/Users/MICET/Desktop")
> getwd ()
[1] "C:/Users/MICET/Desktop"
```

Various data formats can be accommodated by R, including values, letters, vectors, and matrices. R permits users to store the information in several formats such as SPSS, SAS , Comma Delimited, Stata, and Excel. Importing the stored data into the R environment can be performed by employing the right call for the stored data format. A `.csv` (Comma delimited) is considered the simplest way; this option can be found in Excel. The `.csv` (comma delimited) files can be imported into the R environment by employing the function `read.csv ()` as shown below.

```
read.csv (file = " file name. csv" , header = TRUE)
```

Reading the stored data by the function `read.csv()` requires two points, the first point is the name of the stored data in `.csv`, and the second is the header (the header is the first row, which represents the names of the chosen parameters (columns, variables). We have two choices for the header: either TRUE if the data file has names for the variables or FALSE if the stored data does not have column headers. The data have been called and loadeded into the R Consol for the R software or in the R script for Rstudio. Statistical analysis of the data can be started as the next step.

Note: Rstudio offers built-in functions to import the stored data on the upper-right part of the screen (Environment and Workspace or Environment and history). The first step is to click on the Environment button, and then click on the Import Dataset. Six options will show; `From Text (base)` is the first choice, the second choice is `From Text (readr)`, the third choce is `From Excel`, and `Stata` is the last choice. The user can select the choice that suits the stored data, and import it to the R environment.

Further Reading

Alkarkhi, A.F.M., Alqaraghuli, W.A.A., 2019. Easy Statistics For Food Science With R, first ed. Academic Press.

Chi, Y. 2009. R Tutorial, An R Introduction To Statistics [Online]. Available: Http://Www.R-Tutor.Com/R-Introduction [Accessed 15 December 2018].

Computing, T.R.D.C.T.-R.F.F.S., 2008. R: A Language And Environment For Statistical Computing [Online]. Available, Http://Softlibre.Unizar.Es/Manuales/Aplicaciones/R/Fullrefman.Pdf. [(Accessed 5 January 2019)].

Cran. 2019 The Comprehensive R Archive Network [Online]. The R Foundation. Available: Http://Ftp.Heanet.Ie/Mirrors/Cran.R-Project.Org/ [Accessed 5 January 2019].

Cran. 2019. Documentation: Document Collections, Journals And Proceedings [Online]. The R Foundation. Available: Www.R-Project.Org/Other-Docs.Html [Accessed 5 January 2019].

Endmemo. 2016. R Plot Function [Online]. Endmemo. Available: Http://Www.Endmemo.Com/Program/R/Plot.Php [Accessed 5 January 2019].

Kabacoff, R. I. 2017. Quick-R [Online]. Datacamp. Available: Https://Www.Statmethods.Net/ [Accessed 5 January 2019].

Mccown, F. 2016. Producing Simple Graphs With R [Online]. Available: Https://Www.Harding.Edu/Fmccown/R/ [Accessed 5 January 2019].

Paradis, E. 2005. R For Beginners [Online]. Institut Des Sciences De L'evolution, Universit´E Montpellier Ii F-34095 Montpellier C´Edex 05 France. Available: Https://Cran.R-Project.Org/Doc/Contrib/Paradis-Rdebuts_En.Pdf [Accessed 5 January 2019].

R Development Core Team. 2019. The R Manuals [Online]. Available: Https://Cran.R-Project.Org/Manuals.Html [Accessed 5 January 2019].

Rstudio. 2018. Rstudio: Take Control Of Your R Code [Online]. Dmca Trademark Privacy Policy Eccn. Available: Https://Www.Rstudio.Com/Products/Rstudio/ [Accessed 5 January 2019].

The R Foundation For Statistical Computing, C. O. I. F. S. A. M 2018. R: Software Development Life Cycle A Description Of R's Development, Testing, Release And Maintenance Processes [Online]. Available: Https://Www.R-Project.Org/Doc/R-Sdlc.Pdf [Accessed 5 January 2019].

Venables, W. N., Smith, D. M., The R Core Team. 2018. An Introduction To R - Notes On R: A Programming Environment For Data Analysis And Graphics [Online]. The R Development Core Team. Available: Https://Cran.R-Project.Org/Doc/Manuals/R-Release/R-Intro.Pdf [Accessed 5 January 2019].

William, R., Jason A. French 2019. Using R For Psychological Research: A Simple Guide To An Elegant Language [Online]. Available: Http://Personality-Project.Org/R/ [Accessed 5 January 2019].

CHAPTER

3

Statistical Notions

LEARNING OBJECTIVES

After careful consideration of this chapter, you should be able:

- *To understand the significance of statistics.*
- *To describe the meaning of data.*
- *To explain the concept of variable.*
- *To understand the classes of variables.*
- *To distinguish between qualitative and quantitative variables.*
- *To distinguish between continuous and discrete variables.*
- *To compare the meaning of two concepts: population and sample.*
- *To explain the methods of data collection.*
- *To comprehend the sampling mechanism.*

3.1 INTRODUCTION

Statistical techniques are useful during research activities, covering designing, gathering data, testing the data, summarizing the outcomes, and eventually assisting in making smart decisions. Therefore, statistics is widely used by researchers to comprehend the behavior of the chosen variables (parameters) under investigation. Thus, professionals should study statistics to understand the common idea and related statistical terms that are needed for their research to save effort and time and reduce expenses, in addition to implementing forecasts. Moreover, scientists should understand the principles of statistics to comprehend the behavior of the variables treated by different statistical methods to get a clear picture of the research project and to obtain a deeper view of the project.

In summary, choosing a suitable statistical test during the work stages will help in providing smart decisions employing minimal time, effort and expense.

The concept of statistics and the two branches of statistics (descriptive and inferential) are presented in this chapter; moreover, the basic concepts of data, sample, population, random variable, and the techniques for gathering information (data) are given in this chapter as well.

3.2 THE CONCEPT OF STATISTICS

It is well recognized that professionals are unable to undertake worthy studies without employing statistical techniques for designing, testing, and extracting useful inferences. Thus, professionals should comprehend the concept of statistics to be able to perform high-quality studies and gain excellent outcomes.

Various definitions of statistics have been given by various writers. Thus, we can define statistics as the science of designing projects (studies), gathering data, arranging, summarizing, testing, explaining, and drawing useful inferences based on the gathered information (data).

In general, statistics can be divided into two main areas: the first area is called descriptive statistics and the second, inferential statistics.

Applied Statistics for Environmental Science with R
https://doi.org/10.1016/B978-0-12-818622-0.00003-4

Descriptive statistics represents the area of statistics that includes description of the methods of gathering, arranging, summarizing and displaying the gathered information (data).

Inferential statistics represents the area of statistics that includes description of hypothesis testing, methods of estimation, and the investigation of the connections (relationships) between chosen variables under study and prediction.

3.3 COMMON CONCEPTS

It is better to define several terms that are helpful and important to comprehend the subjects in this book.

A population refers to a set of observations (objects, individuals, and members) of interest to the scientists; the observations could be any subject under study such as measurements, humans, or other subjects. We can recognize two types of populations: limited and unlimited. A limited population is, for example, kinds of treatment in the landfill leachate (limited number), and unlimited population, for example, the number of bacteria in a landfill leachate.

A sample refers to a portion (subset) of objects (observations, item, individuals, and members) drawn (chosen) from a larger group called the population. Scientists in all areas can investigate the characteristics of populations using representative samples, because samples are easier to examine than the larger group (population). Moreover, samples are employed to save effort and time and reduce expenses. For example, investigating the heavy metal concentrations in sediment can be performed by choosing several points from a particular river (the population represents the river in the research region). The number of points in this river form a sample and have properties similar to the sediment in the river.

Data refers to a group of observations (measurements) that have been produced and gathered from an experiment; for example, survey returns, measurements, and test outcomes. We can use a random variable to describe data according to the type of data.

Random Variable refers to any characteristic of interest that can be measured and counted. The value of the variable is likely to change or vary from one test to another or may change in value over time. A variable is usually represented by an uppercase letter, such as X, Y, etc., and the value of the variable is usually represented by a lowercase letter, such as x, y, etc.

In summary, we can identify two main kinds of random variables: qualitative variables and quantitative variables.

3.3.1 Qualitative Variables

The qualitative variable is also known as a categorical or attribute variable (that are not numerical). This kind of variable assumes values that are tags, rubrics, or brands that are not helpful for computing various measures in statistics, such as the mean or variance, and others that employ numerical values. However, we can give numbers to appear numeric when we key in the data for identification aims, but the numbers are meaningless (identification objective). Of the kind of oxidation processes employed for stabilized leachate treatment, seven oxidation processes are chosen, including O3, Fenton, Fenton followed by O3, persulfate, persulfate followed by O3, simultaneous O3/Fenton, and O3/persulfate, and are considered as examples of qualitative variable. The second model that represent a qualitative variable includes the outcomes of an inspection classified either negative or positive.

3.3.2 Quantitative Variables

A quantitative variable refers to a measureable quantity (takes numerical values) and can be arranged in order. For example, measures of heavy metals concentration in sediment, such as cadmium (Cd), iron (Fe), copper (Cu), and chromium (Cr), represent models of quantitative variables.

Quantitative variables are classified into two types: discrete variable and continuous variable.

3.3.2.1 *Discrete Variable*

A variable that can take on a countable or finite number of distinct values is called a discrete variable. In other words, discrete variables are numeric variables that can assume values such as 0, 1, 2, etc., (there is a gap between the values) and can be counted; for example, the number of sampling points in a river, number of visits to a landfill leachate.

3.3.2.2 *Continuous Variable*

A numeric variable that can assume any value within any two particular values (interval) is called a continuous variable. The values of this variable are gained by measuring, for example, pressure, temperature, or any readings obtained from a machine.

3.4 DATA GATHERING

Scientists are required to perform studies to gain information that helps them to respond to a research problem to get a better picture or to find out something new. Thus, the objectives of the project and the type of data required should be understood by researchers, how the data is to be collected, and when the data is to be collected will help scientists to accomplish their goal and extract intelligent results. The data needed for the project relies on the type of variables in the study, and the origin.

3.4.1 Approaches for Gathering Data

Collecting data can be carried out using several approaches. Four common approaches can be recognized in the environmental field:

1. Experimental work method
2. Archives method
3. Survey method
4. Automated instrument method

We usually use a sample to collect the required data, and the type of data imposes the approach of choosing the individuals (items, sample data).

1. The experimental work method
   The experimental work method is widely used in the field of environmental science. We carry out experiment in the field or laboratories to collect the required information that helps us recognize the attitude of the chosen variable involved in the investigation. For example, we analyze water samples obtained from a river for their physiochemical parameters such as chemical oxygen demand (COD), biochemical oxygen demand (BOD), and total suspended solids (TSS); this test requests a sample (water) to be chosen and then examined in the laboratory for physiochemical parameters. Or, a scientist may wish to investigate the concentration of heavy metals such as lead (Pb), zinc (Zn), cadmium (Cd), copper (Cu), mercury (Hg), and chromium (Cr) in sediment gained from two rivers. Samples are gathered from each river and analyzed for the chosen heavy metals.
2. The historical records
   Historical records are records and statistics such as archived reports, statistics, studies, and indices that are saved (stored) in the archives of offices or any other bureau. In general, historical data is inexpensive and provides a description of past events.
3. The survey approach
   The survey approach is a non-experimental approach to collect data about individuals (observations, subjects) in a population. This approach demands a proper questionnaire to include the required data to be collected for the research; data are collected from a group of people or a community. The questions should be comprehensible, suitable, and linked to the topic under investigation. We can collect data by several methods, such as mailed questionnaires, telephone interviews, personal interviews, and online interviews.
4. The automated instrument method
   The last approach for collecting data is the automated instrument approach. This approach requires providing instruments (tools) to record the chosen variables and collect the desired information. By using this method, the data is produced automatically without varying the setting of the instruments.

3.5 SAMPLING METHODS

Most projects are performed employing a sample (subset) to gather facts (data) regarding the chosen variables under study. It is well recognized that collecting data employing samples gives rise to reduced time, cost, and effort, and helps the scientist to gain all the needed data regarding the behavior of the chosen variable as the population is investigated. Thus, a sampling method has been defined as the method employed by scientists to choose the individuals (observations) from the population under study to be contained in the representative sample. There are various methods to choose a sample from the population of concern that meets the study target. We must select representative and random samples to guarantee independence between different individuals (observations). Each individual must be given an equal chance to be chosen in the sample.

The sampling methods studied in this book are systematic sampling, simple random sampling, cluster sampling, and stratified sampling.

3.5.1 Simple Random Sampling

Simple random sampling is an essential method and is one of the most commonly applied sampling methods, granting each individual (item, observation) in the population the same opportunity of being chosen in the sample. Individuals in the sample are chosen by assigning a number to every individual in the population; then, numbered cards (each card indicates a defined individual) are put in a box and shuffled, and the desired number of individuals are chosen. Alternatively, we can use random-number tables or a computer to generate random numbers to choose the individuals of the sample.

EXAMPLE 3.1 CHECKING PRODUCTS

Companies should test their product before shipping to customers, and they should check whether the output is ready for market (the produced units are similar). The ultimate output in the store of an environmental company represents the population of concern. Checking the appropriateness of the output requires choosing units randomly to represent the population of the company; these units form the sample to be checked before shipping, and a decision about the entire output is made based on the data delivered by the sample. The product is sent to market only if the analysis of the sample shows it to be suitable. Otherwise, the product should not be sent to market.

3.5.2 Systematic Sampling

Systematic sampling is the second type of sampling method; we should choose the individuals (observations, units) of the sample based on a random starting point selected from N/n, (where N refers to the population size and n refers to the sample size), and then each mth unit is chosen according to a fixed periodic interval until we get the required sample size.

EXAMPLE 3.2 VARIOUS BATCHES

An environmental company needs to choose a sample of their product and check it for capability. The output in the store contains various batches, and all batches generated by the company should be represented in the sample. We should select a systematic sample; a number should be assigned to all units in the sequence, and then a sample is chosen according to the interval between various numbers. As an example, if 10,500 units are generated in various batches in the store (the output is stored based on the batches), a sample of 250 units is required to be examined. The first unit should be selected randomly according to $(10{,}500/250)=42$; then, for every 42*m*th ($m=1, 2$, etc., up to 250) units, one unit is chosen. If a random start is chosen from the first forty two units (between 1 and 42)—say, number 8 is selected to be the first unit in the sample—then add 42 to 8 to obtain the second unit (number 50) and so on until the last unit that constitutes the sample is chosen. In this case, the sample would contain the units whose number is (8, 50, 92, etc.).

3.5.3 Stratified Sampling

Stratified sampling is applied when the population under study is heterogeneous. The observations (individuals, units) of the chosen population should be separated into different layers, called strata, based on some major (main) properties. The observations of each layer must be similar and distinct from other layers. A simple random sample should be drawn from each layer (stratum) to be included in the sample.

EXAMPLE 3.3 PERFORMANCE OF A COMPANY

An environmentalist tries to get better performance of a company. She proposes a modern plan (schedule) that helps to improve performance. She needs to understand how employees feel about the proposed plan. It is well known that various classes of employees work in the company, such as technicians, engineers, administrators, and so on. The sample should cover all classes in the company. Thus, we used stratified sampling to cover all classes, which requires choosing a random sample from each class (stratum) to be included in the sample. The chosen random samples were combined into one sample to represent the required sample that enables the environmentalist to study the ideas from all classes.

3.5.4 Cluster Sampling

The forth sampling technique is called cluster sampling, which is employed when the population is partitioned into parts (sections), called clusters, for geographic reasons. Clusters are usually chosen randomly and then the entire observations (units, individuals) are employed in the chosen clusters. Cluster sampling method is different from stratified sampling because in cluster sampling the whole cluster is used, while we select a random sample in stratified sampling.

EXAMPLE 3.4 LANDFILL LEACHATE

There are many landfill leachate locations distributed around Malaysia. An environmentalist wishes to investigate the behavior of the employees toward a certain issue, and he is unable to survey all landfills. Thus, cluster sampling should be employed at choosing landfills randomly, say four centers, and meeting all the employees at the selected landfills.

Further Reading

Alkarkhi, A.F.M., Alqaraghuli, W.A.A., 2019. Easy Statistics for Food Science with R, first ed. Academic Press.
Alkarkhi, A.F.M., Low, H.C., 2012. Elementary Statistics For Technologist. Universiti Sains Malaysia Press, Pulau Pinang.
Allan, G.B., 2007. Elementary Statistics: A Step by Step Approach. Mcgraw-Hill.
Donald, H.S., Robert, K.S., 2000. Statistics: A First Course. Mcgraw-Hill.
Mario, F.T., 2004. Elementary Statistics. Pearson.

CHAPTER

4

Measures of Center and Variation

LEARNING OBJECTIVES

After careful consideration of this chapter, you should be able:

- *To understand the concept of measures of central tendency.*
- *To understand the concept of measures of variation.*
- *To describe how and where to apply measures of central tendency and measure of variation.*
- *To understand the meaning of covariance and correlation.*
- *To describe the concept of correlation and covariance between various variables.*
- *To know how to produce a scatter plot for two variables and a scatter plot matrix.*
- *To explain the idea of distance.*
- *To know how to compute and use distances in multivariate cases.*
- *To employ R's commands and built-in functions to compute the measures of central tendency and variation.*
- *To describe how to use R functions beyond the built-in functions to gain certain details or to impose a certain idea on a graph.*
- *To comprehend and translate the results of R regarding measures of central tendency and variation.*
- *To extract smart inferences regarding measures of central tendency and variation.*

4.1 INTRODUCTION

Scientists generally develop a common opinion regarding the gathered data by computing several measurements to determine the middle and recognize the shape of the data by considering measures of central tendency and variation. A speedy opinion regarding the center of a data set can be accomplished by employing measures of the central tendency that represent the center of the data by a single number, such as the average, the midrange, the median, or the mode, which are applied to represent data by a single value.

However, measures of central tendency are not sufficient to characterize a data set because information on the homogeneity of the values is not given. The dispersion of the values in a data set gives an opinion regarding the homogeneity and can be measured by employing measures of variation (dispersion).

The center of the data can be located by the measures of central tendency such as the mean, and the spread of the data around the mean can be measured by the measures of variation such as the variance. The two measures give a clear picture of the center and the shape of the data set. The correlation between various variables is of concern to the professionals; it assists them in comprehending the manner of every variable in the presence of other variables. This chapter covers measures of central tendency and variation for one variable (univariate) and several variables (multivariate), along with a comparison between univariate and multivariate distributions. Also, the relationship between various variables is explored through correlation analysis, representing the bivariate data pictorially employing a scatter diagram as well as the distance between two points.

Applied Statistics for Environmental Science with R
https://doi.org/10.1016/B978-0-12-818622-0.00004-6

4.2 MEASURES OF CENTER AND DISPERSION IN R

Various built-in functions are offered by R statistical software to perform descriptive statistics covering measures of central tendency and variation. R supplies many functions for calculating measures of location and dispersion such as the mode, median, mean, variance, standard deviation, and others in various packages such as the Hmisc package, pastecs package, and psych package. Several statistical procedures are part of a base installation of R and labeled *recommended*, while others can be installed by the user.

We can carry out descriptive statistics in R employing various commands to order built-in functions to perform a particular job.

1. The mean value for a set of observations can be calculated by employing the built-in function `mean ()` in R.

```
mean (data frame)
```

2. The sum of all of the observations (values) can be found by employing the built-in function `sum ()`.

```
sum (data frame)
```

3. The number of observations (values) in a vector can be calculated by using the built-in function `length ()`.

```
length (data frame)
```

4. Performing the same operation (function) on a number of rows or columns of a data set can be carried out by using the built-in function `sapply ()`.

```
sapply (data frame, required measure)
```

5. The variance of a data set can be computed by using the built-in function `var ()`.

```
var (data frame)
```

6. The standard deviation of a data set can be calculated by using the built-in function `sd ()`.

```
SD (data frame)
```

7. The covariance between two variables can be calculated by using the built-in function `cov ()`.

```
cov (X1, X2)
```

   where `X1` and `X2` represent the first and second variables of interest respectively.

8. The correlation between two variables can be calculated by employing the built-in function `cor ()`. This correlation is called Pearson's correlation.

```
cor (X1, X2)
```

   where `X1` and `X2` represent the first and second variables of interest, respectively.

9. The coefficient of the correlation and also the significance level of the correlations can be computed by employing the built-in function `corr.test ()`.

```
corr.test (data frame)
```

10. The Euclidean distance can be computed by using the function `dist ()`.

```
dist (data frame)
```

11. **The range, median, minimum, maximum, and quantile values for a variable can be calculated by using** the built-in functions `range ()`, `median ()`, `min ()`, `max ()`, and **`quantile ()`, respectively.**

```
Range ()
Median ()
Min ()
Max ()
Quantile ()
```

12. A scatter plot matrix for k variables can be created by using the built-in function `pairs ()`.

```
pairs (~ X1 + X2 +... + Xk, data = data frame)
```

where `X1, X2,..., Xk` are the k variables.

13. A summary of descriptive statistics such as the quartiles, median, mean, and others is offered by the packages pastecs, psych, and Hmisc. We can install the packages from the library, as presented below.

```
install.packages ("pastecs")
library (pastecs)
install.packages ("psych")
library (psych)
install.packages ("Hmisc")
library (Hmisc)
```

4.3 MEASURES OF CENTER

Measures of center are employed to recognize the middle of a set of observations (values) and represent the whole data by one value. The idea of the average, also called the mean, covering the formulas and computations for one variable and several variables are delivered in the next sub-sections with several samples employing R built-in functions and commands to clarify the step-by-step analysis and explanation of the R output.

4.3.1 The Arithmetic Mean for a Single Variable

The value that is computed by dividing the sum of all data values (observations) by the total number of observations (data values) is called the average or the mean. The average can be computed employing the formula in (4.1).

$$\overline{X} = \frac{\sum_{i=1}^{n} X_i}{n} = \frac{X_i + X_i + \ldots + X_n}{n} \tag{4.1}$$

where

$\overline{X}$ is the mean;
$\sum$ (Uppercase Greek Sigma) is the sum of all observations (data values);
X is the variable employed to represent the data values; and
n is the total number of observations (data values, subjects) in the sample.

EXAMPLE 4.1 MEAN OF THE PH OF SURFACE WATER

An environmentalist wishes to investigate the value of the pH of surface water in the Juru River in the Penang state of Malaysia as a physicochemical parameter. The data for the pH for ten different samples obtained from Juru River are presented in Table 4.1.

TABLE 4.1 The Data for pH of Surface Water Obtained from Ten Sampling Points

7.88	7.92	7.41	7.84	7.86
7.89	7.4	7.41	7.25	7.18

R statistical software can easily compute the average value of the pH of surface water in the Juru River employing the built-in function `mean ()`.

```
> mean (data frame)
```

Three options are suggested to calculate the average value of the pH of surface water employing R built-in functions, as given below:

- Option 1:
  We should call the stored data (Example4_1) for the pH value of surface water into an R frame either by employing the command read.csv () to read and load the stored data for pH values of surface water in the Juru River or by importing the data employing a built-in function from the upper-right hand, by clicking "Environment"—six options will appear, choose the option "Import Dataset." Here, the data have been called and loaded into the R environment, and all of the requested computations can be performed.

```
mean (Example4_1 $ pH)
```

The function mean () is employed to compute the average value of the pH. One can observe in the R command for the mean the character $ which can be employed to call a variable (pH) from a file (data frame). The output of employing built-in functions in R to compute the average value of the pH of surface water is presented below.

```
> mean (Example4_1 $ pH)
[1] 7.604
```

- Option 2:
  The second approach to compute the average value is to represent the data values in a vector form by employing built-in functions and commands as given below.

```
pH <- c (7.88, 7.92, 7.41, 7.84, 7.86, 7.89, 7.4, 7.41, 7.25, 7.18)
mean (pH)
```

The output of performing the R built-in functions and commands to compute the average value is shown below.

```
> pH <- c (7.88, 7.92, 7.41, 7.84, 7.86, 7.89, 7.4, 7.41, 7.25, 7.18)
> mean (pH)
[1] 7.604
```

One can observe that the first row shows the operator <-, which is known as the assignment operator, pH is the variable name, and the function c () is employed to represent the data values in a vector form.

- Option 3:
  If the sum and number of observations are required, R software offers facilities for researchers to carry out (implement) their opinions. The R commands and built-in functions employed to compute the number of observations, the sum, and the mean value are presented below.

```
pH <- c (7.88,7.92,7.41,7.84,7.86,7.89,7.4,7.41,7.25,7.18)
s = sum (pH)
s                        #print the sum (s)
n = length (pH)
n                        #print number of values (n)
Average = s/n
Average                  #print the Average
```

The variable pH was represented in a vector form as appeared in the first row. Then, a call is made to calculate the sum of all of the pH data values. Two functions were employed; the function length () and the function sum () to compute the number of pH values and the sum of all of the pH values. The character # can be employed to insert remarks into a program without performing them or to stop implementing a command. The command Average=S/n is a call to compute the average value of the pH, and the last command is the Average to print the result. The outputs for the number of values, the sum, and the average value of the pH of surface water in the Juru River are shown below.

```
> pH <- c (7.88,7.92,7.41,7.84,7.86,7.89,7.4,7.41,7.25,7.18)
> s = sum (pH)
> s                      #print the sum (s)
[1] 76.04
> n = length (pH)
> n                      #print number of values (n)
[1] 10
```

```
> Average = s/n
> Average                    #print the mean
[1] 7.604
```

In summary, the average value of the pH (7.604) provides a general view regarding the middle of the pH values, which shows that the value of the pH of surface water in the Juru River on average is 7.604.

EXAMPLE 4.2 MEAN OF THE SOLID WASTE IN A PALM OIL MILL

An investigator wishes to examine the average amount of solid waste generated in a palm oil mill. The amounts of solid waste in tons generated over 12 months are given in Table 4.2.

TABLE 4.2 The Amounts of Solid Waste in Tons Generated in a Palm Oil Mill Over 12 Months

2550	2540	2509	2662	2642	2540
2553	2540	2504	2531	2491	2537

Three options can be studied to compute the mean value of the solid waste generated in a palm oil mill employing R commands and built-in functions as in Example 4.1.

- Option 1:
  The function `mean ()` can be employed to compute the average amount of solid waste generated over 12 months in a palm oil mill. The data had been stored as. CSV `(Example4_2)` format; we should import the stored data into an R environment.

The result of employing the function `mean ()` in R to compute the average value of the solid waste is shown below

```
mean (Example4_2 $ Solid)
[1] 2549.917
```

- Option 2:
  The second approach for computing the average value of the solid waste is to represent the data set in a vector form, as shown below.

```
> Solid <- c (2550, 2540, 2509, 2662, 2642, 2540, 2553, 2540, 2504, 2531, 2491, 2537)
> mean (Solid)
[1] 2549.917
```

- Option 3:
  The third approach can be used if the sum and number of data values are needed. The R commands for performing the necessary calculations are shown below.

```
> Solid <- c (2550, 2540, 2509, 2662, 2642, 2540, 2553, 2540, 2504, 2531, 2491, 2537)
> s = sum (Solid)
> s                     #print s
[1] 30599
> n = length (Solid)
> n                     #print n
[1] 12
> Average = s / n
> Average                     #print Average
[1] 2549.917
```

In summary, the average value of the amount of solid waste generated over 12 months in a palm oil mill is 2549.917. The mean value supplies input on the middle of the data set but does not provide information on the spread of the values to help us build a common view regarding the consistency of the data set.

4.3.2 The Mean Vector (Multivariate)

The idea of the mean vector in the case of several variables is the same as that of the single variable concerning computations and explanation. In the case of several variables, a vector will be employed to represent the mean value of each variable under study. Consider there are k variables $X_1, X_2, \ldots, X_k$ and n observations, the mean vector for the k variables is shown in (4.2).

$$\overline{X} = \frac{\sum X_i}{n} = \begin{bmatrix} \overline{X}_1 \\ \overline{X}_2 \\ \cdot \\ \cdot \\ \cdot \\ \overline{X}_k \end{bmatrix} = \frac{1}{n} X' j \tag{4.2}$$

where j is a vector of 1's, $j = \begin{bmatrix} 1 \\ 1 \\ \cdot \\ \cdot \\ \cdot \\ 1 \end{bmatrix}$, and $X_i = \begin{bmatrix} X_{i1} \\ X_{i2} \\ \cdot \\ \cdot \\ \cdot \\ X_{ik} \end{bmatrix}$.

$\overline{X}$ is a $k \times 1$ vector of means; and
n is the total number of data values linked with each variable.

EXAMPLE 4.3 CONCENTRATION OF HEAVY METALS IN SEDIMENT (mg/L)

An environmentalist wishes to measure three heavy metal concentrations (copper (Cu), zinc (Zn), cadmium (Cd)) in sediment (mg/L wet weight) at ten sampling locations in the Juru River in the Penang state of Malaysia. The data for the copper (Cu), zinc (Zn), and cadmium (Cd) are presented in Table 4.3.

TABLE 4.3 The Concentration of Selected Heavy Metals in Sediments of the Juru River (mg/L)

Location	Cu	Zn	Cd
1	0.63	1.89	1.95
2	0.73	1.79	1.99
3	0.35	0.90	1.94
4	0.76	2.19	1.98
5	0.60	1.74	1.94
6	0.36	3.38	1.95
7	0.63	3.12	1.98
8	0.52	3.42	1.93
9	0.55	2.86	1.97
10	0.47	3.71	1.92

Because the concentration of three heavy metals (copper (Cu), cadmium (Cd), and zinc (Zn)) were measured in sediment, this case is considered as a multivariate study. Thus, the average value for each measured heavy metal (variable) is computed and placed in a vector. The formula in (4.1) was employed to calculate the average value for the selected heavy metals in sediment. The average value for Cu is given below.

$$\overline{X} = \frac{\sum_{i=1}^{n} X_i}{n} = \frac{0.63 + 0.73 + \ldots + 0.47}{10} = 0.56$$

Similarly, the average values for Zn and Cd are 2.50 and 1.955 = 1.96, respectively. The computed averages for all of the measured responses are positioned in a vector form similar to (4.2).

$$\overline{X} = \begin{bmatrix} \overline{X}_1 = 0.56 \\ \overline{X}_2 = 2.50 \\ \overline{X}_3 = 1.96 \end{bmatrix}$$

We can easily compute the mean vector for the concentration of chosen heavy metals in sediment employing the built-in function `sapply ()` in R.

One should call the stored data (`.CSV` (`Example4_3`)) for the measured responses Cu, Cd, and Zn into an R environment, then the average values for the concentration of chosen heavy metals in sediment are computed employing the built-in function `sapply ()` in R.

```
> Average <- sapply (Example4_3, mean)
> Average
   Cu    Zn    Cd
0.560 2.500 1.955
```

The average values for Cu, Zn and Cd are presented in the last row of the output of the function `sapply ()`. One can explain the results of the chosen variables as follows: the values of Cu oscillated close to the average value of Cu (`0.560` = 0.56) (i.e., these values were higher or lower than 0.56, but nearby it), the values of Zn oscillated close to the average value of Zn (`2.500` = 2.50), and the values of Cd oscillated close to the average value of Cd (`1.955` = 1.96). In conclusion, the average values of the measured heavy metals in sediment are `Cu=0.56`, `Zn=2.50`, and `Cd=1.956`.

EXAMPLE 4.4 PHYSIOCHEMICAL PARAMETERS OF SURFACE WATER (mg/L)

The concentration of five physiochemical parameters (pH, dissolved oxygen (DO), biological oxygen demand (BOD), chemical oxygen demand (COD), and total suspended solids (TSS) were obtained from ten sampling points in the Juru River in the Penang state of Malaysia. The data for the selected parameters obtained from Table 1.2 are reproduced in Table 4.4.

TABLE 4.4 Physiochemical Parameters of Surface Water of the Juru River (mg/L)

Sampling points	pH	DO	BOD	COD	TSS
1	7.88	6.73	10.56	1248	473.33
2	7.92	6.64	10.06	992.5	461.67
3	7.41	5.93	6.01	1265	393.34
4	7.84	6.23	14.57	1124	473.34
5	7.86	6.29	7.36	1029.5	528.34
6	7.89	6.36	5.27	775	458.33
7	7.4	5.67	4.81	551	356.67
8	7.41	5.26	4.96	606	603.33
9	7.25	4.18	6.61	730	430
10	7.18	5.14	5.11	417	445

The built-in function `sapply ()` is employed to compute the average values for the stored data. CSV `Example(4_4)` of the measured physiochemical parameters of surface water for the Juru River.

```
> Average <- sapply (Example4_4, mean)
> Average
   pH     DO    BOD      COD      TSS
7.604 5.843 7.532 873.800 462.335
```

The results of the built-in function `sapply ()` is the average values for the measured physicochemical parameters of surface water. The average values for pH, DO, BOD, COD, and TSS computed by the `sapply ()` function are as follows: `pH=7.604`, `DO=5.843`, `BOD=7.532`, `COD=873.800` and `TSS=462.335`.

In summary, the values for the concentration of measured physicochemical parameters of surface water measured from various samples oscillated close to the computed mean values of the measured variables.

4.4 MEASURE OF VARIATION

Measures of variation are of great value to any group of data values because the shape of the data is related to the values of these measures. The standard deviation is widely employed to characterize the spread of the values. In this section, the standard deviation and variance for a single variable (univariate) and several variables (multivariate) distributions will be discussed, including the definitions, computations, and formulas.

4.4.1 Variance and Standard Deviation for a Single Variable

The value that is computed by dividing the squared deviations of the data values from the average by the total number of observation (total number of observation minus 1 for the sample) is called the variance. The formula for computing the sample variance is presented in (4.3).

$$S^2 = \frac{\sum_{i=1}^{n} (X_i - \overline{X})^2}{n-1} \tag{4.3}$$

or

$$S^2 = \frac{n\left(\sum_{i=1}^{n} X_i^2\right) - \left(\sum_{i=1}^{n} X_i\right)^2}{n(n-1)}$$

where S^2 represents the sample variance.

The standard deviation (S) for a sample is:

$$S = \sqrt{S^2} = \sqrt{\frac{\sum_{i=1}^{n} (X_i - \overline{X})^2}{n-1}}$$

or

$$S = \sqrt{\frac{n\left(\sum_{i=1}^{n} X_i^2\right) - \left(\sum_{i=1}^{n} X_i\right)^2}{n(n-1)}}$$

EXAMPLE 4.5 VARIANCE OF THE PH OF SURFACE WATER

Compute the standard deviation and variance for the data presented in Example 4.1 concerning the pH values measured from the Juru River as a physicochemical parameter.

The mean value of the pH was computed in Example 4.1 to be 7.604. The standard deviation and variance for the pH values of surface water are computed employing the formula presented in (4.3), and they are 0.298 and 0.089, respectively.

The two built-in functions `var. ()` and `sd ()` in R can be employed to compute the variance and standard deviation for a set of values, respectively, as given below.

```
var (data frame)
```

The built-in function for the standard deviation is:

```
sd (data frame)
```

The outcomes of employing the two built-in functions in R to compute the variance and standard deviation for the stored data values of the pH of surface water. CSV (Example4_5) are presented below.

```
> var (Example4_5 $ pH)
[1] 0.08900444
> sd (Example4_5 $ pH)
[1] 0.2983361
```

One can observe that the commands `var. (Example4_5 $ pH)` and `sd (Example4_5 $ pH)` are two calls to compute the variance and standard deviation for the stored data of the pH values. The output of employing the two functions are `0.08900444`=0.089 and `0.2983361` =0.298, as appeared in the second and forth rows; the two values are for the variance and standard deviation for pH values, respectively. The consistency of various values of pH can be measured by employing the value of the standard deviation, which measures the oscillation of the values around the average value, which shows that every value of the pH is more or less than the average value by 0.298. The oscillation is considered low between various records of pH of surface water, which indicates that the pH values of surface water obtained from various sampling points are consistent.

EXAMPLE 4.6 VARIANCE OF THE SOLID WASTE IN A PALM OIL MILL

Compute the standard deviation and variance for the data presented in Example 4.2 for the amount of the solid waste in tons generated in a palm oil mill over 12 months.

The values of the variance and standard deviation for the stored data of the amount of the solid waste in a palm oil mill. CSV (Example4_6) were computed employing R built-in functions. The results of employing built-in functions for the variance and standard are presented below.

```
> var (Example4_6 $ Solid)
[1] 2647.72
> sd (Example4_6 $ Solid)
[1] 51.456
```

The value of standard deviation is `51.456` as appeared in the results of employing the built-in function `sd ()`, which represents the oscillation of the values around the average value of the solid waste. The oscillation of the solid waste is not high, which shows that the variation in the solid waste between various months is not too high. The oscillation in the solid waste values could be due to the amount of fresh fruit bunches (FFB) processed daily. Furthermore, the amounts of solid waste generated in a mill depend on the capacity of the mill.

4.5 THE CONCEPT OF COVARIANCE

The concept of variance for a single random variable can be extended to include two random variables and measure the joint variability between the two random variables. The linear association between any two variables can be measured by the sample covariance. Consider a bivariate case when two variables measured for each research unit, taking X and Y as the two random variables; a measure of the joint variability between the two variables X and Y can be achieved by employing (4.4) to compute the sample covariance. The formula for computing the covariance between X and Y is presented in (4.4).

$$S_{XY} = \frac{1}{n-1}\sum_{i=1}^{n}(X_i - \overline{X})(Y_i - \overline{Y}) \tag{4.4}$$

where S_{XY} represents the sample covariance.

- S_{XY} will be positive if the greater values of one variable would correspond to the greater values of the other variable, or vice versa.
- S_{XY} will be negative if the lesser values of one variable would correspond with the greater values of the other variable.
- S_{XY} will be approximately zero if there is no relationship between the values of the two variables.

4.5.1 Covariance Matrices (Multivariate)

Consider there are k chosen variables included under investigation, the covariance between every two distinct variables of the k chosen variables can be placed in a matrix form called a variance and covariance matrix, as presented in (4.5), which represents the variances and covariances of the k chosen variables.

$$S = (S_{ij}) = \begin{bmatrix} S_{11} & S_{12} & \cdots & S_{1k} \\ S_{21} & S_{22} & \cdots & S_{2k} \\ \vdots & \vdots & & \vdots \\ S_{k1} & S_{k2} & \cdots & S_{kk} \end{bmatrix} \tag{4.5}$$

where $S = (S_{ij})$ represents the sample covariance matrix.

The variances of various variables are represented by the main diagonal, and the covariances are represented by the off-diagonal. The matrix $S = (S_{ij})$ is a symmetric matrix.

S_{ii} refers to the variance S_i^2; S_{22} refers to the variance of the second variable, while S_{ij} refers to the covariance. For instance, S_{2k} refers to the covariance between variable 2 and variable k.

4.6 CORRELATION ANALYSIS

The correlation can be defined as a statistical measure employed to test the relationship between two variables or can be defined as how the two variables are correlated. Researchers usually measure the strength and the trend of the linear relationship between any two variables through correlation analysis.

The linear relationship between any two quantitative variables can be measured by employing Pearson's product–moment correlation coefficient; the value does not depend on the scale of measurement. The formula for computing the correlation coefficient between any two variables is presented in (4.6).

$$r_{XY} = \frac{S_{XY}}{S_X S_Y} = \frac{\sum_{i=1}^{n}(X_i - \overline{X})(Y_i - \overline{Y})}{\sqrt{\sum_{i=1}^{n}(X_i - \overline{X})^2 \sum_{i=1}^{n}(Y_i - \overline{Y})^2}} \tag{4.6}$$

The covariance between the two variables is represented by the numerator of (4.6); while the product of the variance of the two variables is represented by the denominator.

Note:

- The value of the coefficient of the correlation (r) varies between -1 and 1, $-1 \leq r \leq 1$.
- The coefficient of the correlation (r) measures the strength of the linear correlation between two variables.
- If $r = 0$, this indicates that no relationship exists between the two variables.
- The direction of the relationship is indicated by the sign of r.

4.6.1 Correlation Matrices

Consider there are k chosen variables included under investigation, the correlation between any two distinct variables can be represented in a matrix form called a correlation matrix. The correlation matrix for k variables is shown in (4.7).

$$R = (r_{ij}) = \begin{bmatrix} 1 & r_{12} & . & . & . & r_{1k} \\ r_{21} & 1 & . & . & . & r_{2k} \\ . & . & & & & . \\ . & . & & & & . \\ . & . & & & & . \\ r_{k1} & r_{k2} & . & . & . & 1 \end{bmatrix} \tag{4.7}$$

One can observe that the correlation matrix is similar to the covariance matrix with correlations in place of covariances.

EXAMPLE 4.7 HEAVY METALS CONCENTRATION IN SEDIMENT (mg/L)

The concentration of eight heavy metals in sediment was obtained from ten sampling points in the Jejawi River in the Penang state of Malaysia. The measured heavy metals were copper (Cu), zinc (Zn), cadmium (Cd), chromium (Cr), iron (Fe), lead (Pb), mercury (Hg), and manganese (Mn). The data for the selected parameters obtained from Table 1.5 are reproduced in Table 4.5.

TABLE 4.5 The Concentration of Heavy Metals in Sediments for the Jejawi River

	Cu	Zn	Cd	Cr	Fe	Pb	Hg	Mn
Jejawi	0.38	3.27	1.60	0.13	38.72	0.16	0.17	1.63
Jejawi	0.43	2.47	1.69	0.12	38.66	0.18	0.23	1.68
Jejawi	0.39	1.59	1.62	0.17	38.94	0.15	0.15	1.75
Jejawi	0.30	2.12	1.64	0.20	37.74	0.18	0.15	1.75
Jejawi	0.49	2.28	1.66	0.28	39.52	0.28	0.56	1.90
Jejawi	0.45	2.24	1.68	0.11	38.49	0.28	0.24	1.67
Jejawi	0.48	1.76	1.76	0.16	38.22	0.15	0.17	1.71
Jejawi	0.46	2.51	1.75	0.23	39.01	0.42	0.22	1.77
Jejawi	0.23	1.80	1.71	0.21	39.61	0.41	0.24	1.85
Jejawi	0.20	1.45	1.73	0.17	38.80	0.15	0.18	1.70

The formulas given in (4.4) and (4.6) can easily be employed to compute the variance–covariance matrix and the correlation matrix for the concentration of chosen heavy metals in sediment.

The variance–covariance matrix between various heavy metals were computed as given in (4.4) employing the built-in functions in R. The results of employing the built-in function `cov ()` to compute the variance–covariance matrix for the stored data (`.CSV` (`Example4_7`)) of the heavy metals are presented below with a step-by-step demonstration.

```
cov (first variable, second variable)

cov (Example4_7)

round (data frame, digits = Number of digits)

> round (cov (Example4_7), digits=3)
      Cu     Zn     Cd     Cr     Fe    Pb    Hg     Mn
Cu  0.011  0.022  0.000  0.000 -0.003 0.000 0.005  0.000
Zn  0.022  0.288 -0.012 -0.005 -0.011 0.006 0.008 -0.014
Cd  0.000 -0.012  0.003  0.000  0.002 0.002 0.000  0.001
Cr  0.000 -0.005  0.000  0.003  0.014 0.003 0.004  0.004
Fe -0.003 -0.011  0.002  0.014  0.312 0.034 0.039  0.029
Pb  0.000  0.006  0.002  0.003  0.034 0.011 0.004  0.005
Hg  0.005  0.008  0.000  0.004  0.039 0.004 0.015  0.007
Mn  0.000 -0.014  0.001  0.004  0.029 0.005 0.007  0.007
```

The results of employing the function `cov ()` to compute the covariances for the heavy metals in sediment are rounded up to three digits employing the built-in function `round ()` in R.

One can observe that the main diagonal of the results of using R built-in functions refer to the variances of selected variables (`0.011, 0.288, 0, 003, 0.003, 0.312, 0.011, 0.015` and `0.007`), and the off-diagonal values refer to the covariances between the various variables; for instance, the covariance between mercury (Hg) and iron (Fe) is **0.039**.

The formula in (4.6) is employed to compute the correlation coefficient between various variables. The built-in function `cor ()` in R is usually employed to compute the correlation matrix between various heavy metals concentration in sediment and to arrange the outputs in a matrix form. The outputs of the correlations between all possible combinations of two variables for the heavy metals concentration in sediment are shown below.

```
cor (Example4_7)
```

The output of using the built-in function `cor ()` are rounded up to three digits as follows:

```
> round (cor (Example4_7), digits=3)
       Cu     Zn     Cd     Cr     Fe    Pb     Hg     Mn
Cu  1.000  0.402  0.019  0.014 -0.049 0.043  0.382 -0.012
Zn  0.402  1.000 -0.402 -0.161 -0.036 0.108  0.129 -0.304
Cd  0.019 -0.402  1.000  0.138  0.067 0.363 -0.026  0.122
Cr  0.014 -0.161  0.138  1.000  0.484 0.495  0.623  0.907
Fe -0.049 -0.036  0.067  0.484  1.000 0.576  0.582  0.618
Pb  0.043  0.108  0.363  0.495  0.576 1.000  0.347  0.565
Hg  0.382  0.129 -0.026  0.623  0.582 0.347  1.000  0.683
Mn -0.012 -0.304  0.122  0.907  0.618 0.565  0.683  1.000
```

The main diagonal of the correlation matrix is **1**, as appeared in the correlation matrix generated by R built-in function `cor ()` for all of the selected variables, and the off-diagonal values refer to the correlations between any two selected heavy metals. For instance, the correlation coefficient between mercury (Hg) and iron (Fe) is **0.582**.

Note:

- The same results can be obtained by employing another call to calculate the correlation matrix between various variables.

```
cor (data frame, use = "complete.obs")
```

- The correlations between various heavy metals concentration have been calculated without providing input on whether the correlation between various heavy metals is significantly different from zero or not. R offers another built-in function to compute the correlation with the *p*-value of every correlation between various variables. The built-in function is `corr.test ()` to calculate the correlation and the p-value as well.

Two packages, **psych** and **GPArotation**, should be installed and loaded before employing the built-in function `corr.test ()` for checking the significance of the correlation coefficients. This function will provide the correlation matrix and the significance levels (p-value) of the correlations for all of the possible combinations of two parameters (variables).

```
install.packages("psych")
library(psych)
install.packages("GPArotation")
library(GPArotation)
corr.test(Example4_7)
```

The built-in function `corr.test ()` is employed to examine the significance of the correlation for the chosen heavy metals concentration in sediment.

```
> corr.test (Example4_7)
Call:corr.test(x = Example4_7)
Correlation matrix
      Cu    Zn    Cd    Cr    Fe   Pb    Hg    Mn
Cu  1.00  0.40  0.02  0.01 -0.05 0.04  0.38 -0.01
Zn  0.40  1.00 -0.40 -0.16 -0.04 0.11  0.13 -0.30
Cd  0.02 -0.40  1.00  0.14  0.07 0.36 -0.03  0.12
Cr  0.01 -0.16  0.14  1.00  0.48 0.50  0.62  0.91
```

```
Fe -0.05 -0.04  0.07  0.48  1.00 0.58  0.58  0.62
Pb  0.04  0.11  0.36  0.50  0.58 1.00  0.35  0.56
Hg  0.38  0.13 -0.03  0.62  0.58 0.35  1.00  0.68
Mn -0.01 -0.30  0.12  0.91  0.62 0.56  0.68  1.00
Sample Size
[1] 10
Probability values (Entries above the diagonal are adjusted for multiple tests.)
     Cu   Zn   Cd   Cr   Fe   Pb   Hg   Mn
Cu 0.00 1.00 1.00 1.00 1.00 1.00 1.00 1.00
Zn 0.25 0.00 1.00 1.00 1.00 1.00 1.00 1.00
Cd 0.96 0.25 0.00 1.00 1.00 1.00 1.00 1.00
Cr 0.97 0.66 0.70 0.00 1.00 1.00 1.00 0.01
Fe 0.89 0.92 0.85 0.16 0.00 1.00 1.00 1.00
Pb 0.91 0.77 0.30 0.15 0.08 0.00 1.00 1.00
Hg 0.28 0.72 0.94 0.05 0.08 0.33 0.00 0.79
Mn 0.97 0.39 0.74 0.00 0.06 0.09 0.03 0.00

 To see confidence intervals of the correlations, print with the short=FALSE option
```

The output of employing the built-in function `corr.test ()` is the correlation matrix between various heavy metals concentration in sediment, and a table of *p*-values that refers to the significance level (alpha) is presented at the bottom of the results. The table of p-values provides the significance level for every correlation coefficient between various heavy metals, which relies on the lower-triangular of p-values. For instance, the significance level between copper (Cu) and the other heavy metals is presented in the first column, and the significance level between zinc (Zn) and the other heavy metals excluding copper is presented in the second column, and similarly for the other columns.

The probability values in the first three columns did not show significant relationships between Cu, Zn, Cd, and other heavy metals, as the p-values are high (p-value >0.05). Moreover, a significant positive correlation was exhibited between Cr and Mn (p-value <0.00), and a strong positive correlation at p-value <0.05 was exhibited between the Cr and Hg values. Furthermore, a significant positive correlation was indicated at a p-value <0.10 between Fe and Pb, and Hg (p-value <0.08) and Mn (p-value <0.06), while Hg was significantly correlated with Mn (p-value <0.03). The correlation between the other variables did not exhibit significant relationships, as the p-values are high. A significant relationship indicates that the variables have strong association and share a common source with them. Other results can be interpreted in the same manner.

EXAMPLE 4.8 PHYSIOCHEMICAL PARAMETERS OF LANDFILL LEACHATE TREATMENT

A researcher wants to investigate the covariance matrix and correlation matrix for eight physiochemical parameters (chemical oxygen demand (COD), biochemical oxygen demand (BOD), total dissolved solids (TDS), total suspended solids (TSS), electrical conductivity (EC), pH, ammoniacal-N (NH_3-N), and dissolved oxygen (DO)) measured from each sample obtained from the collection pond of a landfill leachate.

The data for the selected physicochemical parameters obtained from Table 1.3 are reproduced in Table 4.6.

TABLE 4.6 The Results of Physiochemical Parameters of Collection Pond in a Landfill Leachate

pH	EC	TDS	TSS	COD	BOD	NH3H	DO
7.19	6242.00	4530.00	43.26	944.67	9.57	81.66	6.90
7.36	6062.00	4392.00	29.45	845.33	8.77	61.33	6.69
7.41	6164.00	4476.00	31.52	897.00	9.21	71.00	6.48
7.92	6710.33	4710.00	39.36	810.00	8.66	52.43	6.85
7.84	6690.00	4709.83	36.46	686.67	9.66	55.80	6.85
7.83	6600.33	4654.83	37.00	1365.00	10.33	52.80	7.23
8.46	6893.00	4849.00	47.33	692.00	8.47	110.67	2.34
8.47	6850.00	4860.00	50.67	967.00	9.27	93.00	2.82
8.48	6873.00	4859.00	49.00	833.00	8.87	103.00	2.67

The results of employing the built-in function to compute the variance–covariance matrix for the stored data (`.CSV (Example4_8)`) of the physiochemical parameters of the collection pond are presented below, the results for the variance–covariance matrix are computed as given in (4.5) and the results were rounded up to two digits.

```
> round (cov (Example4_8), digits = 2)
          pH        EC      TDS     TSS      COD    BOD     NH3H      DO
pH      0.25    151.89    82.80    2.93   -17.54  -0.09     6.68   -0.91
EC    151.89 104975.75 55522.30 1903.75 -9851.39 -26.18  3164.80 -461.20
TDS    82.80  55522.30 30223.24 1129.51 -5839.91 -16.80  2175.26 -280.54
TSS     2.93   1903.75  1129.51   57.82  -152.55  -0.82   131.42  -13.01
COD   -17.54  -9851.39 -5839.91 -152.55 40834.85  86.83 -1398.02  131.55
BOD    -0.09    -26.18   -16.80   -0.82    86.83   0.34    -5.93    0.61
NH3H    6.68   3164.80  2175.26  131.42 -1398.02  -5.93   499.19  -43.17
DO     -0.91   -461.20  -280.54  -13.01   131.55   0.61   -43.17    4.51
```

The correlation matrix for all possible combinations of the various physiochemical parameters are computed by employing the formula for calculating the correlation coefficient presented in (4.6). The correlation matrix between the various physiochemical parameters of the collection pond of the landfill leachate are produced by employing the function `cor ()` in R.

```
> round (cor (Example4_8), digits = 2)
        pH    EC   TDS   TSS   COD   BOD  NH3H    DO
pH    1.00  0.93  0.95  0.77 -0.17 -0.30  0.60 -0.85
EC    0.93  1.00  0.99  0.77 -0.15 -0.14  0.44 -0.67
TDS   0.95  0.99  1.00  0.85 -0.17 -0.16  0.56 -0.76
TSS   0.77  0.77  0.85  1.00 -0.10 -0.18  0.77 -0.81
COD  -0.17 -0.15 -0.17 -0.10  1.00  0.73 -0.31  0.31
BOD  -0.30 -0.14 -0.16 -0.18  0.73  1.00 -0.45  0.49
NH3H  0.60  0.44  0.56  0.77 -0.31 -0.45  1.00 -0.91
DO   -0.85 -0.67 -0.76 -0.81  0.31  0.49 -0.91  1.00
```

The `corr.test ()` function in R was employed to compute the correlation coefficients with p-values to examine the significance of the computed correlations between the various physicochemical variables.

```
> corr.test (Example4_8)
Call:corr.test(x = Example4_8)
Correlation matrix
        pH    EC   TDS   TSS   COD   BOD  NH3H    DO
pH    1.00  0.93  0.95  0.77 -0.17 -0.30  0.60 -0.85
EC    0.93  1.00  0.99  0.77 -0.15 -0.14  0.44 -0.67
TDS   0.95  0.99  1.00  0.85 -0.17 -0.16  0.56 -0.76
TSS   0.77  0.77  0.85  1.00 -0.10 -0.18  0.77 -0.81
COD  -0.17 -0.15 -0.17 -0.10  1.00  0.73 -0.31  0.31
BOD  -0.30 -0.14 -0.16 -0.18  0.73  1.00 -0.45  0.49
NH3H  0.60  0.44  0.56  0.77 -0.31 -0.45  1.00 -0.91
DO   -0.85 -0.67 -0.76 -0.81  0.31  0.49 -0.91  1.00
Sample Size
[1] 9
Probability values (Entries above the diagonal are adjusted for multiple tests.)
       pH   EC  TDS  TSS  COD  BOD NH3H   DO
pH   0.00 0.01 0.00 0.30 1.00 1.00  1.0 0.09
EC   0.00 0.00 0.00 0.30 1.00 1.00  1.0 0.77
TDS  0.00 0.00 0.00 0.08 1.00 1.00  1.0 0.32
TSS  0.02 0.01 0.00 0.00 1.00 1.00  0.3 0.19
COD  0.66 0.70 0.67 0.80 0.00 0.42  1.0 1.00
BOD  0.43 0.72 0.67 0.63 0.02 0.00  1.0 1.00
NH3H 0.09 0.24 0.12 0.01 0.42 0.22  0.0 0.02
DO   0.00 0.05 0.02 0.01 0.42 0.18  0.0 0.00
```

```
To see confidence intervals of the correlations, print with the short=FALSE option
```

A strong positive relationship was exhibited between pH and electrical conductivity (EC) at *p*-value <0.00, as appeared in the probability values of the pH column, between pH and total dissolved solids (TDS) (p-value <0.00), between pH and total suspended solids (TSS) (p-value <0.02), and a negative relationship between pH and DO (*P*-value <0.00). Moreover, there were positive relationships between EC and TDS (p-value <0.00), and a negative relationship between EC and dissolved oxygen (DO) (p-value <0.05), positive relationship between TDS and TSS, negative relationship between TDS and DO (p-value <0.02), positive relationship between chemical oxygen demand (COD) and biochemical oxygen demand (BOD), and negative relationship between ammoniacal-N (NH_3-N) and DO (p-value <0.00). Correlated parameters negatively or positively refer to sharing the origin source.

Note:

- The correlations between any two variables can be computed by employing the built-in function `cor.test()` in R as an alternative command. It works easily by providing the two variables *X* and *Y*.

```
cor.test (x,y)
```

4.7 SCATTER DIAGRAM

A scatter diagram (also called a scatter plot) is a graphical method using cartesian coordinates to represent the data values for two variables (*X*, *Y*). One variable will represent the horizontal axis and the second variable will represent the vertical axis. A scatter diagram can give a common idea of the correlation between two variables.

EXAMPLE 4.9 SCATTER DIAGRAM FOR THE CHROMIUM (Cr) AND MANGANESE (Mn) IN SEDIMENT

The data for two heavy metals, chromium (Cr) and manganese (Mn), in sediment of the Jejawi River are given in Table 4.7,

TABLE 4.7 The Results of Chromium (Cr) and Manganese (Mn) in Sediments of the Jejawi River

Cr	Mn
0.13	1.63
0.12	1.68
0.17	1.75
0.20	1.75
0.28	1.90
0.11	1.67
0.16	1.71
0.23	1.77
0.21	1.85
0.17	1.70

build a scatter plot for Cr and Mn.

R statistical software offers built-in functions to produce various graphs for different statistical objectives. Moreover, R permits us to put a particular opinion on any produced plot, we can select the desired color, the shape and the size of the marks, the legend, and other matters related to plots as well. Furthermore, text, points, and lines can be placed in the graph. The function `plot ()` can be employed to produce various plots.

```
Plot (x,y)
```

The function `plot ()` is employed to produce a scatter diagram for the stored data. CSV `(Example4_9)` of chromium and manganese data values.

```
plot (Example4_9 $ Cr, Example4_9 $ Mn, xlab = "", ylab = "",
    cex=0.9, pch = 15, col = "darkgreen", cex.axis = 0.9, xaxt = "n", yaxt ="n")
```

where `ylab=""`refers to the y-axis and `xlab=""`refers to the x-axis.

The function `pch=?` is usually employed to assign the characters to be employed for plotting points. The function `col.=""`is employed to assign the desired color.

Fig. 4.1 shows the scatter diagram for the chromium and manganese. One can observe that there is obvious direction for the correlation between the chromium and the manganese. Furthermore, a positive correlation is noted between them because increasing the value of the chromium would increase the manganese. This correlation includes a common inference regarding the behavior of chromium and the manganese. However, a scatter diagram does not show the strength of the correlation; thus, we should investigate it through the correlation coefficient.

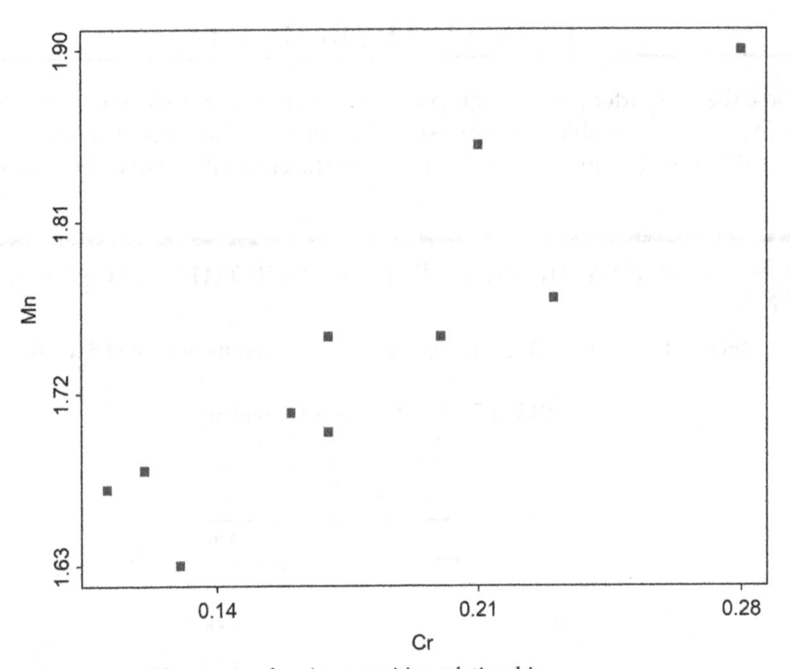

FIG. 4.1 Scatter plot of chromium versus. Manganese, showing a positive relationship.

Note:

- Scatter diagrams can have various patterns other than the one presented in Fig. 4.1. Fig. 4.2 presents other types of scatter diagrams. We can use the same function for a scatter diagram with different values to generate the scatter diagrams.

4.7.1 The Scatter Diagram Matrix

In the case of several variables, the scatter diagram matrix is usually employed to display and organize scatter diagrams for all possible combinations of two variables to be within one vision. Various functions are offered by R for generating a scatter diagram matrix. The function `pairs ()` is employed to produce a scatter diagram

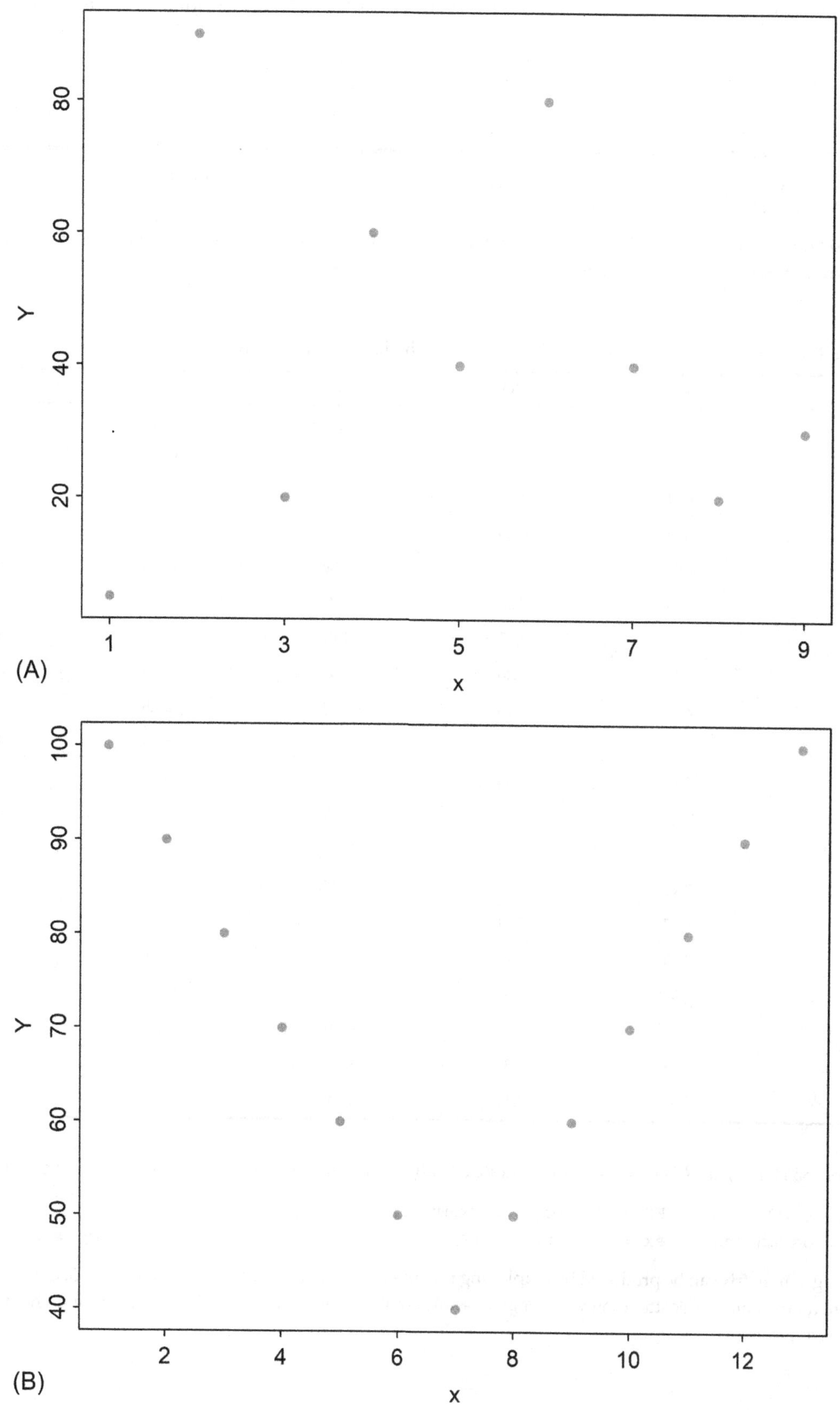

FIG. 4.2 Scatter plots for different types of relationships.

matrix. Consider that there are $Y_1, Y_2, \ldots, Y_k$, then the structure for the call to build a scatter-plot matrix in R is given below.

```
pairs (~ Y1+ Y2 +... + Yk, data = data frame)
```

EXAMPLE 4.10 SCATTER GRAPH MATRIX FOR THE PHYSIOCHEMICAL PARAMETERS OF WATER

Generate a scatter graph matrix for the water quality parameters of five physiochemical parameters selected from two rivers, namely the Juru and Jejawi Rivers. The sample data are given in Table 4.8.

TABLE 4.8 The Results of Physiochemical Parameters of the Juru and Jejawi Rivers

Location	pH	DO	BOD	EC	TSS
Juru	7.88	6.73	10.56	42.45	473.33
Juru	7.92	6.64	10.06	42.75	461.67
Juru	7.41	5.93	6.01	29.47	393.34
Juru	7.84	6.23	14.57	42.35	473.34
Juru	7.86	6.29	7.36	40.70	528.34
Juru	7.89	6.36	5.27	42.50	458.33
Juru	7.40	5.67	4.81	29.51	356.67
Juru	7.41	5.26	4.96	28.45	603.33
Juru	7.25	4.18	6.61	26.25	430.00
Juru	7.18	5.14	5.11	25.79	445.00
Jejawi	7.40	4.96	36.65	29.55	815.00
Jejawi	7.53	5.86	37.20	21.80	858.34
Jejawi	7.45	5.52	41.50	27.55	823.34
Jejawi	7.41	5.27	36.75	16.45	821.67
Jejawi	7.27	5.71	37.95	21.10	736.67
Jejawi	7.36	5.16	34.85	11.50	373.34
Jejawi	7.37	5.08	37.05	13.35	385.00
Jejawi	7.40	5.23	37.10	10.85	403.34
Jejawi	7.38	5.30	38.30	13.90	550.00
Jejawi	7.36	5.17	38.40	12.65	363.34

The function `pairs ()` in R is employed to generate a scatter diagram for all possible combination of two variables.

```
pairs (~ pH + DO + BOD + EC +TSS, data = Example4_10,
    col ="darkgreen", cex = 0.9, pch = 12, cex.axis = 0.9, tck = -0.03, mgp = c(3, .3, 0))
```

A scatter diagram matrix can be produced by employing the function `pairs ()` as presented in Fig. 4.3. One can see that the scatter diagram matrix contains scatter diagrams for all possible combinations of two variables for the physiochemical variables selected from two rivers.

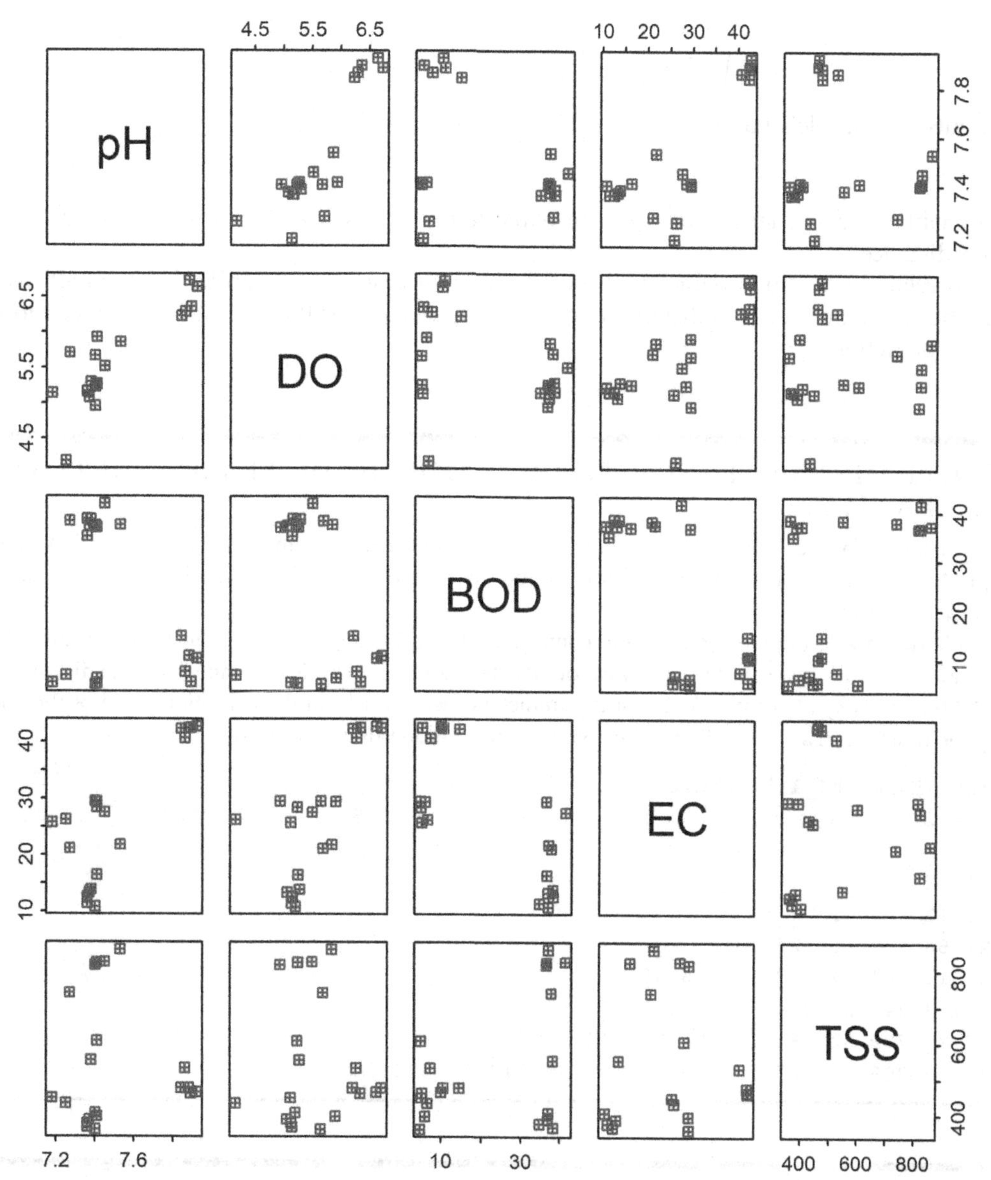

FIG. 4.3 A scatter plot matrix for physiochemical parameters of surface water.

4.8 EUCLIDEAN DISTANCE

Consider there are two points, $A=(X_1, X_2, \ldots, X_k)$ and $B=(Y_1, Y_2, \ldots, Y_k)$, the Euclidean distance in either the plane or 3-dimensional space that connect between the two points A and B is calculated employing the formula presented in (4.8).

$$d=\sqrt{(X_1-Y_1)^2+(X_2-Y_2)^2+\ldots+(X_k-Y_k)^2} \tag{4.8}$$

The formula in (4.8) does not consider the differences in variances of the variables; thus, we can develop a formula that considers the differences and the correlation or covariances as well. The contribution of the coordinates will be influenced by the variation and the correlation between various variables.

The distance that considers the variances of the selected variables and their covariances or correlations is called a statistical distance, i.e., the variation and the correlation (or covariances) between various variables is considered by the statistical distance. The formula for the statistical distance between the two points A and B is presented in (4.9).

$$\sqrt{\frac{(X_1 - Y_1)^2}{S_{11}} + \frac{(X_2 - Y_2)^2}{S_{22}} + \ldots + \frac{(X_k - Y_k)^2}{S_{kk}}} \quad (4.9)$$

where S_{ii} represents the sample's variance.

Note:

- If a random variable showed a larger variance compared to the others, it will receive less weight (it contributes less to the squared distance).
- Two slightly correlated random variables will contribute more than two variables that are highly correlated.
- The difference between the statistical distance and Euclidean distance is that statistical distance formula is divided by the standard deviation.

EXAMPLE 4.11 DISTANCES FOR THE PHYSIOCHEMICAL PARAMETERS OF LANDFILL LEACHATE

Compute the distances for the data of physiochemical parameters measured from the collection pond of a landfill leachate used in Example 4.8. The data represents the first nine samples extracted from Table 1.3 and eight physiochemical variables measured from each sample.

The distances between all sampling points were computed by the function `dist ()` in R and results of distances are produced in a matrix form to represent the distances for all selected physiochemical parameters. The distance matrix for the stored data. CSV (`Example4_11`) of physiochemical parameters measured from the collection pond of the landfill leachate is produced by the function `dist ()` in R and the results were rounded up to two digits.

```
> round (dist (Example4_11), digits = 2)
       1      2      3      4      5      6      7      8
2 248.83
3 107.35 142.22
4 520.33 723.11 601.01
5 548.02 721.56 613.06 125.08
6 567.04 793.14 664.64 568.49 686.45
7 768.30 962.13 845.24 264.74 252.50 761.43
8 692.29 925.33 789.82 261.62 358.24 514.45 279.14
9 720.68 937.07 809.20 227.73 282.06 633.83 142.98 136.34
```

EXAMPLE 4.12 DISTANCES FOR THE HEAVY METALS IN SEDIMENT

Compute the distance for the eight selected heavy metals presented in Table 4.5. The data represent samples for eight heavy metals measured for each sample; the samples were selected form Juru and Jejawi Rivers.

The distances between all of the selected heavy metals were computed by employing the function `dist ()` in R and then generate a distance matrix for the stored data. CSV (`Example4_12`) of heavy metals in sediment. The entries of the distance matrix for the selected heavy metals in sediments are rounded up to two digits as shown below.

```
> round (dist (Example4_12), digits = 2)
      1    2    3    4    5    6    7    8    9
2  0.81
3  1.70 0.93
4  1.52 1.00 1.32
5  1.38 0.99 1.02 1.85
6  1.07 0.30 0.82 0.80 1.12
7  1.60 0.84 0.76 0.64 1.48 0.58
8  0.89 0.45 0.98 1.37 0.69 0.63 1.13
9  1.76 1.22 0.78 1.92 0.66 1.25 1.45 0.96
10 1.84 1.06 0.30 1.26 1.23 0.90 0.72 1.15 0.94
```

Further Reading

Alkarkhi, A.F.M., Ahmad, A., Ismail, N., Easa, A.M., Omar, K., 2008. Assessment of surface water through multivariate analysis. J. Sustain. Develop. 1, 27–33.

Alkarkhi, A.F.M., Alqaraghuli, W.A.A., 2019. Easy Statistics for Food Science with R, first ed. Academic Press.

Alkarkhi, A.F.M., Ismail, N., Ahmed, A., Easa, A.M., 2009. Analysis of heavy metal concentrations in sediments of selected estuaries of Malaysia—a statistical assessment. Environ. Monit. Assess. 153, 179–185.

Alkarkhi, A.F.M., Low, H.C., 2012. Elementary Statistics For Technologist. Universiti Sains Malaysia Press, Pulau Pinang.

Allan, G.B., 2007. Elementary Statistics: A Step by Step Approach. Mcgraw-Hill.

Ben, B. 2010. The Many Flavors Of Apply [Online]. Available: http://ms.mcmaster.ca/~bolker/classes/m2e03/labs/lab1X.html [Accessed 26 May 2018].

Black, K. 2015. R Tutorial, Basic Operations And Numerical Descriptions [Online]. Datacamp. Available: https://www.cyclismo.org/tutorial/R/basicOps.html [Accessed 28 May 2018].

Bryan, F.J.M., 1991. Multivariate Statistical Methods: A Primer. Chapman & Hall, Great Britain.

Chard, N.C., Abdul Rahman, N.N.N., Alkarkhi, A.F.M., Ahmad, A., Abdel Kadir, M.O., 2010. Assessment of major solid wastes generated in palm oil mills. Int. J. Environ. Technol. Manage. 13, 245–252.

Chi, Y. 2018. R Tutorial, An R Introduction To Statistics [Online]. Available: http://www.r-tutor.com/elementary-statistics/numerical-measures/mean [Accessed 26 May 2018].

Daniel & Hocking, 2013. Blog Archives, High Resolution Figures In R [Online]. R-Bloggers. Available, https://www.r-bloggers.com/author/daniel-hocking.

Fanara, C., 2016. Tutorial On The R Apply Family, The Apply Functions As Alternatives To Loops. [Online]. Datacamp Available: https://www.datacamp.com/community/tutorials/r-tutorial-apply-family#Gs.Rqdg=Dw. [(Accessed 26 May 2018)].

Johnson, R.A., Wichern, D.W., 2002. Applied Multivariate Statistical Analysis. Prentice Hall, New Jersey.

Kabacoff, R. I. Quick-R, Scatterplots [Online]. Available: https://www.statmethods.net/graphs/scatterplot.html [Accessed 26 May 2018].

Mark, G., Using R For Statistical Analyses—Introduction. [Online]. Available: http://www.gardenersown.co.uk/education/lectures/r/. [(Accessed 28 May 2018)].

Rencher, A.C., 2002. Methods of Multivariate Analysis. J. Wiley, New York.

Wickham, H., Advanced R, Functionals. [Online]. Available: http://adv-r.had.co.nz/functionals.html. [(Accessed 26 May 2018)].

William, B.K., 2016. Tutorials [Online]. Available, http://ww2.coastal.edu/kingw/statistics/r-tutorials/index.html. [(Accessed 30 May 2018)].

Further Reading

CHAPTER

5

Statistical Hypothesis Testing

LEARNING OBJECTIVES

After careful consideration of this chapter, you should be able:

- *To explain the concept of statistical hypothesis testing.*
- *To write the null and alternative hypotheses in mathematical equations.*
- *To perform statistical hypothesis testing regarding a mean value for a single population.*
- *To perform statistical hypothesis testing about a mean vector for a single population.*
- *To conduct a comparison for two population means (with a single variable).*
- *To conduct a comparison for two multivariate population means.*
- *To describe the difference between univariate and multivariate tests.*
- *To apply R commands and built-in functions to carry out statistical hypothesis testing.*
- *To comprehend and explain R output to help in making intelligent decisions.*
- *To report useful conclusions about the problem under study.*

5.1 INTRODUCTION

An essential aspect of any test is to make inferences regarding the matter under consideration to draw conclusions concerning a population. Scientists must have sufficient and appropriate data concerning a subject to decide properly. We commonly make a decision regarding any issue using the sample data gathered from a population of interest. Statistical hypothesis testing is one of the best methods to lead scientists to make the right decision regarding any matter and in any area. Statistical hypothesis testing is easier to implement on single variable studies than several variables studies because multivariate data need techniques that take into account the relationship between various variables. Statistical hypothesis testing with regard to a single variable and the corresponding tests for several variables (the multivariate case) will be presented in this chapter.

5.2 STATISTICAL HYPOTHESIS TESTING IN R

A variety of built-in functions are offered by R statistical software in the basic R installation to include all of the popular statistical methods. Moreover, more built-in functions have been offered by many supplementary packages and beyond the standard packages (base installation). We provided the structure of each built-in function with detailed explanation.

1. The built-in function `t.test ()` can be employed to calculate a t-test statistic for one sample.

```
t.test (data frame) # H0:mu = 0
t.test (data frame, mu = value) # H0:mu = Claimed (proposed) value of the average
t.test (data frame, mu = value, conf.level = alpha, alternative = "less")
```

`alternative= "greater"` instead of `"less"` can be set for the alternative hypothesis.

Applied Statistics for Environmental Science with R
https://doi.org/10.1016/B978-0-12-818622-0.00005-8

2. Consider there are two numeric variables, y_1 and y_2, a *t*-test statistic for two independent samples can be computed by employing the built-in function `t.test ()`.

```
t.test (y1,y2)
```

   The option `alternative="less"` or `"greater"`, `mu=value` can be included, and an equal variances option can be added as well (`vari.equal=TRUE`).
3. Consider there are two variables, *y* (numeric variable) and *x* (binary variable), a *t*-test statistic for two independent samples can be computed by employing the built-in function `t.test ()`.

```
t.test (y~x)
```

4. Consider there are two numeric variables, y_1 and y_2, a *t*-test statistic for two dependent samples can be computed by employing the built-in function `t.test ()`.

```
t.test (y1, y2, paired = TRUE)
```

5. The function `sqrt ()` can be employed to compute the square root.

```
sqrt (data frame)
```

6. The function `t ()` can be employed to generate the transpose of a matrix.

```
t (data frame)
```

7. The function `qt ()` can be employed to compute the quantiles and critical values of a t distribution.

```
qt (data frame)
```

8. The probability of the normal distribution curve can be calculated by employing the built-in function `dnorm ()`.

```
dnorm (x, mean = 0, sd = 1, log = FALSE)
```

   The mean and standard deviation for a standard normal distribution are set to 0 and 1, respectively. The probability for other values of the mean and standard distribution can be computed; `x` marks that a variable is being rated.
9. The cumulative probability to the left of the `q` value (the area below a given value of "X" can be computed employing the function `pnorm ()`.

```
pnorm (q, mean = 0, sd = 1,lower.tail = TRUE,log.p = FALSE)
```

   The option `lower.tail=TRUE` means compute the area to the left of `q`, whereas the option `FLASE` means calculate the area to the right of `q`.
10. The inverse of the function `pnorm ()` is the function `qnorm ()`, which can be employed to get quantiles or `"critical values"`; `qnorm ()` deals by default with areas below the given boundary value.

```
qnorm (p, mean = 0,sd = 1,lower.tail =TRUE, log.p = FALSE)
```

   where `p` refers to the area under the standard normal distribution curve.
11. The inverse matrix can be calculated by employing the built-in function `solve ()`.

```
solve (data frame)
```

5.3 COMMON STEPS FOR HYPOTHESIS TESTING

We should provide several remarkable concepts and expression that are important to statistical hypothesis testing before talking about the common steps for hypothesis testing.

5.3.1 The Concept of Null and Alternative Hypotheses

A statistical hypothesis refers to any assumption concerning a population parameter; this assumption may or may not be true.

We can recognize two opposing statistical hypotheses: the first hypothesis is known as the *null hypothesis* (denoted by H_0), while the second hypothesis is the opposite of the null hypothesis and is known as the *alternative hypothesis* (denoted by H_1 or H_a). The *null hypothesis* states that the difference between a population parameter (such as the average or standard deviation) and the proposed value is equal to zero. Or the null hypothesis states that the distinction between the two values (a population parameter and the proposed values) is usually attributed to coincidence (no real difference).

The *alternative hypothesis* states that there is a significant (real) distinction between the two values (a population parameter and the proposed values). In other words, it says that the difference between the two values exists and is real. This result would suggest that the population parameter has a value that differs from the proposed value.

EXAMPLE 5.1 REPRESENT THE NULL AND ALTERNATIVE HYPOTHESES

Consider the three situations below to represent the null and alternative hypotheses concerning the iron (Fe) mg/L value in wastewater.

Situation 1: The mean value of the Fe in the wastewater is 1.2 (mg/L).
Situation 2: The mean value of the Fe in the wastewater is >1.2 (mg/L).
Situation 3: The mean value of the Fe in the wastewater is <1.2 (mg/L).

Three different cases will be investigated for formulating the null and alternative hypotheses.

Situation 1: The mean value of the Fe in the wastewater is 1.2 (mg/L).

The null hypothesis states that the mean value of the Fe in the wastewater is exactly 1.2 (mg/L) (the mean value is equal to 1.2), the null hypothesis can be expressed mathematically as given below.

$$H_0: \mu = 1.2 \quad \text{(claim)}$$

The exact value of iron in the wastewater (1.2) is covered under the null hypothesis, while the complement to this direction (more or <1.2) will form the alternative hypothesis, which is the opposite direction of the null hypothesis. The alternative hypothesis can be expressed mathematically as given below:

$$H_1: \mu \neq 1.2.$$

The two hypotheses H_0 and H_1 are stated in such a way as to be mutually exclusive. That is, if one is correct, the other must be false, and vice versa.

Situation 2: The mean value of the Fe in the wastewater is >1.2 (mg/L).

The assumption covers only one side (>1.2) without an equality sign; thus, the assumption will be attached to the alternative hypothesis. The complement of this assumption will be either equal to 1.2 or less than1.2. The two options equal to 1.2 and <1.2 will be attached to the null hypothesis because of the equality direction.

$$H_0: \mu \leq 1.2 \quad \text{vs} \quad H_1: \mu > 1.2 \quad \text{(claim)}$$

Situation 3: The mean value of the Fe in the wastewater is <1.2 (mg/L).

The assumption covers only one side (<1.2) without an equality sign, thus, the assumption will be attached to the alternative hypothesis. The complement of this assumption will be either equal to 1.2 or more than1.2. The two options equal to 1.2 and >1.2 will be attached to the null hypothesis because of the equality direction.

$$H_0: \mu \geq 1.2 \quad \text{vs} \quad H_1: \mu < 1.2 \quad \text{(claim)}$$

Three different cases of statistical hypotheses were discussed. We must use one case for testing any assumption.

We can summarize the three cases of the statistical hypothesis, one-sided test and two-sided test as presented below.

Two-sided test	One-side test	
	Right-sided test	Left-sided test
$H_0: \mu = \mu_0$ $H_1: \mu \neq \mu_0$	$H_0: \mu \leq \mu_0$ $H_1: \mu > \mu_0$	$H_0: \mu \geq \mu_0$ $H_1: \mu < \mu_0$

where, μ_0 is a given value.

Note:

- The equality sign is always used with the null hypothesis, H_0.
- We always use either <, or >, or ≠ with the alternative hypothesis, H_1.

5.3.2 Basic Concepts

- *Test statistic*:
  Every test has a mathematical formula used to calculate a value using the sample data; the produced value is called the test statistic. The test statistic value is usually employed to reject or support the null hypothesis.
- *Rejection region*:
  Also known as critical region, this is the range of the values of the test statistic for which the null hypothesis should be rejected.
- *Nonrejection region*:
  Also known as noncritical region, this is the range of values of the test statistic for which the null hypothesis should not be rejected.
- *Critical value*:
  The value that breaks the area under the distribution curve into two regions (critical region and noncritical region) is called the critical value.
- *Significance level*:
  The probability of rejecting the null hypothesis when it is true is called the *significance level and* denoted by α. Scientists generally select α to be 0.05 or 0.01.
- *p-Value*:
  If the null hypothesis is true, then the probability of gaining a more extreme value of the test statistic than the observed value is called the p-value, which is employed in hypothesis testing to decide whether to support or reject the null hypothesis. The range of the *p*-value varies between 0 and 1. The null hypothesis should be rejected if the *p*-value is small, while the null hypothesis should not be rejected if the *p*-value is large.

EXAMPLE 5.2 REPRESENT REJECTION AND NONREJECTION REGIONS

Consider the three situations below to represent the rejection and nonrejection regions. Choose a significance level of $\alpha = 0.01$ to carry out the analysis. Assume that the distribution is normal.

Situation 1: $H_0: \mu = 10$ vs $H_1: \mu \neq 10$.

Situation 2: $H_0: \mu \leq 10$ vs $H_1: \mu > 10$.

Situation 3: $H_0: \mu \geq 10$ vs $H_1: \mu < 10$

Situation 1: The alternative hypothesis is $H_1: \mu \neq 10$, the rejection region (critical region) is distributed on both sides of the normal distribution, where each side includes an area of $(0.01/2 = 0.005)$. The Z critical value is ±2.58, as displayed in Fig. 5.1A.

(A) The critical and noncritical regions for $H_1: \mu \neq 10$.

Situation 2: The alternative hypothesis is $H_1: \mu > 10$, the rejection region is in the right side of the normal curve and has a right-side area of 0.01. The Z critical value is 2.326, as displayed in Fig. 5.1B.

(B) The critical and noncritical regions for $H_1: \mu > 10$.

Situation 3: The alternative hypothesis is $H_1: \mu < 10$, the rejection region is in the left side of the normal curve and has a left-side area of 0.01. The Z critical value is −2.326, as displayed in Fig. 5.1C.

(C) The critical and noncritical regions for $H_1: \mu < 10$.

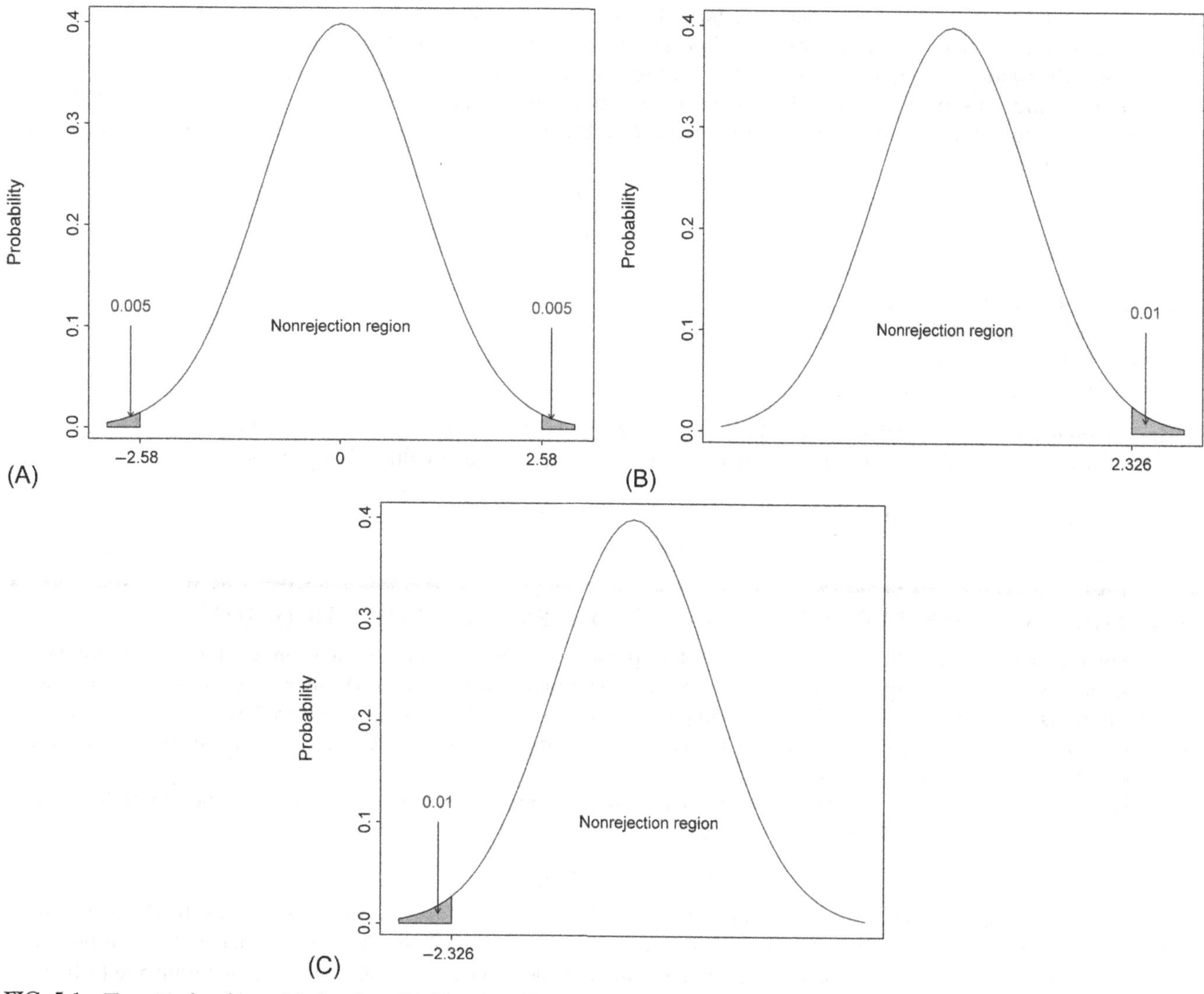

FIG. 5.1 The critical and noncritical regions (A) $H_1:\mu \neq 10$, (B) $H_1:\mu > 10$, (C) $H_1:\mu < 10$.

5.4 HYPOTHESIS TESTING FOR A MEAN VALUE

This section discusses hypothesis testing concerning the average value of a single variable (univariate) and several variables (multivariate).

5.4.1 Hypothesis Testing for One Population Mean

Hypothesis testing for the average value in the case of one measured variable covers two situations, depending on whether the population variance is given (known) or not given (unknown). The two situations are shown below.

Situation 1: When population standard deviation σ is known

Two-tailed test will only be covered for hypothesis testing concerning an average value of a variable when the population standard deviation is given (known) because we cannot generalize the alternative hypothesis in the case of multivariable tests for other cases. The scientists become concerned in examining the hypothesis presented in (5.1), to make a decision with confidence regarding the average value of any process besides the one-tailed test.

$$H_0:\mu=\mu_0 \text{ vs } H_1:\mu\neq\mu_0 \tag{5.1}$$

The average value of the variable is equal to μ_0 (specific value) as stated by the null hypothesis, whereas the complement of the null hypothesis is stated under the alternative hypothesis (H_1), which states that the average value differs from μ_0 (specific value), which indicates that the average value is either less or more than μ_0.

Consider a random sample of size n is selected from a population that has normal distribution. The test statistic employed to examine the null hypothesis in (5.1) regarding the average value is called the Z-test. The Z-test formula is presented in (5.2).

$$Z = \frac{\overline{Y} - \mu_0}{\sigma / \sqrt{n}} \tag{5.2}$$

where

$\overline{Y}$ refers to the average value;
μ_0 refers to a specific (given) value;
σ refers to the standard deviation, and
n refers to the chosen sample size.

The calculated value of Z-test and the critical value of Z are used to make a decision regarding the null hypothesis. We reject the null hypothesis (H_0) if $|Z = \text{calculated value}| \geq Z_{\frac{\alpha}{2}} = \text{critical value}$; this guides us to believe that the average value does not equal the μ_0 value.

EXAMPLE 5.3 THE PARTICULATE MATTER PM_{10} IN THE AIR ($\mu G/M^3$)

An environmentalist claims that the average value of the particulate matter (PM_{10}) in the air of an equatorial coastal location is 77 ($\mu g/m^3$). Seventy-five samples are chosen to examine the environmentalist's claim. The average value of the particulate matter (PM_{10}) is 65 ($\mu g/m^3$) as calculated from the sample data. The standard deviation of the population is 44 ($\mu g/m^3$). Is there enough evidence to support the environmentalist's claim? Use a significance level of $\alpha = 0.05$ to carry out the analysis. Assume that the population is normally distributed.

The claim concerning the average value of the particulate matter in the air can be placed in a mathematical formula to represent the null and alternative hypotheses as given below.

$$H_0: \mu = 77 \quad \text{vs} \quad H_1: \mu \neq 77$$

Testing the hypothesis regarding the average value of the particulate matter (PM_{10}) in the air can be achieved by employing the Z-test. However, the Z-test is not covered in the default R packages. Thus, we should use R commands and built-in functions to carry out a Z-test about the average value. To clarify the computations of the Z-test in R employing the formula in (5.2), we should provide the sample size, average value, proposed average value, and standard deviation. The R commands and build-in functions with the results for testing the average value of the particulate matter (PM_{10}) in the air using the Z-test are presented below.

```
average = 65
m0 = 77
s = 44
n = 75
z = (average - m0) / (s / sqrt(n))
z
alpha = 0.05
z1 = round (qnorm (1 - alpha/2), digits = 2)
c.v = c(-z1, z1)
c.v
```

The required data to employ the Z formula were provided in the first four rows of R commands, including the average value (65), proposed average value (77), population standard deviation (44), and sample size (75). The outputs generated by R built-in functions and commands for a two-sided Z-test for the particulate matter (PM_{10}) in the air are presented below.

```
> average = 65
> m0 = 77
> s = 44
> n = 75
> z = (average - m0)/(s/sqrt(n))
```

```
> z
[1] -2.361887
> alpha = 0.05
> z1 = round (qnorm (1 - alpha/2), digits = 2)
> c.v = c(-z1, z1)
> c.v
[1] -1.96 1.96
```

One can observe that the input data (information) appeared in the first four rows of the R results. The Z-test formula and the result: `-2.361887`$=-2.36$ appeared in next two rows of the outputs. The function `qnorm ()` was employed to compute the critical values (`z1`) choosing the significance level alpha to be 0.05. The two critical values for a two-tailed Z-test are `-1.96` and `1.96` as appeared in the last row of the output. We can compare the absolute computed value of the Z-test statistic $|-2.36|$ with the Z critical value of 1.96, the null hypothesis should be rejected because $|-2.36|>1.96$. Alternately, one can use a normal curve to identify the critical (rejection) and noncritical (nonrejection) regions as presented in Fig. 5.2. We use Fig. 5.2 to make a decision based on the results of the Z-test for the PM_{10} value in the air and the critical values; obviously, we should reject the null hypothesis because the test statistic value -2.36 falls in the critical (rejection) region. The desired plot with the summary statistics for PM_{10} in the air is presented in Fig. 5.2.

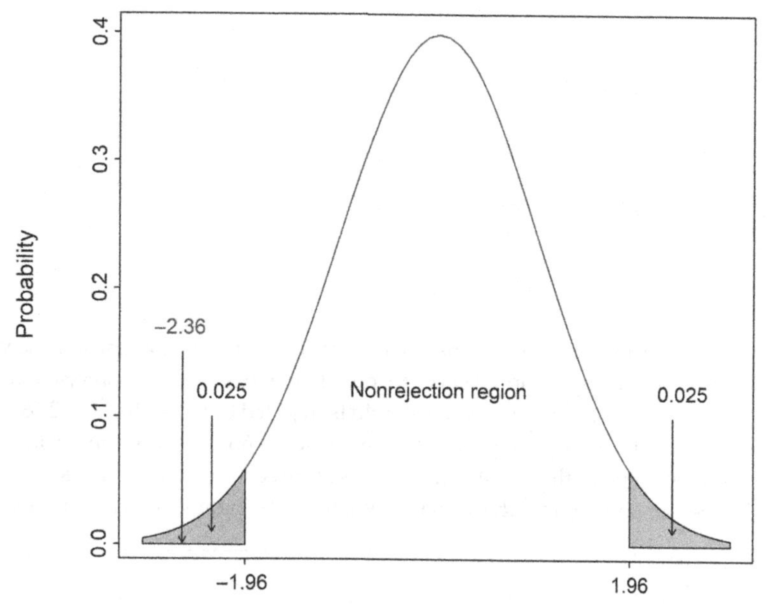

FIG. 5.2 The critical and noncritical regions for the PM_{10} value in the air.

In summary, one can believe that there is enough evidence to reject the environmentalist's claim, "the mean value of PM_{10} value in the air is 77" and be confident that the average value of the PM_{10} in the air differs from 77.

EXAMPLE 5.4 THE CONCENTRATION OF TOTAL SUSPENDED SOLIDS (TSS) IN THE WATER (MG/L)

An environmentalist wishes to verify that the average value of the concentration of the total suspended solids (TSS) in the water of a certain river is 450 mg/L. Forty samples are selected and examined for the concentration of TSS. The results show that the mean concentration of TSS is 465 mg/L. The variance of the population is 85 mg/L as reported by previous study. Is there enough evidence to support the claim? Assume that the population is normally distributed. Use a significance level of $\alpha=0.01$ to carry out the analysis.

The hypothesis regarding the average concentration of TSS in the water (450 mg/L) can be stated mathematically as given below.

$$H_0: \mu = 450\ \text{(Claim)} \quad \text{vs} \quad H_1: \mu \neq 450$$

The commands and built-in functions provided by R were employed to calculate the Z-test statistic value for the concentration of TSS in the water. The steps and commands employed to examine the hypothesis are presented below.

```
average = 465
m0 = 450
v = 85
s = sqrt(v)
n = 40
z = (average - m0)/(s / sqrt(n))
z
alpha = 0.01
z1 = round (qnorm (1 - alpha/2), digits = 2)
c.v = c(-z1, z1)
c.v
```

The outputs of employing R built-in functions and commands for examining the claim that the average concentration of the total suspended solids (TSS) in the water is 450 mg/L are given below.

```
> average = 465
> m0 = 450
> v = 85
> s = sqrt(v)
> n = 40
> z = (average - m0)/(s / sqrt(n))
> z
[1] 10.28992
> alpha = 0.01
> z1 = round (qnorm (1 - alpha/2), digits = 2)
> c.v = c(-z1, z1)
> c.v
[1] -2.58 2.58
```

There are two ways to make a decision, either by comparing the absolute computed Z-test statistic value (`10.28992`=10.29) with the critical value of 2.58, or by employing a normal curve to identify the rejection and nonrejection regions. The null hypothesis concerning the average concentration of TSS in the water should be rejected because $|10.29| > 2.58$. Thus, there is enough evidence to believe that, on average, the total suspended solids in the water is not 450. The same conclusion can be achieved by employing a normal curve with the value of the Z-test statistic, critical (rejection), critical values, and noncritical (nonrejection) regions as shown in Fig. 5.3. One can observe that the value 10.29 falls in the right side, which represents a rejection region.

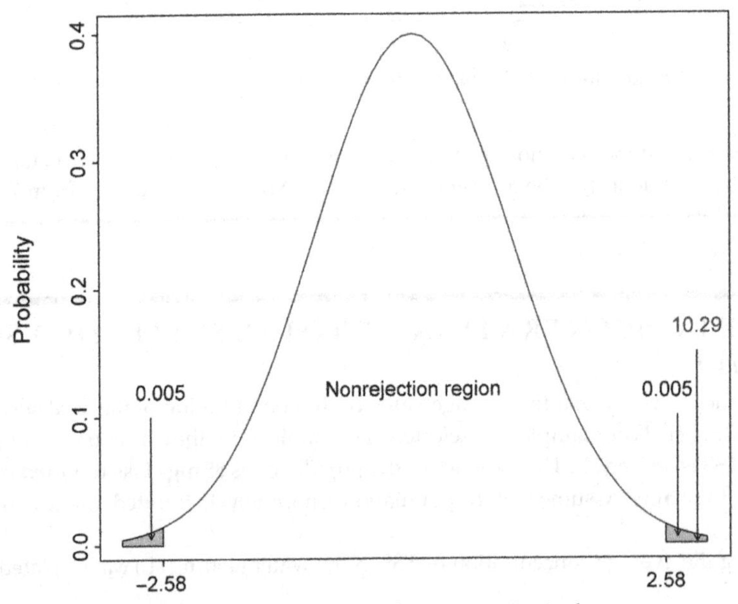

FIG. 5.3 The critical and noncritical regions for the total suspended solids concentration in the water.

Situation 2: When σ is unknown

Claims about the average value of a variable when the standard deviation of the population is given (known) was discussed employing the Z-test. This test cannot be employed if the variance of the population is unknown. We should employ another procedure to deal with the situation of unknown variance. Consider a random sample of size n is selected from a population that has normal distribution. The test statistic employed to examine the null hypothesis regarding the average value is called the t-test. The t-test formula is presented in (5.3).

$$t = \frac{\overline{Y} - \mu_0}{S/\sqrt{n}}, \text{with } n-1 \text{ degrees of freedom} \tag{5.3}$$

where

$\overline{Y}$ refers to the average value;
μ_0 refers to a specific (given) value;
n refers to the sample size; and
S refers to the sample's standard deviation.

We reject the null hypothesis (H_0) if the absolute value of the computed t-test statistic value is greater than or equal to the critical value, $|t| \geq t_{\alpha/2,\ n-1}$, where $n-1$ is the degree of freedom.

Note:

- $t_{\alpha/2,\ n-1}$ refers to the critical value obtained from the t table for critical values.
- The null and alternative hypotheses are the same as those presented in (5.1).
- If the selected sample size is <30 ($n<30$), we employ t-test instead of Z-test.
- R built-in functions and commands can be employed to generate the t curve for various degrees of freedom. The curve for t distribution is presented in Fig. 5.4 (see the appendix for R codes).

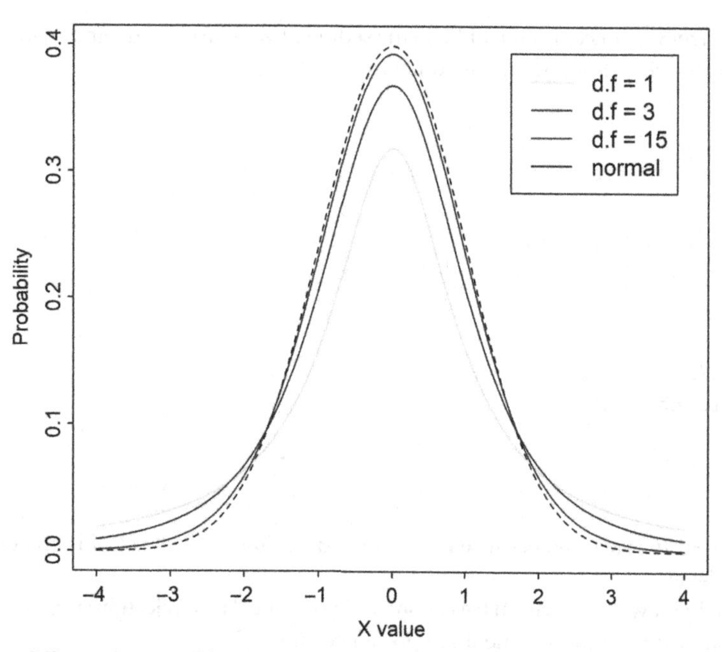

FIG. 5.4 T distribution curve at different degrees of freedom.

EXAMPLE 5.5 THE CONCENTRATION OF LEAD (PB) IN THE SEDIMENT (MG/L)

An environmentalist claims that the mean concentration of lead (Pb) in the sediment is 0.16 (mg/L). Eleven samples are chosen to test the environmentalist's claim. The sample data revealed that the mean concentration of Pb in the sediment is 0.23 and the standard deviation is 0.03. A significance level of $\alpha = 0.05$ was chosen to examine the researcher's claim. Assume that that the population is normally distributed.

The mean concentration of lead in the sediment is 0.16 (mg/L), which represents the claim and should be placed under the null hypothesis, while the complement of the equality is that the mean concentration of lead is more or <0.16 (mg/L), which represents the alternative hypothesis. The two hypotheses concerning the average concentration of lead can be expressed mathematically as given below.

$$H_0 : \mu = 0.16\,(\text{claim}) \quad \text{vs} \quad H_1 : \mu \neq 0.16$$

The built-in function `t.test ()` in R can be employed for the actual data while this example supplied a summary of the recorded values for the concentration of Pb in the sediment including the mean, sample size, and standard deviation; thus, we cannot employ this function here. We should write scripts employing built-in functions and commands in R to carry out the *t*-test. R built-in functions and commands for a two-tailed *t*-test are presented below.

```
Average = 0.23
m0 = 0.16
s = 0.03
n = 11
t = (Average - m0)/(s / sqrt(n))
t
alpha = 0.05
c.v = qt (1 - alpha/2, df = n-1)
t1 = round (c.v, digits = 2)
c.v = c(-t1, t1)
c.v
```

A two-tailed *t*-test for the average concentration of Pb in the sediment was carried out employing the commands and built-in functions, the outputs regarding the *t*-test are presented below.

```
> Average = 0.23
> m0 = 0.16
> s = 0.03
> n = 11
> t = (Average - m0)/ (s / sqrt(n))
> t
[1] 7.738791
> alpha = 0.05
> c.v = qt (1 - alpha/2, df = n-1)
> t1 = round (c.v, digits = 2)
> c.v = c(-t1, t1)
> c.v
[1] -2.23 2.23
```

The null hypothesis $H_0 : \mu = 0.16$ is rejected because the computed *t* value `7.7387` $= 7.74$ is greater than the *t* critical value of 2.23 using 0.05 significance level.

One can generate a normal plot with the critical (rejection) and noncritical (nonrejection) regions. The critical and noncritical regions for the concentration of lead in the sediment are presented in Fig. 5.5.

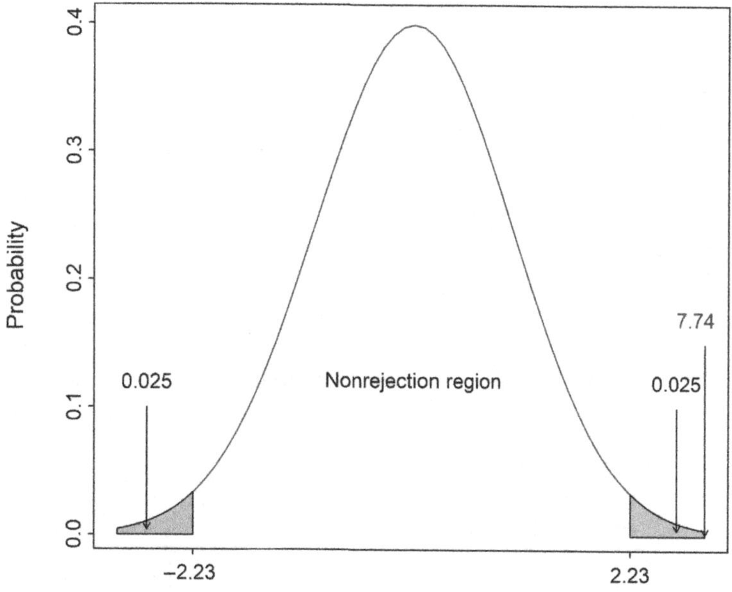

FIG. 5.5 The critical and noncritical regions for the Pb concentration in the sediments.

One can observe that the *t*-test statistic value of 7.74 falls in the rejection region as presented in Fig. 5.5. Thus, the null hypothesis regarding the average concentration of lead in the sediment should be rejected. This conclusion shows that the average value of lead in the sediment is not equal to 0.16 mg/L.

EXAMPLE 5.6 THE CONCENTRATION OF ZINC (ZN) IN THE WATER (MG/L)

A scientist in environmental science claims that the average value of the concentration of zinc (Zn) in the water of a certain river is 0.02 (mg/L). Eight sites are chosen and tested for the concentration of zinc (Zn). The sample data shows that the mean concentration of Zn is 0.092 and the standard deviation is 0.014. Use a significance level of $\alpha = 0.10$ to examine the researcher's claim, assuming that the population is normally distributed.

We should express the scientist claim in terms of mathematical form employing the null and alternative hypotheses.

$$H_0: \mu = 0.02 \text{ (claim)} \quad \text{vs} \quad H_1: \mu \neq 0.02$$

The *t*-test statistic for examining the mean concentration of zinc (Zn) in the water was computed employing R built-in functions and commands in R. The outputs produced by employing various commands are presented below.

```
> Average = 0.092
> m0 = 0.02
> s = 0.014
> n = 8
> t = (Average - m0) / (s / sqrt (n))
> t
[1] 14.5462
> alpha = 0.10
> c.v = qt (1 - alpha / 2, df = n-1)
> t1 = round (c.v, digits = 2)
> c.v = c (-t1, t1)
> c.v
[1] -1.89 1.89
```

We reject the null hypothesis concerning the average concentration of zinc in the water because the *t*-test statistic value `14.5462` =14.55 is greater than the *t* critical value of 1.89, or one can reach the same conclusion by employing the *t* curve with the rejection and nonrejection regions as presented in Fig. 5.6. There is enough evidence to conclude that the mean concentration of zinc (Zn) in the water of the river differs from the claimed value 0.02 mg/L.

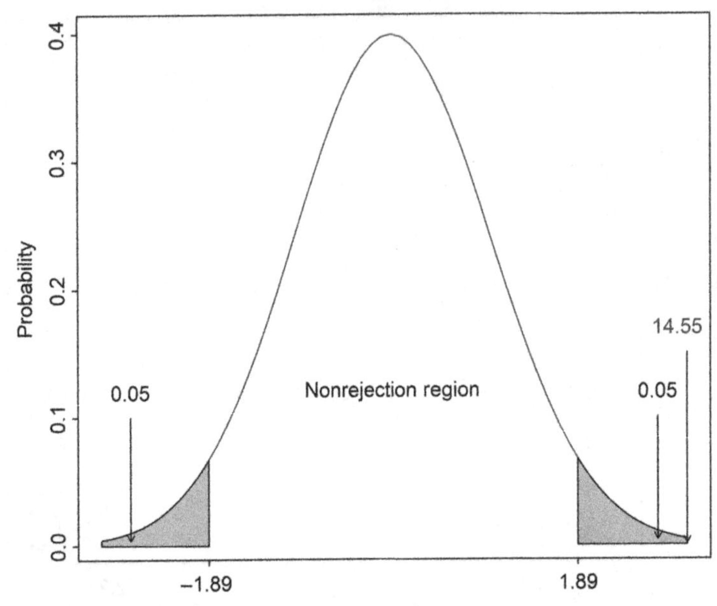

FIG. 5.6 The critical and noncritical regions for the concentration of Zn in the water.

Note:

- If the actual data of the research are employed, then the built-in function `t.test ()` is employed to perform the test.

```
t.test (data frame)
```

- The built-in function `t.test ()` can also be employed if the null hypothesis equals a given value. We can employ the command given below in this case.

```
t.test (data frame, mu = value) # Ho: mu = Claimed value
t.test (data frame, mu = value, alternative = "two.sided")
```

- The command for *t*-test can provide the same output without using the option `"two.sided"` since this option is the default. We can set the alternative hypothesis as `"less"` or `"greater"`. For example:

```
t.test (data frame, mu = value, alternative = "less")
```

EXAMPLE 5.7 THE CONCENTRATION OF CHROMIUM (CR) IN THE WATER (MG/L)

The concentration of Cr in the water of the Juru River in the Penang state of Malaysia was studied. Ten sampling points were chosen and tested for the concentration of Cr. The sample data are presented in Table 5.1.

Use a significance level of $\alpha=0.05$ to conduct the analysis and verify the four hypotheses given below concerning the mean concentration of Cr in the water.

Situation 1: $H_0:\mu=0$ vs $H_1:\mu\neq0$
Situation 2: $H_0:\mu=0.12$ vs $H_1:\mu\neq0.12$
Situation 3: $H_0:\mu\geq0.12$ vs $H_1:\mu<0.12$
Situation 4: $H_0:\mu\leq0.12$ vs $H_1:\mu>0.12$

TABLE 5.1 The Concentration of Chromium (Cr) mg/L in the Water Obtained from Ten Sampling Points

0.08	0.03	0.10	0.09	0.13	0.03	0.11	0.07	0.12	0.15

The built-in function `t.test ()` will be employed to verify the given hypotheses for the mean concentration of Cr in the water.

Situation 1

In the first situation, $\mu=0$ is equal to zero, the mean concentration of Cr in the water is equal to zero, while the other side is either less or >0. Thus, the two hypotheses for this case in terms of mathematical form are given below.

$$H_0:\mu=0 \quad \text{vs} \quad H_1:\mu\neq0$$

Thus, we should import the stored data .CSV (Example5_7) for the concentration of Cr in the water to R environment. Then, we can carry out the steps to verify the null hypothesis $H_0: \mu = 0$.

```
t.test (Example5_7)
```

The outputs of employing the built-in function `t.test ()` to perform a *t*-test to decide whether the mean concentration of Cr in the water equals or not equal to 0 are given below.

```
> t.test (Example5_7)

    One Sample t-test

data: Example5_7
t = 7.2218, df = 9, p-value = 4.965e-05
alternative hypothesis: true mean is not equal to 0
95 percent confidence interval:
 0.06249522 0.11950478
sample estimates:
mean of x
    0.091
```

One can observe that the first command appearing in the R outputs is the function `t.test (Example5_7)` to import the stored data in the file `Example5_7` and perform a *t*-test. The second row shows a notice "`one-sample t-test`" to tell us the type of the test. The computed *t*-test statistic value (`t`), degrees of freedom (`df`), and `p-value` are shown below the third row "`data: Example5_7`" of the outputs. The statement regarding the "`alternative hypothesis of:true mean is not equal to 0`" informs us that the test is a two-tailed test. Moreover, the two statements "`95 percent confidence interval`" and (`mean of x`) are presented as the last two results.

The null hypothesis is rejected because the *p*-value is very small (`p-value=4.965e-05`), which is less than $\alpha = 0.05$. We can conclude that that the mean concentration of Cr in the water is not 0.

Situation 2

In the second situation, $\mu = 0.12$, the mean concentration of Cr in the water is equal to a particular value (0.12), thus the null and alternative hypotheses in a mathematical form are:

$$H_0: \mu = 0.12 \quad \text{vs} \quad H_1: \mu \neq 0.12$$

The function `t.test ()` can also be employed to carry out the analysis for this situation.

```
t.test (Example5_7, mu = 0.12)
```

The built-in function `t.test()` produced the outputs shown below.

```
> t.test (Example5_7, mu = 0.12)

    One Sample t-test

data: Example5_7
t = -2.3015, df = 9, p-value = 0.04689
alternative hypothesis: true mean is not equal to 0.12
95 percent confidence interval:
 0.06249522 0.11950478
sample estimates:
mean of x
    0.091
```

The explanation of the R outputs is similar to the first situation when $\mu = 0$.

The null hypothesis should be rejected since the `p-value=0.04689`, which is less than $\alpha = 0.05$. It can be said that the mean concentration of Cr in the water differs from the given value of 0.12.

Situation 3

In the third situation, $\mu \geq 0.12$, the mean concentration of Cr in the water is more than or equal to a particular value (0.12), thus the null and alternative hypotheses in a mathematical form are:

$$H_0: \mu \geq 0.12 \quad \text{vs} \quad H_1: \mu < 0.12$$

The hypothesis of "greater than or equal" can be examined by employing the function `t.test ()`. Similar steps to those used for situations 1 and 2 can be employed to verify the hypothesis of greater than or equal, and choose the option "`less`" for the alternative hypothesis.

```
t.test (Example5_7, mu = 0.12, alternative = "less")
> t.test (Example5_7, mu = 0.12, alternative = "less")

    One Sample t-test

data: Example5_7
t = -2.3015, df = 9, p-value = 0.02344
alternative hypothesis: true mean is less than 0.12
95 percent confidence interval:
      -Inf 0.1140985
sample estimates:
mean of x
    0.091
```

One can see that the results of R are similar in meaning to the earlier two situations.

We reject the null hypothesis because of small p-value `0.02344`, which says that the mean concentration of Cr in the water is greater than or equal to 0.12.

Situation 4

In the fourth situation, $\mu \leq 0.12$, the mean concentration of Cr in the water is less than or equal to a particular value (0.12), thus the null and alternative hypotheses in mathematical form are:

$$H_0: \mu \leq 0.12 \quad \text{vs} \quad H_1: \mu > 0.12$$

The same structure of the R commands can be employed to verify the analysis; we should choose the option of `"greater"` for the alternative hypothesis.

```
t.test (Example5_7, mu = 0.12, alternative = " greater")
```

The output of employing the built-in function `t.test()` is similar to the other t-test result.

```
> t.test (Example5_7, mu = 0.12, alternative = "greater")

    One Sample t-test

data: Example5_7
t = -2.3015, df = 9, p-value = 0.9766
alternative hypothesis: true mean is greater than 0.12
95 percent confidence interval:
0.06790148        Inf
sample estimates:
mean of x
    0.091
```

We fail to reject the null hypothesis as the p-value (`p-value=0.9766`) is greater than 0.05. This conclusion indicates that the mean concentration of Cr in the water is less than or equal to 0.12.

It can be seen that the same built-in function `t.test` in R can be employed to carry out a t-test for various situations, and the only difference is to change the alternative choice.

5.4.2 Hypothesis Testing for a Mean Vector for one Sample

Hypothesis testing concerning an average value of the data set has been discussed for two tests, the Z-test when the sample size is large and the variance of the population is given, and the t-test when the sample size is small and the variance is unknown. Inferences regarding a mean vector in the case of several variables (k variables) will be discussed. Researchers are usually concerned in examining the hypotheses presented below in a vector form.

$$H_0: \boldsymbol{\mu} = \boldsymbol{\mu_0} \quad \text{vs} \quad H_1: \boldsymbol{\mu} \neq \boldsymbol{\mu_0}$$

$$H_0: \begin{bmatrix} \mu_1 \\ \mu_2 \\ \cdot \\ \cdot \\ \cdot \\ \mu_k \end{bmatrix} = \begin{bmatrix} \mu_{01} \\ \mu_{02} \\ \cdot \\ \cdot \\ \cdot \\ \mu_{0k} \end{bmatrix} \quad \text{vs} \quad H_1: \begin{bmatrix} \mu_1 \\ \mu_2 \\ \cdot \\ \cdot \\ \cdot \\ \mu_k \end{bmatrix} \neq \begin{bmatrix} \mu_{01} \\ \mu_{02} \\ \cdot \\ \cdot \\ \cdot \\ \mu_{0k} \end{bmatrix}$$

The average value of each selected variable equals a specified value, as stated by the null hypothesis (H_0), and the complement of the null hypothesis states that at least one average value does not equal its specified value, which represents the alternative hypothesis (H_1).

Hotelling's test is the test statistic employed for examining a hypothesis in the case of several variables. Consider a random sample of size n is selected from a population that has normal distribution and k variables measured for each sampling unit. The formula for the Hotelling's test (T^2) for examining k variables is presented in (5.4).

$$T^2 = n(\overline{Y} - \mu_0)' S^{-1} (\overline{Y} - \mu_0) \tag{5.4}$$

where $\overline{Y}$ refers to a mean vector and μ_0 is a vector of the specified values.

If the computed test statistic value is greater than the critical value of the Hotelling's test (T^2), the null hypothesis should be rejected ($T^2 > T^2_{\alpha,\, k,\, n-1}$), where $T^2_{\alpha,\, k,\, n-1}$ is the critical value (theoretical value).

EXAMPLE 5.8 THE CONCENTRATION OF HEAVY METALS IN THE SEDIMENT (MG/KG)

A scientist wants to study the concentrations of heavy metals in sediment. An experiment is carried out to measure the concentrations of five heavy metals, cadmium (Cd), iron (Fe), copper (Cu), zinc (Zn), and lead (Pb), at eight stations distributed across six rivers in the Penang State of Malaysia. The regions were Air Hitam River (stations 1–3), Dondang River (4), Air Puteh River (5), Air Terjun River (6), Jelutong River (7), and Pinang River (8). Three samples were selected from each station. Assume that the data are normally distributed. The sample data are presented in Table 5.2. Is there sufficient evidence to state that $\mu_1 = 0.67$, $\mu_2 = 15$, $\mu_3 = 102$, $\mu_4 = 27$, $\mu_5 = 1450$ for the concentrations of Cd, Pb, Zn, Cu and Fe, respectively? Use a significance level of $\alpha = 0.05$ to carry out the analysis.

TABLE 5.2 The Results of Heavy Metals Concentration in Sediments mg/kg

Region	Cd	Pb	Zn	Cu	Fe	Region	Cd	Pb	Zn	Cu	Fe
1	0.15	3.45	76.39	9.40	1658.83	13	0.35	0.01	34.45	4.50	1015.90
2	0.70	11.65	101.29	17.25	1680.33	14	0.95	0.01	28.85	0.40	1558.84
3	0.01	9.30	120.09	19.75	1674.83	15	0.50	0.02	48.25	1.10	1321.87
4	0.35	6.50	75.84	20.45	1294.37	16	0.55	0.01	85.39	9.95	1648.84
5	0.01	0.10	52.04	8.40	1711.83	17	1.95	1.90	59.14	8.15	1650.33
6	0.02	10.40	72.34	46.25	1605.34	18	1.10	0.01	166.58	18.20	1589.84
7	0.02	20.05	58.14	43.90	1790.32	19	0.01	0.10	63.29	2.65	1585.84
8	0.01	22.85	67.09	19.35	1705.83	20	0.01	0.01	77.59	58.94	2069.29
9	0.01	23.95	120.54	15.00	1481.85	21	1.20	124.44	470.25	81.09	1947.81
10	1.40	10.50	26.55	14.40	1713.33	22	3.90	83.79	259.67	59.49	2000.30
11	1.10	0.10	68.14	40.70	1492.35	23	2.60	24.30	183.93	43.10	1930.81
12	0.60	0.10	28.15	9.05	1674.83	24	2.45	36.05	298.67	53.29	1983.80

The mean concentration of heavy metals in the sediment should be placed in a vector form to represent the null and alternative hypotheses as given below.

$$H_0: \begin{bmatrix} \mu_1 \\ \mu_2 \\ \mu_3 \\ \mu_4 \\ \mu_5 \end{bmatrix} = \begin{bmatrix} 0.67 \\ 15 \\ 102 \\ 27 \\ 1450 \end{bmatrix} \quad \text{vs} \quad H_1: \begin{bmatrix} \mu_1 \\ \mu_2 \\ \mu_3 \\ \mu_4 \\ \mu_5 \end{bmatrix} \neq \begin{bmatrix} 0.67 \\ 15 \\ 102 \\ 27 \\ 1450 \end{bmatrix}$$

Because there are five variables (heavy metals) measured for each sampling unit, this case is considered as a multivariate case and Hotelling's test (T^2) should be employed for testing the null hypothesis for the equality of the mean vector of the heavy metals. We should calculate the mean value for each variable and define the means in a vector form, the variance–covariance matrix, and then compute the Hotelling's test statistic value (T^2).

Hotelling's test T^2 can be employed to examine the hypothesis regarding the mean vector of the concentrations of heavy metals in the sediment. Thus, one should employ the commands and built-in functions in R to carry out Hotelling's test T^2. The commands and built-in functions in R were employed to carry out Hotelling's test for the stored data of the heavy metals concentrations in the sediment `.CSV (Example5_8)`, including the computations of the variance–covariance matrix, mean vector, inverse matrix, transpose, and the value of Hotelling's test statistic.

```
# Compute the mean vector
Ms <- round (sapply (Example5_8, mean), digits = 3)
Ms
Mp <- c (0.67,15,102,27,1450)
# Compute the difference
D = Ms - Mp
D
#Compute the transpose
Dt = t(D)
Dt
# Compute the number of observations
n <- length (Example5_8 $ Cd)
n
# Calculate the covariance matrix
cov = round (cov (Example5_8), digits = 3)
cov
# Compute the inverse
INV = solve (cov)
INV
#Compute Hotelling's test value
TSQ = Dt %*% INV %*% D * n
TSQ
```

The built-in function `sapply ()` can be employed when a similar job must be performed multiple times for all of the columns of an array or list. Applying R commands and built-in functions for the concentrations of heavy metals in the sediment produced detailed results regarding the calculation of the Hotelling's test statistic.

```
> # Compute the mean vector
> Ms <- round (sapply (Example5_8, mean), digits = 3)
> Ms
    Cd        Pb        Zn       Cu       Fe
 0.831    16.233   110.111   25.198  1657.813
> Mp <- c (0.67,15,102,27,1450)
> # Compute the difference
> D = Ms - Mp
> D
    Cd       Pb       Zn        Cu        Fe
 0.161    1.233    8.111    -1.802    207.813
> #Compute the transpose
> Dt = t(D)
> Dt
        Cd     Pb     Zn      Cu       Fe
[1,] 0.161  1.233  8.111  -1.802  207.813
> # Compute the number of observations
> n <- length (Example5_8 $ Cd)
```

```
> n
[1] 24
> # Calculate the covariance matrix
> cov = round (cov (Example5_8), digits = 3)
> cov
          Cd        Pb         Zn       Cu        Fe
Cd     1.025    14.543     54.127    9.007   104.794
Pb    14.543   872.514   2713.982  483.172  3511.601
Zn    54.127  2713.982  10708.011 1712.913 13164.684
Cu     9.007   483.172   1712.913  497.764  3567.803
Fe   104.794  3511.601  13164.684 3567.803 57229.676
> # Compute the inverse
> INV = solve (cov)
> INV
            Cd            Pb            Zn            Cu
Cd  1.418548473 -7.484346e-03 -4.877454e-03  0.0102419129
Pb -0.007484346  5.812084e-03 -1.239927e-03 -0.0014929695
Zn -0.004877454 -1.239927e-03  5.030978e-04 -0.0003963994
Cu  0.010241913 -1.492969e-03 -3.963994e-04  0.0062571633
Fe -0.001654811  3.537434e-05 -6.003726e-06 -0.0002260441
              Fe
Cd -1.654811e-03
Pb  3.537434e-05
Zn -6.003726e-06
Cu -2.260441e-04
Fe  3.380609e-05
> #Compute Hotelling's test value
> TSQ = Dt %*% INV %*% D * n
> TSQ
         [,1]
[1,] 38.09285
```

The results of employing R commands and built-in functions showed the computed mean values for the selected heavy metals (MS) and the proposed mean values (MP), then the difference between computed (observed) and the proposed means (D) and the transpose of the difference (DT) are computed and presented. Then, the built-in functions are employed to compute the sample size (n), the variance–covariance matrix, and its inverse as well. The Hotelling's test value is presented in the last row of the results (38.09285=38.09).

The calculated value of the Hotelling's test statistic is 38.09 (TSQ=38.09) and the critical value at 0.05 significance level is 16.585 ($T^2_{0.05,\,3,\,5}=16.585$). The decision regarding the concentrations of heavy metals in the sediment is to reject the null hypothesis because the computed value (38.09) is greater than the critical value (16.585) ($38.09>16.585$) This conclusion marks that at least one of the selected heavy metals does not equal the proposed value. The values for Cd, Pb, Zn, Cu, and Fe are $\mu_1=0.67$, $\mu_2=15$, $\mu_3=102$, $\mu_4=27$, and $\mu_5=1450$ respectively.

EXAMPLE 5.9 THE THE CONCENTRATION OF HEAVY METALS IN THE WATER

The concentrations of three heavy metals in the water were measured in twenty samples obtained from the Juru River in the Penang state of Malaysia. The heavy metals are chromium (Cr), arsenic (As), and cadmium (Cd). The sample data are presented in Table 5.3.

TABLE 5.3 The Concentrations of Three Heavy Metals mg/L in the Water Obtained from the Juru River

Cd	Cr	As	Cd	Cr	As
0.14	0.08	5.98	0.17	0.13	2.92
0.12	0.03	1.77	0.19	0.14	3.51
0.13	0.10	2.88	0.21	0.14	3.02
0.13	0.09	3.80	0.18	0.15	3.64
0.12	0.13	3.27	0.17	0.14	3.21
0.14	0.03	1.67	0.21	0.15	3.80
0.13	0.11	3.83	0.28	0.15	3.59
0.12	0.07	5.28	0.27	0.15	3.28
0.15	0.12	2.20	0.25	0.12	3.47
0.16	0.15	3.87	0.12	0.13	2.73

Is there enough evidence to conclude that $\mu_1=0.25$, $\mu_2=0.10$, and $\mu_3=4.50$? Use a significance level of $\alpha=0.05$ to carry out the analysis.

The mean concentration of selected heavy metals in the water can be placed in a vector form to represent the null and alternative hypotheses in a mathematical form.

$$H_0:\begin{bmatrix}\mu_1\\ \mu_2\\ \mu_3\end{bmatrix}=\begin{bmatrix}0.25\\ 0.10\\ 4.50\end{bmatrix} \quad \text{vs} \quad H_1:\begin{bmatrix}\mu_1\\ \mu_2\\ \mu_3\end{bmatrix}\neq\begin{bmatrix}0.25\\ 0.10\\ 4.50\end{bmatrix}$$

The outputs of applying the necessary R built-in functions and commands to calculate the Hotelling's test statistic (T^2) for the stored data .CSV (Example5_9) of the heavy metals concentrations in the water are presented below.

```
> #calculate the mean vector (observed)
> Ms <- round (sapply (Example5_9, mean), digits = 3) # Compute the mean vector (observed)
> Ms                                   # print the mean vector
   Cd    Cr    As
0.170 0.116 3.386
> # proposed (Hypothesized) mean vector
> Mp<- c(0.25,0.10,4.50)
> #Compute the difference
> D = Ms - Mp
> D
    Cd    Cr     As
-0.080 0.016 -1.114
> # Compute the transpose for the difference
> Dt = t(D)
> Dt
        Cd    Cr     As
[1,] -0.08 0.016 -1.114
> # Compute the number of values
> n <- length (Example5_9 $ Cd)
> n
[1] 20
> # Ccompute the covariance matrix
> cov = round (cov (Example5_9), digits = 3)
> cov
      Cd    Cr    As
Cd 0.003 0.001 0.002
Cr 0.001 0.001 0.006
```

```
As 0.002 0.006 1.014
> # computee the inverse
> IN = solve(cov)
> IN
              Cd          Cr           As
Cd    504.123711  -516.494845   2.061856
Cr   -516.494845  1565.979381  -8.247423
As      2.061856    -8.247423   1.030928
> # Ccompute Hotelling's test value
> TSQ = Dt %*% IN %*% D * n
> TSQ                              # Print Hotelling's test value
         [,1]
[1,] 137.8079
```

Because the Hotelling's test statistic value as it appeared in the last row of the R outputs (`TSQ=137.8079` =137.81) and the critical value at 0.05 significance level is 10.117 ($T^2_{0.05,2,20}=10.117$), we should reject the null hypothesis. This result indicates that the mean concentration of at least one of the selected heavy metals (Cd, Cr and As) differs from the proposed (hypothesized) values of $\mu_1=0.25$, $\mu_2=0.10$, $\mu_3=4.50$.

Note:

- In some situations, the conclusion of a univariate test does not agree with the conclusion of a multivariate test. However, we should prefer the multivariate conclusion.

5.5 HYPOTHESIS TESTING FOR TWO POPULATION MEANS

The idea of hypothesis testing for a single population can be improved to cover the two population means, including a single variable and several variables. Scientists are usually concerned in exploring the differences between two sets of data regarding the same variable (variables) measured in each experimental unit to get a clear and inclusive view of the variable (variables) under investigation.

5.5.1 Hypothesis Testing for Two Population Means

Suppose there are two normally distributed populations. A sample is chosen from each population, with n_1 to represent the sample size selected from population 1 and n_2 to represent the sample size selected from population 2. Assume that the variance of the population 1 is equal to the variance of population 2; $\sigma_1^2=\sigma_2^2=\sigma^2$.

The scientists are usually concerned in examining the hypotheses for equality of two means, as given below:

$$H_0:\mu_1=\mu_2 \quad \text{vs} \quad H_1:\mu_1\neq\mu_2$$

or

$$H_0:\mu_1-\mu_2=0 \quad \text{vs} \quad H_1:\mu_1-\mu_2\neq 0$$

The two populations have the same mean values as stated by the null hypothesis (H_0), which indicates that there is no difference between the average values of the two populations, while the alternative hypothesis indicates (H_1) that a difference between the average values of the two populations exists.

The test statistic employed to examine the hypothesis of equality of the two means is called the *t*-test for two independent samples. Consider that two samples were selected, with n_1 to represent the sample size for the first population, n_2 to represent the sample size for the second population, and a common standard deviation called pooled standard deviation S_p as defined in (5.5).

$$Sp=\sqrt{\frac{(n_1-1)S_1^2+(n_2-1)S_2^2}{n_1+n_2-2}} \tag{5.5}$$

Consider there are two numeric variables y_1 and y_2, the *t*-test statistic formula for two independent samples is presented in (5.6).

$$t = \frac{\overline{Y}_1 - \overline{Y}_2}{Sp\sqrt{\frac{1}{n_1} + \frac{1}{n_2}}} \quad (5.6)$$

where $d.f = n_1 + n_2 - 2$.

The null hypothesis should be rejected if the critical value is less than or equal to the absolute value of the computed *t*-test value: $t_{\frac{\alpha}{2}, n_1+n_2-2} \le |t|$, or one can employ a *t* distribution curve to recognize the rejection and nonrejection regions; we reject the null hypothesis if the *t*-test statistic value falls in the critical (rejection) region.

We can to compute the value of the *t*-test statistic for two independent samples easily employing built-in functions offered by *R*. The command for performing the *t*-test for two independent samples employing the function `t.test ()` is given below:

```
t.test (y1, y2)
```

where y_1 and y_2 are numeric variables.

Note:

- The same function `t.test ()` is employed for two variables including one numeric variable and one binary variable.

```
t.test (y ~ x)
```

where y is numeric and x is binary.

EXAMPLE 5.10 THE CONCENTRATION OF CADMIUM (CD) IN THE SEDIMENT (MG/L)

A researcher is concerned in comparing the concentration of cadmium (Cd) in the sediment obtained from the Juru and Jejawi Rivers in the Penang state of Malaysia. Ten sampling points are chosen from each river, and the Cd concentration is measured. Assume that the two populations are normally distributed. The sample data are presented in Table 5.4. Use a significance level of $\alpha = 0.05$ to carry out the analysis regarding the mean concentration of Cd in the two rivers.

The null and alternative hypotheses for the difference in the concentration of Cd in the sediment for the two rivers in a math-

TABLE 5.4 The Concentration of Cadmium (Cd) mg/L in Sediment from the Juru and Jejawi Rivers

Juru	Jejawi	Juru	Jejawi
1.91	1.60	1.93	1.68
1.92	1.69	1.94	1.76
1.95	1.62	2.00	1.75
2.02	1.64	2.00	1.71
1.90	1.66	1.95	1.73

ematical form are:

$$H_0: \mu_1 = \mu_2 \quad \text{vs} \quad H_1: \mu_1 \neq \mu_2$$

There is no difference in the concentration of the cadmium in the sediment for the two rivers as stated by the null hypothesis, this means that the concentration of the cadmium in the sediment of Juru River (μ_1) is equal to the concentration of the cadmium of Jejawi River (μ_2), whereas the alternative hypothesis stated that the concentration of cadmium in sediment of Juru and Jejawi Rivers is different.

Testing the hypothesis for the stored data. CSV (`Example5_10`) about the concentration of Cd in the two rivers can be performed employing the built-in function `t.test (y1, y2)`.

```
t.test (Example5_10 $ Juru, Example5_10 $ Jejawi)
```

The sample data for the concentration of Cd in the sediment recorded from Juru and Jejawi Rivers were analyzed employing a *t*-test for two independent samples. The results of employing the built-in function `t.test(y1, y2)` are given below.

```
> T.v <- t.test (Example5_10 $ Juru, Example5_10 $ Jejawi)
> T.v

      Welch Two Sample t-test

data: Example5_10$Juru and Example5_10$Jejawi
t = 12.406, df = 16.789, p-value = 7.047e-10
alternative hypothesis: true difference in means is not equal to 0
  95 percent confidence interval:
   0.2223791 0.3136209
sample estimates:
mean of x mean of y
    1.952          1.684
```

One can observe that the first command appeared in the first row is the function `t.test ()` to perform a *t*-test for the stored data for the two rivers (`Example5_10$Juru and Example5_10$Jejawi`). Then, the information regarding the *t*-test is provided immediately after performing the built-in function `t.test()`, such as `Welch Two Sample t-test, data: Example5_10$Juru and Example5_10$Jejawi`, the test statistic value (`t`), `df`, and `p-value`.

The null hypothesis should be rejected because the *p*-value is very small: the `p-value <7.047e-10`, which is <0.05. One can conclude that the concentration of Cd in the sediment of Juru and Jejawi Rivers is different. This difference could be due to the differences in the surrounding area of each river, such as different industrial activities or residential areas.

This conclusion can be presented in a plot. R built-in functions and commands can be employed to generate a *t* curve with the computed *t*-test statistic values, critical values, critical (rejection) region, and noncritical (nonrejection) regions, as presented in Fig. 5.7.

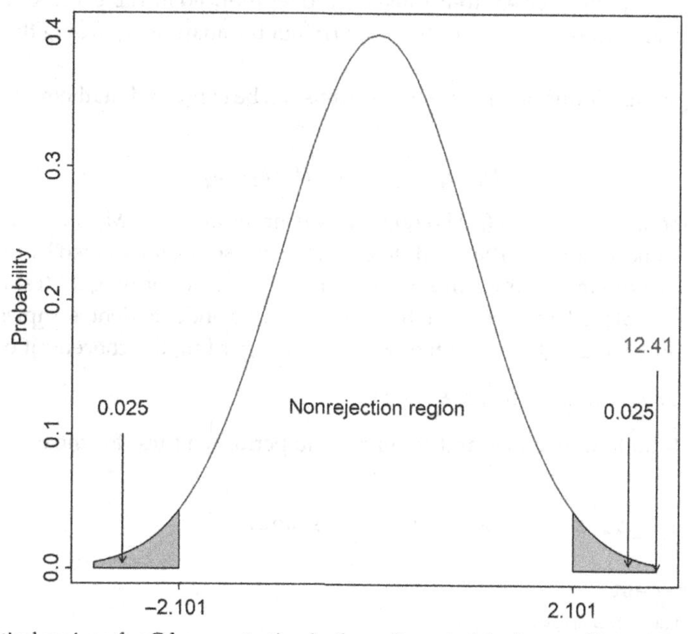

FIG. 5.7 The critical and noncritical regions for Cd concentration in the sediment of the Juru and Jejawi Rivers.

EXAMPLE 5.11 THE PARTICULATE MATTER PM_{10} IN THE AIR FOR TWO SEASONS ($\mu G/M^3$)

A researcher wants to study the concentration of particulate matter (PM_{10}) in the air of an equatorial urban coastal location. Air pollution levels are studied during the summer (dry) and winter monsoon (wet) seasons using high-volume sampling

TABLE 5.5 Particulate Matter PM_{10} $\mu g/m^3$ in the Air During the Summer and Winter Monsoon Seasons

Summer (Dry)	Winter (Wet)	Summer (Dry)	Winter (Wet)
37.71	82.87	123.40	62.28
48.03	138.25	72.39	65.87
67.87	82.67	51.85	122.83
39.01	46.08	77.59	50.18
38.33	15.33	30.30	60.19
29.70	30.78	100.40	28.28
53.66	155.00	132.98	42.50
132.28	61.75	126.38	49.68
66.31	88.90	31.82	39.91
69.20	38.47	110.53	37.94
78.17	24.46	38.41	27.97
31.63	14.29	124.26	45.34
66.73	65.52	53.62	58.45
113.56	63.33	23.30	43.30

techniques. Atomic absorption spectrophotometry is used to collect PM_{10} samples with an average time of 24h. Twenty-eight records from the dry and wet seasons are selected from Table 1.4 and reproduced in Table 5.5. Assume that the two populations are normally distributed. Use a significance level of $\alpha=0.05$ to carry out the analysis regarding the concentration of PM_{10} in the dry and wet seasons.

The concentrations of PM_{10} in the air during dry and wet seasons can be expressed mathematically to represent the null and alternative hypotheses:

$$H_0:\mu_1=\mu_2 \quad \text{vs} \quad H_1:\mu_1\neq\mu_2$$

The concentration of PM_{10} in the summer season (μ_1) is equal to the concentration of PM_{10} in the winter monsoon (μ_2), as stated by the null hypothesis, and the concentration of PM_{10} is different in the two seasons, as stated by the alternative hypothesis. The equality of the means of PM_{10} in both the summer and winter was tested by performing a t-test for two samples. The built-in function `t.test ()` can be employed to carry out the t-test for two independent samples for the stored data `.CSV` (`Example5_11 $ Dry, Example5_11 $ Wet`) of the two seasons regarding the concentration of PM_{10} in the air.

```
t.test (Example5_11 $ Dry, Example5_11 $ Wet)
```

The outputs of employing R built-in functions and commands to perform a t-test for the concentration of PM_{10} in the two seasons are shown below.

```
> T.v <- t.test (Example5_11 $ Dry, Example5_11 $ Wet)
> T.v
    Welch Two Sample t-test
data: Example5_11$Dry and Example5_11$Wet
t = 1.2416, df = 53.868, p-value = 0.2198
alternative hypothesis: true difference in means is not equal to 0
95 percent confidence interval:
 -7.181198 30.538341
sample estimates:
mean of x mean of y
 70.33643  58.65786
```

The output generated by using R's built-in functions to examine the concentration of PM_{10} in the dry and wet seasons showed that one cannot reject the null hypothesis, as the `p-value` of `0.2198` is large (>0.05). R commands and built-in functions enable users to display the decision in graphical form, including the critical (rejection) and noncritical (nonrejection) regions, with the critical values and computed t-test statistic value, as presented in Fig. 5.8.

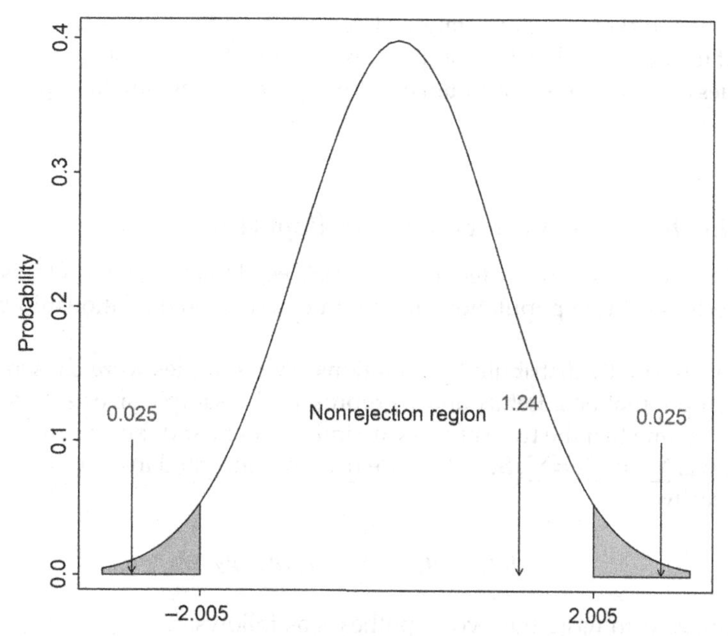

FIG. 5.8 The critical and noncritical regions for the concentration of PM_{10} in the air.

In summary, one can conclude that different seasons do not influence the concentration of PM_{10} in the air.

Note:

- The built-in function `t.test ()` in R cannot be employed to carry out the *t*-test when brief data are provided including the average, variance, and sample size. However, we still can employ built-in functions and commands to compute the *t*-test statistic value. Use the information presented in Example 5.11 to calculate the *t*-test statistic value for the two independent samples.

```
> #Input information
> MA = 70.336
> MB = 58.658
> n1 = 28
> n2 = 28
> df = n1 + n2 - 2
> M = MA - MB
> sA = 36.0569
> sB = 34.313
> pv = sqrt (((n1 - 1)*(sA ^ 2) + (n2 - 1)*(sB ^ 2)) /df)
> pv
[1] 35.19575
> A = sqrt ((1/n1) + (1/n2))
> T.v = M /(pv * A)
> T.v
[1] 1.241487
> alpha = 0.05
> c.v = qt(1 - alpha/2, df= n1 + n2 - 2)
> t = round (c.v, digits = 3)
> c.vs = c(-t, t)
> c.vs
[1] -2.005 2.005
```

The information concerning the average value, standard deviation, and number of observations for each sample are presented in the first part of the output, while the computations related to the t-test are presented in the second part of the output. The critical values (±2.005) were computed employing the built-in function `qt ()` and appeared in the last row of the results.

5.5.2 Hypothesis Testing for Mean Vectors for Two Populations

Suppose that there are k response variables (dependent variables). Hotelling's test (T^2) is usually employed to test the equality of the mean vectors for two populations taking into account the relationship between various response variables.

Consider two multivariate normally distributed populations. Two samples were chosen, one sample from each of the two multivariate normal populations. Let n_1 and n_2 represent the sample sizes selected from population 1 and population 2, respectively. Assume that the two samples are independent and the variance of population 1 is the same as the variance of population 2. $\sum_1 = \sum_2 = \sum$. Scientists are usually concerned in examining the hypotheses presented in a vector form mathematically.

$$H_0: \boldsymbol{\mu}_1 = \boldsymbol{\mu}_2 \quad \text{vs} \quad H_1: \boldsymbol{\mu}_1 \neq \boldsymbol{\mu}_2$$

The vector form can be employed to write the two hypotheses as follows:

$$H_0: \begin{bmatrix} \mu_{11} \\ \mu_{21} \\ \cdot \\ \cdot \\ \cdot \\ \mu_{k1} \end{bmatrix} = \begin{bmatrix} \mu_{12} \\ \mu_{22} \\ \cdot \\ \cdot \\ \cdot \\ \mu_{k2} \end{bmatrix} \quad \text{vs} \quad H_1: \begin{bmatrix} \mu_{11} \\ \mu_{21} \\ \cdot \\ \cdot \\ \cdot \\ \mu_{k1} \end{bmatrix} \neq \begin{bmatrix} \mu_{12} \\ \mu_{22} \\ \cdot \\ \cdot \\ \cdot \\ \mu_{k2} \end{bmatrix}$$

Let $\overline{\mathbf{Y}}_1$ and $\overline{\mathbf{Y}}_2$ represent the mean vectors for the sample 1 and sample 2 respectively, with a common variance called pooled variance (Sp), computed as given in (5.7).

$$S_p = \frac{(n_1 - 1)\,\mathbf{S}_1 + (n_2 - 1)\,\mathbf{S}_2}{n_1 + n_2 - 2} \tag{5.7}$$

where $\mathbf{S}_1$ and $\mathbf{S}_2$ are the variance–covariance matrices for the sample 1 and sample 2, respectively.

The Hotelling's test (T^2) statistic formula for comparing two multivariate populations means is given in (5.8).

$$T^2 = \frac{n_1 n_2}{n_1 + n_2}\left(\overline{\mathbf{Y}}_1 - \overline{\mathbf{Y}}_2\right)' Sp^{-1}\left(\overline{\mathbf{Y}}_1 - \overline{\mathbf{Y}}_2\right) \tag{5.8}$$

distributed as $T_{\alpha,\ k,\ n_1+n2-2}$.

The null hypothesis (the equality of two mean vectors) should be rejected if $T^2 \geq T^2_{\alpha,\ k,\ n_1+n_2-2}$.

Note:

- It is necessary that $n_1 + n_2 - 2 > k$.

EXAMPLE 5.12 THE CONCENTRATIONS OF HEAVY METALS IN TWO RIVERS (MG/L)

A scientist wishes to compare the concentrations of five heavy metals in water samples obtained from the Juru and Jejawi Rives in the Penang state of Malaysia. Ten sampling points are chosen from each river and examined for copper (Cu), zinc (Zn), iron (Fe), mercury (Hg), and lead (Pb) concentrations (mg/L). Assume that the two populations are normally distributed. The sample data are presented in Table 5.6. Use a significance level of $\alpha = 0.05$ to carry out the analysis concerning equality of the mean concentrations of the selected heavy metals in the Juru and Jejawi Rivers.

TABLE 5.6 Selected Heavy Metals Concentrations in Water mg/L

Location	Cu	Zn	Fe	Hg	Pb
Juru	0.069	0.394	18.510	9.415	0.498
Juru	0.030	0.355	22.400	7.950	0.126
Juru	0.088	0.440	23.120	9.315	0.809
Juru	0.049	0.583	23.100	9.240	0.723
Juru	0.048	0.696	22.580	7.535	0.644
Juru	0.091	0.559	23.730	12.40	0.138
Juru	0.393	1.745	24.270	9.375	0.534
Juru	0.284	2.099	29.160	10.88	0.479
Juru	0.323	2.724	25.720	13.100	0.499
Juru	0.233	2.131	25.870	9.370	0.406
Jejawi	0.031	0.452	1.025	11.230	0.061
Jejawi	0.011	0.182	1.665	11.430	0.209
Jejawi	0.019	0.197	1.160	19.050	0.105
Jejawi	0.013	0.157	0.907	9.780	0.073
Jejawi	0.019	0.199	1.501	12.970	0.229
Jejawi	0.096	0.211	1.303	14.620	0.054
Jejawi	0.005	0.192	1.542	19.080	0.286
Jejawi	0.012	0.166	0.916	16.020	0.181
Jejawi	0.007	0.237	0.976	15.460	0.081
Jejawi	0.017	0.225	1.163	13.120	0.139

The concentrations of heavy metals in the water of Juru River are the same as those in the Jejawi River—this statement represents the null hypothesis, which means that the difference is 0, while the complement of the null hypothesis indicates that at least one chosen heavy metal concentration differs in the two rivers. The two hypotheses can be expressed mathematically to represent the concentrations of the selected heavy metals.

$$H_0 : \mu_1 = \mu_2 \quad \text{vs} \quad H_1 : \mu_1 \neq \mu_2$$

$$H_0 = \begin{bmatrix} 0.161 \\ 1.173 \\ 23.850 \\ 9.858 \\ 0.468 \end{bmatrix} = \begin{bmatrix} 0.023 \\ 0.222 \\ 1.216 \\ 14.276 \\ 0.142 \end{bmatrix} \quad \text{vs} \quad H_1 = \begin{bmatrix} 0.161 \\ 1.173 \\ 23.850 \\ 9.858 \\ 0.468 \end{bmatrix} \neq \begin{bmatrix} 0.023 \\ 0.222 \\ 1.216 \\ 14.276 \\ 0.142 \end{bmatrix}$$

The mean vector, variance–covariance matrix for the stored data (`.CSV Example5_12`) for the two rivers with Hotelling's test (T^2) are computed employing the commands and built-in functions in R. The R commands and built-in functions for examining the mean vectors of the chosen heavy metals in the water of the two rivers are shown below.

```
#compute the mean vector
MV1<- sapply(Example5_12[1:10,2:6], mean)
MV1
MV2 <- sapply (Example5_12[11:20,2:6], mean)
MV2
#Compute the variance-covariance matrix
round (cov (Example5_12[1:10,2:6]), digits = 3) #Juru River
round (cov (Example5_12[11:20,2:6]), digits = 3) #Jejawi River
# Calculate the number of observations
```

```
n1 = length (Example5_12[1:10,2:6] $ Cu)
n2 = length (Example5_12[11:20,2:6] $ Cu)
#Calculate the pooled variance
spv = (((n1 - 1)*cov(Example5_12[1:10,2:6])) + ((n2 - 1)*cov(Example5_12[11:20,2:6])))/(n1 +
n2 - 2)
round (spv, digits = 3)
D = MV1 - MV2                 # Compute the difference
D1 = t(D)        # Compute the transpose
INV = solve(spv) # Compute the inverse
#Compute the value of the Hotelling's test statistic
TSQ = ((n1*n2) / (n1 + n2)) * (D %*% INV %*% D)
TSQ                 #print the result of Hotelling's test
```

The results of employing R built-in functions and commands to carry out Hotelling's test for the concentrations of chosen heavy metals in the water are presented in detail:

```
> #compute the mean vector
> MV1<- sapply(Example5_12[1:10,2:6], mean)
> MV1
     Cu      Zn      Fe      Hg      pb
 0.1608  1.1726 23.8460  9.8580  0.4856
> MV2 <- sapply (Example5_12[11:20,2:6], mean)
> MV2
     Cu      Zn      Fe      Hg      pb
 0.0230  0.2218  1.2158 14.2760  0.1418
> #Compute the variance-covariance matrix
> round (cov (Example5_12[1:10,2:6]), digits = 3)
      Cu     Zn     Fe     Hg     pb
Cu 0.018  0.107  0.241  0.110  0.001
Zn 0.107  0.808  1.888  0.827  0.001
Fe 0.241  1.888  7.699  2.104 -0.030
Hg 0.110  0.827  2.104  3.155 -0.093
pb 0.001  0.001 -0.030 -0.093  0.049
> round (cov (Example5_12[11:20,2:6]), digits = 3)
       Cu     Zn     Fe     Hg     pb
Cu  0.001  0.000  0.000 -0.006 -0.001
Zn  0.000  0.007 -0.005 -0.068 -0.003
Fe  0.000 -0.005  0.076  0.100  0.015
Hg -0.006 -0.068  0.100 10.118  0.082
pb -0.001 -0.003  0.015  0.082  0.006
> # Calculate the number of observations
> n1 = length (Example5_12[1:10,2:6] $ Cu)
> n2 = length (Example5_12[11:20,2:6] $ Cu)
> #Calculate the pooled variance
> spv = (((n1 - 1)*cov(Example5_12[1:10,2:6])) + ((n2 - 1)*cov(Example5_12[11:20,2:6])))/(n1 +
n2 - 2)
> round (spv, digits = 3)
      Cu     Zn     Fe     Hg     pb
Cu 0.009  0.054  0.121  0.052  0.000
Zn 0.054  0.408  0.942  0.380 -0.001
Fe 0.121  0.942  3.887  1.102 -0.007
Hg 0.052  0.380  1.102  6.637 -0.006
pb 0.000 -0.001 -0.007 -0.006  0.028
>
> D = MV1 - MV2                 # Compute the difference
> D1 = t(D)        # Compute the transpose
```

```
> INV = solve(spv)     # Compute the inverse
> #Compute the value of the Hotelling's test statistic
> TSQ = ((n1*n2) / (n1 + n2)) * (D1 %*% INV %*% D)
> TSQ                  #print the result of Hotelling's test
         [,1]
[1,] 1317.652
```

The mean vectors for the heavy metals concentrations in Juru and Jejawi Rivers and the variance–covariance matrices are computed employing the two built-in functions `sapply ()` and `cov ()` for the mean and covariance, respectively. We can compute the value of the Hotelling's test statistic to test the equality of the mean vectors of heavy metals concentrations. Because the critical value at $\alpha=0.05$ significance level $T^2_{0.05,\,5,\,22}=19.017$ of the Hotelling's test is less than the computed Hotelling's test statistic value `TSQ=1317.652`, the null hypothesis should be rejected. One can conclude that the mean vectors of the two rivers are different for at least one heavy metal concentration in the water.

EXAMPLE 5.13 COMPARING THE PHYSIOCHEMICAL PARAMETERS OF TWO RIVERS (MG/L)

An assessment of water quality in the Juru and Jejawi Rivers in the Penang State of Malaysia was conducted by monitoring seven parameters. The electrical conductivity (EC) was measured employing a HACH portable pH meter, and dissolved oxygen (DO) was measured with a YSI 1000 DO meter. The biochemical oxygen demand (BOD), chemical oxygen demand (COD), and turbidity were measured employing a nephelometer. The total suspended solids (TSS) was analyzed gravimetrically in the laboratory. The APHA Standard Methods for the Examination of Water and Wastewater were applied to analyze the concentration of the abovementioned water quality parameters. The data obtained from ten different sites are given in Table 5.7. Use a significance level of $\alpha=0.05$ to carry out the analysis concerning the equality of the mean concentration of the selected physiochemical parameters of the Juru and Jejawi samples.

TABLE 5.7 Selected Water Quality Parameters mg/L for the Juru and Jejawi Rivers

River	PH	DO	BOD	COD	TSS	EC	Turbidity
Juru	7.88	6.73	10.56	1248.00	473.33	42.45	13.05
Juru	7.92	6.64	10.06	992.50	461.67	42.75	14.11
Juru	7.41	5.93	6.01	1265.00	393.34	29.47	25.95
Juru	7.84	6.23	14.57	1124.00	473.34	42.35	22.00
Juru	7.86	6.29	7.36	1029.50	528.34	40.70	12.36
Juru	7.89	6.36	5.27	775.00	458.33	42.50	17.90
Juru	7.40	5.67	4.81	551.00	356.67	29.51	27.70
Juru	7.41	5.26	4.96	606.00	603.33	28.45	29.35
Juru	7.25	4.18	6.61	730.00	430.00	26.25	35.35
Juru	7.18	5.14	5.11	417.00	445.00	25.79	27.75
Jejawi	7.40	4.96	36.65	82.35	815.00	29.55	1.01
Jejawi	7.53	5.86	37.20	122.90	858.34	21.80	0.77
Jejawi	7.45	5.52	41.50	46.35	823.34	27.55	1.12
Jejawi	7.41	5.27	36.75	52.65	821.67	16.45	0.46
Jejawi	7.27	5.71	37.95	99.95	736.67	21.10	1.01
Jejawi	7.36	5.16	34.85	36.30	373.34	11.50	0.73
Jejawi	7.37	5.08	37.05	33.50	385.00	13.35	0.71
Jejawi	7.40	5.23	37.10	36.35	403.34	10.85	0.57
Jejawi	7.38	5.30	38.30	30.00	550.00	13.90	0.75
Jejawi	7.36	5.17	38.40	26.15	363.34	12.65	0.62

The two hypotheses can be expressed mathematically to represent the concentrations of the selected physiochemical parameters for the Juru and Jejawi Rivers:

$$H_0: \mu_1 = \mu_2 \quad \text{vs} \quad H_1: \mu_1 \neq \mu_2$$

$$H_0: \begin{bmatrix} 7.604 \\ 5.843 \\ 7.532 \\ 873.800 \\ 462.335 \\ 35.022 \\ 22.552 \end{bmatrix} = \begin{bmatrix} 7.393 \\ 5.326 \\ 37.575 \\ 56.650 \\ 613.004 \\ 17.870 \\ 0.775 \end{bmatrix} \quad \text{vs} \quad H_1: \begin{bmatrix} 7.604 \\ 5.843 \\ 7.532 \\ 873.800 \\ 462.335 \\ 35.022 \\ 22.552 \end{bmatrix} \neq \begin{bmatrix} 7.393 \\ 5.326 \\ 37.575 \\ 56.650 \\ 613.004 \\ 17.870 \\ 0.775 \end{bmatrix}$$

The Hotelling's test was performed for the stored data (`.CSV` files `Example5_13`) to examine the equality of the mean vectors of the water quality parameters of the Juru and Jejawi Rivers.

```
#compute the mean vector
MV1 <- sapply (Example5_13[1:10,], mean)
MV1
MV2 <- sapply (Example5_13[11:20,], mean)
MV2
#Compute the variance-covariance matrix
round (cov(Example5_13[1:10,]), digits = 3)
round (cov(Example5_13[11:20,]), digits = 3)
# Compute the number of observations
n1 = length (Example5_13[1:10,] $ DO)
n1
n2 = length (Example5_13[11:20,] $ DO)
n2
#Compute the pooled variance
spv = (((n1 - 1)*cov(Example5_13[1:10,])) + ((n2 - 1)*cov(Example5_13[11:20,])))/(n1 + n2 - 2)
round (spv, digits = 3)      # Compute the difference
D = MV1 - MV2
D1 = t(D)        # Compute the transpose
INV = solve (spv)     # Compute the inverse
#Compute the value of the Hotelling's test statistic
TSQ = ((n1*n2)/(n1 + n2))*(D1%*%INV%*%D)
TSQ               #print the result of Hotelling's test
```

The above commands and built-in functions were carried out to calculate the Hotelling's test statistic value for the physicochemical parameters of the water samples obtained from the Juru and Jejawi Rivers.

```
> #compute the mean vector
> MV1 <- sapply (Example5_13[1:10,], mean)
> MV1                              #Print m1
      PH        DO       BOD       COD       TSS  Conductivity
   7.604     5.843     7.532   873.800   462.335        35.022
Turbidity
   22.552
> MV2 <- sapply (Example5_13[11:20,], mean)
> MV2
      PH        DO       BOD       COD       TSS Conductivity
   7.393     5.326    37.575    56.650   613.004       17.870
Turbidity
    0.775
> #Compute the variance-covariance matrix
> round (cov(Example5_13[1:10,]), digits = 3)
```

```
                    PH         DO        BOD          COD        TSS  Conductivity  Turbidity
PH               0.089      0.208      0.590       55.326      5.284         2.257     -2.125
DO               0.208      0.631      1.250      149.793      4.106         5.263     -5.759
BOD              0.590      1.250     10.366      622.417     23.173        16.280    -11.420
COD             55.326    149.793    622.417    89890.900    -67.858      1391.275  -1354.542
TSS              5.284      4.106     23.173      -67.858   4640.688       109.394   -120.320
Conductivity     2.257      5.263     16.280     1391.275    109.394        58.148    -53.855
Turbidity       -2.125     -5.759    -11.420    -1354.542   -120.320       -53.855     62.494
> round (cov(Example5_13[11:20,]), digits = 3)
                    PH     DO       BOD       COD       TSS  Conductivity  Turbidity
PH               0.004  0.005     0.019     0.632     6.233         0.138     -0.001
DO               0.005  0.081     0.174     6.366    32.253         0.531      0.017
BOD              0.019  0.174     2.928    -4.082   115.741         4.887      0.177
COD              0.632  6.366    -4.082  1116.571  5278.207       138.684      2.841
TSS              6.233 32.253   115.741  5278.207 47083.713      1227.236     21.222
Conductivity     0.138  0.531     4.887   138.684  1227.236        45.702      1.133
Turbidity       -0.001  0.017     0.177     2.841    21.222         1.133      0.045
> # Compute the number of observations
> n1 = length (Example5_13[1:10,] $ DO)
> n1
[1] 10
> n2 = length (Example5_13[11:20,] $ DO)
> n2
[1] 10
> #Compute the pooled variance
> spv = (((n1 - 1)*cov(Example5_13[1:10,])) + ((n2 - 1)*cov(Example5_13[11:20,])))/(n1 + n2 - 2)
> round (spv, digits = 3)    # Compute the difference
                    PH     DO     BOD        COD        TSS Conductivity Turbidity
PH               0.047  0.107   0.304     27.979      5.758        1.197    -1.063
DO               0.107  0.356   0.712     78.080     18.179        2.897    -2.871
BOD              0.304  0.712   6.647    309.167     69.457       10.583    -5.621
COD             27.979 78.080 309.167  45503.736   2605.174      764.980  -675.851
TSS              5.758 18.179  69.457   2605.174  25862.200      668.315   -49.549
Conductivity     1.197  2.897  10.583    764.980    668.315       51.925   -26.361
Turbidity       -1.063 -2.871  -5.621   -675.851    -49.549      -26.361    31.269
> D = MV1 - MV2
> D1 = t(D) # Compute the transpose
> INV = solve (spv) # Compute the inverse
> #Compute the value of the Hotelling's test statistic
> TSQ = ((n1*n2)/(n1 + n2))*(D1%*%INV%*%D)
> TSQ    #print the result of Hotelling's test
        [,1]
[1,] 3838.936
```

The computed test statistic value (Hotelling's test) TSQ=3838.936 and the critical value is $T^2_{0.05,\ 7,\ 18}=48.715$; the null hypothesis is rejected because the computed value is greater than the critical value 3838.936 > 45.715. Thus, the Juru and Jejawi Rivers have different concentration of selected water quality parameters in the water. One can conclude that at least one parameter of the chosen water quality parameters is different for the Juru and Jejawi Rivers.

Further Reading

Alkarkhi, A.F.M., Ahmad, A., Ismail, N., Easa, A.M., Omar, K., 2008. Assessment of surface water through multivariate analysis. Journal Of Sustainable Development 1, 27–33.

Alkarkhi, A.F.M., Alqaraghuli, W.A.A., 2019. Easy Statistics for Food Science with R, first ed. Academic Press.

Alkarkhi, A.F.M., Ismail, N., Ahmed, A., Easa, A.M., 2009. Analysis of heavy metal concentrations in sediments of selected estuaries of Malaysia—A statistical assessment. Environ. Monit. Assess. 153, 179–185.

Alkarkhi, A.F.M., Low, H.C., 2012. Elementary Statistics For Technologist. Universiti Sains Malaysia Press, Pulau Pinang.

Allan, G.B., 2007. Elementary Statistics: A Step by Step Approach. McGraw-Hill.

Bhabuk, U. R. 2013. R Graph Gallery-Rg#47: Shaded Normal Curve [Online]. Blogger. Available: http://rgraphgallery.blogspot.my/2013/04/Shaded-Normal-Curve.Html [Accessed 28 May 2018].

Bryan, F.J.M., 1991. Multivariate Statistical Methods: A Primer. Chapman & Hall, Great Britain.

Daniel & Hocking, 2013. Blog Archives, High Resolution Figures In R [Online]. R-Bloggers. Available, https://www.r-bloggers.com/Author/Daniel-Hocking/.

Donald, H.S., Robert, K.S., 2000. A First Course. Mcgraw-Hill.

Fanara, C., 2016. Tutorial On The R Apply Family, The Apply Functions As Alternatives To Loops. [Online]. Datacamp Available, Https://Www.Datacamp.Com/Community/Tutorials/R-Tutorial-Apply-Family#Gs.Rqdg=Dw. [(Accessed 26 May 2018)].

Farah, N.M.S., Nik Norulaini, N.A.R., Alkarkhi, A.F.M., 2012. Assessment of Heavy Metal Pollution in Sediments of Pinang River, Malaysis. In: International Conference on Environmental Research And Technology (ICERT 2012). Universiti Sains Malaysia, Pinang, Malaysia.

Johnson, R.A., Wichern, D.W., 2002. Applied Multivariate Statistical Analysis. Prentice Hall, New Jersey.

Rencher, A.C., 2002. Methods of Multivariate Analysis. J. Wiley, New York.

Yusup, Y., Alkarkhi, A.F.M., 2011. Cluster analysis of inorganic elements in particulate matter in the air environment of an equatorial urban coastal location. Chem. Ecol. 27, 273–286.

CHAPTER

6

Multivariate Analysis of Variance

LEARNING OBJECTIVES

After careful consideration of this chapter, you should be able:

- *To understand the concept of the analysis of variance (ANOVA).*
- *To carry out one-way analysis of variance.*
- *To carry out two-way analysis of variance.*
- *To understand the concept of multivariate analysis of variance (MANOVA).*
- *To carry out one-way multivariate analysis of variance.*
- *To carry out two-way multivariate analysis of variance.*
- *To understand the interaction between various variables.*
- *To employ R built-in functions and commands for analysis of variance and multivariate analysis of variance.*
- *To comprehend and explain the R output for analysis of variance and multivariate analysis of variance.*
- *To extract helpful conclusions and report on them.*

6.1 INTRODUCTION

Tests such as the Z-test and *t*-test, which were investigated earlier, are used for testing a single variable in the case of one or two populations, while in a case of multivariate (several variables), Hotelling's test (T^2) is employed. In the case of more than two means, we cannot employ these tests. Thus, another test should be studied for such an objective. In this chapter, we test the equality of means for more than two populations employing new techniques for univariate and multivariate cases. The proposed techniques are an analysis of variance (ANOVA) for one response variable, and a multivariate analysis of variance (MANOVA) for more than one response.

6.2 ANALYSIS OF VARIANCE IN R

A set of built-in functions are offered by R to carry out various comparisons among population means. The built-in functions can be employed to carry out an analysis of variance (ANOVA) for only a single response variable and a multivariate analysis of variance (MANOVA) for more than one response variable. Furthermore, R offers a built-in function to generate interaction graphs between various variables. The built-in functions to perform ANOVA and MANOVA in R are given below, with detailed explanation for each function.

1. A one-way analysis of variance (ANOVA) can be performed by employing the function `aov ()` in R.

```
aov (Dependent variable ~ Independent variables, data = data frame)
aov (y ~ x, data = data frame)
```

where y refers to the response variable and x represent the predictor (independent variable).

Applied Statistics for Environmental Science with R
https://doi.org/10.1016/B978-0-12-818622-0.00006-X

2. More details concerning the analysis can be provided by employing the function `summary ()`, including a complete ANOVA table and a p-value.

```
summary (data frame)
```

3. The function `aov ()` in R can also be employed to carry out two-way ANOVA for both cases where the interaction between various variables in the study exists or does not exist. The R built-in functions for two-way ANOVA are presented below with detailed explanation for each function.

```
aov (Dependent variable ~ Independent variables, data = data frame)
aov (y ~ x1 + x2, data = data frame)
```

Furthermore, the built-in functions for ANOVA, covering two-factor interaction between various independent variables and the main effect as well, are presented below.

```
aov (y ~ x1 + x2 + x1:x2, data = data frame)
or
aov (y ~ x1*x2, data = data frame)
```

where `y` refers to the response variable (dependent variable) and `x1` and `x2` refer to the independent variables.

4. The function `interaction.plot ()` can be employed to produce the two-factor interaction plot between the selected variables.

```
Interaction.plot (x1, x2, Y)
```

The `interaction.plot ()` function requires the `graphics` package to be installed first, and then we can employ the function to produce the two-factor interaction plot.

5. The function `manova ()` can be employed to carry out multivariate analysis of variance (MANOVA) in R. Two packages should be installed and loaded before employing the function `manova ()`, the two packages are, `car` and `mvtnorm`. The three functions `cbind ()`, `manova ()` and `summary ()` are used to perform MANOVA in R.
First, we should employ the function `cbind ()` to call all selected response variables and place them in a new location (this is called `Responses`) while the function `manova (y ~ x)` is used to carry out MANOVA. More information on the analysis can be provided by the function `summary ()`, including the p-value, the approximate *F* value, and the degrees of freedom.

```
Responses <- cbind (y1, y2, ..., yk), #y1,..., yk Represent response variablrs.
MANOVA <- manova (Responses ~ x) # x Represents independent variable (factor)
```

where, `y1, y2, ..., yk` are *k* response variables (dependent variables), and x refers to the independent variable.

```
summary (MANOVA, test = "wilks")
summary (MANOVA, "wilks")
```

The structure of the `summary ()` function includes the test options, which are, `"pillai"`, `"Hotelling-Lawley"`, and `"Roy"` tests.

6. The function `manova ()` can be employed to carry out two-way MANOVA including the interaction option. Consider that `y1, y2, ..., yk` are *k* response variables (dependent variables) and `x1` and `x2` are two independent variables. Thus, the commands for performing the two-way MANOVA are:

```
Responses <- cbind (y1, y2, ..., yk)
MANOVA <- manova (Responses ~ x1*x2)
summary (MANOVA, "wilks")
```

The options of the test could be `"pillai"`, `"Hotelling-Lawley"`, and `"Roy"` tests.

6.3 THE CONCEPT OF ANALYSIS OF VARIANCE

We have investigated the impact of one or several factors (independent variables), each factor at one or two levels (groups, settings, samples) on a response variable by employing the Z-test and *t*-test. These tests cannot be employed when a factor has more than two settings. For example, an investigator wishes to understand the impact of iron (II) sulphate ($FeSO_4$) reagent at three different dosages: 1, 4, and 7 g/L. Here, there are three dosages (levels), and thus, a new method should be employed to investigate the impact of three or more levels (samples).

The variation of a response variable could be due to the impact of one or more independent variables that have various levels. Thus, the method used to analyze the variation of the response is called an analysis of variance (ANOVA). The concept of ANOVA depends on partitioning the total variation into two parts, the first part attributed to one or more independent variables, while the second part is attributed to an unknown origin (source) which is called the residual (error).

Note:

- Analysis of variance gives the same results as the Z or t tests when it is employed for two population means (two levels, two groups).

Assumptions of analysis of variance

Performing an analysis of variance requires three assumptions to be satisfied:

- Normality: the populations under study must be normally or approximately normally distributed.
- Independence: the samples should be randomly chosen and independent.
- Homogeneity: equality of the variances of the populations (called homoscedasticity).

6.3.1 One-Way Analysis of Variance

A one-way analysis of variance is a test that concerns only one independent variable. Let X be an independent variable (called a factor), which has a different levels (settings, groups), and let Y be a response variable (continuous variable). Here, only one independent variable is considered; one-way analysis of variance can be employed to investigate the influence of various setting of the chosen factor on the response variable. For example, a scientist wishes to investigate the effect of iron(II) sulphate ($FeSO_4$) reagent on color removal as a dependent variable; four dosages of $FeSO_4$ are chosen: 1, 4, 7, and 10 g/L. Based on ANOVA, the differences or the variation in the color removal could be due to the influence of $FeSO_4$ or could be caused by an unknown origin. Thus, we can decompose the total sum of squares into two parts (called components), the sum of squares due to the chosen factor $FeSO_4$ (SS_B), which represents between samples (various groups, various levels), and the sum of squares due to error (SS_E). The total sum of squares can be written as the sum of these two parts.

$$SS_{Total} = SS_{Between\ groups} + SS_{Error}$$

$$SS_T = SS_B + SS_E$$

where

SS_T refers to the total sum of squares,
SS_E refers to the sum of squares of error, and
SS_B refers to the sum of squares between different levels (groups).

Corresponding to each component is a number; these are called degrees of freedom.

F test is employed to test the influence of a factor on a response variable to determine whether or not the factor affects the output (response) (statistically significant or insignificant), as presented in (6.1).

$$F = \frac{Variance\ between\ groups}{Variance\ due\ to\ error} = \frac{MS_B}{MS_E} \tag{6.1}$$

The general configuration for a single-factor experiment is presented in Table 6.1.

TABLE 6.1 The Arrangement of a Single-Factor Experiment

Factor (Level)	Replication			
1	Y_{11}	Y_{12}	...	Y_{1n1}
2	Y_{21}	Y_{22}	...	Y_{2n2}
⋮	⋮	⋮	...	⋮
a	Y_{a1}	Y_{a2}	...	Y_{ana}

TABLE 6.2 The Entries of ANOVA Table

Source of variation (S.O.V)	Degrees of freedom (d.f)	Sum of squares (S.S)	Mean sum of squares (M.S)	F-value
Between groups	$a-1$	$SS_B=\sum n_i(\overline{Y}_i-\overline{Y})^2$	$MS_B=\frac{SS_B}{a-1}$	$F=\frac{MS_B}{MS_E}$
Within groups (Error)	$N-a$	$SS_{Error}=SS_E=\sum(n_i-1)S_i^2$	$MS_E=\frac{SS_E}{N-a}$	
Total	$N-1$	$SS_T=\sum(Y_i-\overline{Y})^2$		

Symbols and computational formulas

Below are the symbols employed in the computational formula:

- a refers to the number of levels (groups, settings),
- n_i refers to the number of observations in each level (groups, settings), $i=1, 2,\ldots, a$,
- $N=\sum_i n_i$ refers to the total number of observations,
- Y_{ij} refers to the observation j in the ith group, $j=1, 2,\ldots, n_i$,
- $\overline{Y}=\frac{\sum Y}{N}$ refers to the grand mean, and $\overline{Y}_i$ is the mean of group i.

The formulas for building a one-way analysis of variance are presented in Table 6.2.

6.3.1.1 Hypothesis Testing for a One-Way Analysis of Variance

In a one-way analysis of variance, scientists are concerned with examining the hypotheses given below.

$$H_0:\mu_1=\mu_2=\cdots=\mu_a$$

vs

$$H_1:\mu_i\neq\mu_j \text{ for at lease one pair } (i,j),\ i=1,2,\ldots,a$$

The effect of $FeSO_4$ on color removal can be used as an example for forming the hypothesis. The null and alternative hypotheses for examining four various dosages of $FeSO_4$, 1, 4, 7, and 10, on the color removal are:

$$H_0:\mu_1=\mu_2=\mu_3=\mu_4$$

vs

$$H_1:\mu_i\neq\mu_j \text{ for at least one mean is difference from others, } i=1,2,3,4$$

The color removal at 1, 4, 7, and 10 is the same (no real difference in the color removal between various dosages), as stated by the null hypothesis, whereas the complement of the null hypothesis stated that at least one dosage of $FeSO_4$ out of four dosages will produce a different color removal; this statement represents the alternative hypothesis.

6.3.1.2 Explanation of the Analysis of Variance Results

The explanation of the analysis of variance table starts with the F-value in the last column of Table 6.2. This F-value is called the computed F-value, and it should be compared with the critical value (theoretical). A decision should be made by comparing the computed F-value with the critical value at a specified α (significance level) such as 0.05, 0.01, or other values. This comparison will lead to the rejection of the null hypothesis if the computed F-value is greater than the critical value. Another useful value is provided by computer outputs; this value is called the p-value (probability value), which can be employed to produce a quick decision about the null hypothesis. We should reject the null hypothesis for small values of the p-value; the p-value varies between 0 and 1.

EXAMPLE 6.1 THE CONCENTRATION OF THE TOTAL SUSPENDED SOLIDS (TSS) AT THE BERIS DAM

An environmentalist wishes to investigate the impact of the location on the total suspended solids of the water at the Beris dam. Three locations are of interest: the three locations were before the dam (B), at the dam (D), and after the dam (A). The samples are collected for eight months and analyzed for the total suspended solids (TSS) gravimetrically at the laboratory. Assuming that the data for TSS are normally distributed, use a significance level of $\alpha=0.05$ to carry out the analysis. The sample data are presented in Table 6.3.

TABLE 6.3 The Results of TSS for Three Locations of Beris Dam Water

B	A	D
2	15	4
4	45	6
4	35	8
3	20	3
4	12	4
3	25	7
3	45	8
2	30	5

We wish to investigate the effect of location on the total suspended solids (TSS) values measured at different points of the dam whether there is no impact of the location on the TSS result or the effect exists. The location of the sampling is considered as an independent variable (factor) that affects the TSS values. The decision regarding the effect of various locations on the TSS values (whether or not various locations would give different TSS results) can be achieved by employing hypothesis testing. The null and alternative hypotheses in a mathematical form are presented below.

$$H_0 : \mu_1 = \mu_2 = \mu_3$$

vs

$$H_1 : \mu_i \neq \mu_j \text{ for at least one mean is different from others, } i = 1,2,3$$

The same results of TSS are obtained from all selected locations as stated by the null hypothesis, which indicates that the location does not influence the TSS result; whereas, the complement of the null hypothesis states that at least one location within the chosen locations would show a distinctly different result of the total suspended solids (TSS) compared to the others—this statement represents the alternative hypothesis. Analysis of variance (ANOVA) can be carried out to examine the equality of means (the difference is zero) for the TSS values indicated by various locations. The sample data for the total suspended solids (TSS) gathered from various locations were analyzed. The one-way analysis of variance was used to test the null hypothesis; the entries of the one-way analysis of variance were computed by employing the built-in functions `aov ()` and `summary ()` in R.

```
aov (TSS ~ Location, data = Example6_1)
```

The built-in function `aov ()` produces five values including, `Sum of Squares`, `Deg. of Freedom`, `Location`, `residuals`, and the last value is for `Residual standard error`, without providing information on the significance of the chosen variable (factor), which is location in this example. Information such as the p-value `Pr(>F)`, *F* value (`F value`), and mean sum of squares (`Mean Sq`) can be produced by employing the function `summary ()`.

```
summary ()
```

The function `aov ()` can be employed to carry out analysis of variance (ANOVA) for the stored data `.CSV` (`Example6_1`) for the total suspended solids (TSS) measured from various locations.

```
ANOVA <- aov(TSS ~ Location, data = Example6_1)
ANOVA
Summary (ANOVA)
```

The outputs of employing the two functions `aov ()` and `summary ()` to perform analysis of variance are presented below.

```
> ANOVA <- aov (TSS ~ Location, data = Example6_1)
> ANOVA
Call:
   aov(formula = TSS ~ Location, data = Example6_1)
```

```
Terms:
                Location Residuals
Sum of Squares  3097.000 1158.625
Deg. of Freedom        2        21

Residual standard error: 7.427827
Estimated effects may be unbalanced
> summary(ANOVA)
            Df Sum Sq Mean Sq F value    Pr(>F)
Location     2   3097  1548.5   28.07  1.17e-06 ***
Residuals   21   1159    55.2
---
Signif. codes: 0 '***' 0.001 '**' 0.01 '*' 0.05 '.' 0.1 ' ' 1
```

The outputs of R show that the built-in function `aov ()` is a call (request) to carry out analysis of variance on the TSS gathered data and the output should be stored in a new file called `ANOVA`. Moreover, the built-in function `summary ()` supplies more details on the entries of the analysis of variance, as appeared in the output.

The analysis of variance results clearly show that the location of the sampling point exhibited a real (significant) influence on the total suspended solids values as displayed by the p-value of `1.17e-06`. We should reject the null hypothesis, which stated that the results of TSS is the same at various locations (location does not have effect on the TSS) and believe that the location is an important and influential factor. It can be concluded that at least one location is different compared to the other locations. This is a general conclusion, and we are unable to locate which location gives a different TSS value. The source of variation (the location) can be identified by performing further analysis, such as multiple comparisons technique, to identify the location that gives a different TSS result, if required.

EXAMPLE 6.2 THE LEVEL OF THE PARTICULATE MATTER PM_{10} IN THE AIR OF PALM OIL MILLS

Air monitoring was conducted, and the total particulate matter (PM_{10}) was measured in five palm oil mills in the Penang and Kedah States, Malaysia. Twelve samples were selected from each mill and tested for PM_{10}. Assuming that the data of the particulate matter PM_{10} in the air of palm oil mills are normally distributed, use a significance level of $\alpha=0.05$ to carry out the analysis. The data for the particulate matter PM_{10} in the air of palm oil mills are presented in Table 6.4.

TABLE 6.4 The Data for PM_{10} Collected from Five Mills ($\mu g/m^3$)

Mill				
1	2	3	4	5
487.19	231.92	712.03	709.26	169.85
716.40	593.17	281.55	622.17	346.06
1164.88	173.71	172.97	514.38	165.88
583.62	178.46	175.70	500.10	123.86
731.32	564.43	166.51	444.23	466.57
1013.7	208.3	318.96	396.64	344.92
853.55	135.39	226.63	1080.82	200.81
302.52	76.85	340.34	1049.77	284.14
323.48	122.25	105.00	958.86	287.49
1000.95	118.79	173.42	190.93	101.90
855.60	141.48	434.62	217.64	79.84
708.56	163.28	168.49	818.53	93.54

At significance level $\alpha = 0.05$, we examine the hypothesis that there is no difference between the PM_{10} results measured in various mills.

The location of the mill is considered as an important factor (independent variable) that influences the result of the PM_{10} produced by different mills. To make a decision regarding the effect of location on the PM_{10} produced by different mills, we should employ hypothesis testing to examine the claim regarding the PM_{10} value. The null and alternative hypotheses in a mathematical form are presented below.

$$H_0 : \mu_1 = \mu_2 = \cdots = \mu_5$$

vs

$$H_1 : \mu_i \neq \mu_j \text{ for at least one mean is different from others, } i = 1,2,\ldots,5$$

All mills produce the same level of PM_{10} regardless of the location of the mill, as stated by the null hypothesis, which indicates that the location of the mill does not influence the level of PM_{10}. Whereas, the complement of the null hypothesis states that at least one mill within the chosen mills would show a distinct result of PM_{10} level than the others. Examining the equality of PM_{10} means generated by various mills can be done by performing a one-way analysis of variance. The one-way ANOVA was used to examine the null hypothesis regarding the stored data of the particulate matter PM_{10} in the air of palm oil mills.`CSV (Example6_2)`. The entries of the one-way analysis of variance were computed by employing the built-in functions `aov ()` and `summary ()` in R.

The two functions `aov ()` and `summary ()` were employed to perform a one-way analysis of variance to examine the PM_{10} results produced by various mills (location); the outputs produced by R built in functions are given below.

```
> ANOVA <- aov (PM10 ~ Location, data = Example6_2)
> ANOVA
Call:
  aov(formula = PM10 ~ Location, data = Example6_2)
Terms:
                Location  Residuals
Sum of Squares   2828557    2605953
Deg. of Freedom        4         55
Residual standard error: 217.6717
Estimated effects may be unbalanced
> summary(ANOVA)
           Df  Sum Sq Mean Sq F value   Pr(>F)
Location    4 2828557  707139   14.93 2.55e-08 ***
Residuals  55 2605953   47381
---
Signif. codes: 0 '***' 0.001 '**' 0.01 '*' 0.05 '.' 0.1 ' ' 1
```

The outputs of analysis of variance show that there is a significant difference in the PM_{10} results produced by different mills, which indicates that the location of the mill influences the level of PM_{10}. Thus, the null hypothesis is rejected because the p-value is very small, at `2.55e-08`, and we conclude that the location of the mill is important.

In summary, it can be said that the location of the mill influences the PM_{10} result and should be considered as an influential factor when the level of PM_{10} is measured. The difference in the level of PM_{10} values could be due to the origins (sources) of pollution that might be different in each location.

6.3.2 Two-Way Analysis of Variance

A two-way analysis of variance (two-way ANOVA) is a test that considers only two factors (independent variables), each factor has at least two levels (settings, groups) and one response variable (continuous variable). The aim of using the two-way ANOVA is to test the influence of the chosen factors on the selected response variable. For example, an environmentalist wishes to investigate the effect of two factors on the percentage of turbidity reduction as a response; the two factors are pH at three levels 5, 7, and 9, and the cation concentration at three levels 0.01, 0.03, and 0.05 mM. The differences in the percentage of turbidity reduction could be due to the effect of pH, the effect of the cation concentration, or the interaction between the two factors pH and the cation concentration, which are well-known origins (sources) of variation. The second source of variation (cause difference in the percentage of turbidity reduction) is called the error (residual), which represents the effect caused by any undefined source (not included in the model).

TABLE 6.5 General Arrangement for a Two-Factor Experiment

Factor A		Factor B			
		1	2	...	b
	1	$Y_{111}, Y_{112}, \ldots, Y_{11n}$	$Y_{121}, Y_{122}, \ldots, Y_{12n}$		$Y_{1b1}, Y_{1b2}, \ldots, Y_{1bn}$
	2	$Y_{211}, Y_{212}, \ldots, Y_{21n}$	$Y_{221}, Y_{222}, \ldots, Y_{22n}$	...	$Y_{2b1}, Y_{2b2}, \ldots, Y_{2bn}$
	.	.	.	.	.
	.	.	.	.	.
	.	.	.	.	.
	a	$Y_{a11}, Y_{a12}, \ldots, Y_{a1n}$	$Y_{a21}, Y_{a22}, \ldots, Y_{a2n}$	...	$Y_{ab1}, Y_{ab2}, \ldots, Y_{abn}$

Thus, four components can represent the total sum of squares (SS_T): the first component is the sum of squares due to unknown source (error) (SS_E), the second component is the sum of squares due to the first factor, pH (SS_A), the third component is the sum of squares due to the second factor, cation concentration (SS_B), and the fourth component is due to the interaction between the two chosen factors (pH and cation concentration) (SS_{AB}). If the effect of one of the factors varies for various levels of the other factors this would indicate that the interaction exists.

The total sum of squares (SS_T) can be written as the sum of all the four components.

$$SS_T = SS_{Error} + SS_A + SS_B + SS_{AB}$$

where

SS_T refers to the total sum of squares;
SS_E refers to the sum of squares of error;
SS_A refers to the sum of squares between different levels of the first factor;
SS_B refers to the sum of squares between different levels of the second factor; and
SS_{AB} refers to the sum of squares due to the interaction between the two factors.
Corresponding to each component is a number; these are called degrees of freedom.

The general configuration for a two-factor experiment with factor A at a levels and factor B at b levels, is presented in Table 6.5.

Symbols and computational formulas

- a and b represent the levels of the two selected factors X_1 and X_2 respectively.
- $Y_{...}$ refers to the sum of all the observations: $Y_{...} = \sum_{i=1}^{a}\sum_{i=1}^{b}\sum_{k=1}^{n} Y_{ijk}$
- $Y_{i..}$ refers to the sum of the b observations in row i: $Y_{i..} = \sum_{j=1}^{b}\sum_{k=1}^{n} Y_{ijk}$
- $Y_{.j.}$ refers to the sum of the a observations in column j: $Y_{.j.} = \sum_{i=1}^{a}\sum_{k=1}^{n} Y_{ijk}$
- $Y_{ij.}$ refers to the sum of each cell under the same combination: $Y_{ij.} = \sum_{k=1}^{n} Y_{ijk}$

The formulas for the two-way ANOVA entries are summarized in Table 6.6.

TABLE 6.6 The Formulas for Two-Way ANOVA Entries

Source of variation (S.O.V)	Degrees of freedom (d.f)	Sum of squares (S.S)	Mean sum of squares (M.S)	F-value
A	$a-1$	$SS_A = \frac{1}{bn}\sum_{i=1}^{a} Y_{i..}^2 - \frac{Y_{...}^2}{abn}$	$MS_A = \frac{SS_A}{a-1}$	$F = \frac{MS_A}{MS_E}$
B	$b-1$	$SS_B = \frac{1}{an}\sum_{j=1}^{b} Y_{.j.}^2 - \frac{Y_{...}^2}{abn}$	$MS_B = \frac{SS_B}{b-1}$	$F = \frac{MS_B}{MS_E}$
AB	$(a-1)(b-1)$	$SS_{AB} = SS_{subtotal} - SS_A - SS_B$ $SS_{subtotal} = \frac{1}{n}\sum_{i=1}^{a}\sum_{j=1}^{b} Y_{ij.}^2 - \frac{Y_{...}^2}{abn}$	$MS_{AB} = \frac{SS_{AB}}{(a-1)(b-1)}$	$F = \frac{MS_{AB}}{MS_E}$
Error	$ab(n-1)$	$SS_{Error} = SS_E = SS_T - SS_{AB} - SS_A - SS_B$	$MS_E = \frac{SS_E}{ab(n-1)}$	
Total	$abn-1$	$SS_T = \sum\sum\sum(\overline{Y}_{i..} - \overline{Y}_{...})^2 = \sum_{i=1}^{a}\sum_{j=1}^{b}\sum_{k=1}^{n} Y_{ijk}^2 - \frac{Y_{...}^2}{abn}$		

6.3.2.1 *Hypothesis Testing for a Two-way Analysis of Variance*

Consider there are two factors, factor A at a levels and factor B at b levels. The scientists are concerned in examining the hypotheses given below:

Hypothesis 1

Hypothesis regarding the impact of the factor A (first factor) in a mathematical form is:

$$H_0: \alpha_1 = \alpha_2 = \cdots = \alpha_a = 0$$

Factor A does not affect the response variable (the response value is the same at all levels of factor A) as stated by the null hypothesis, while the complement of the null hypothesis is attached to the alternative hypothesis, which states that factor A has an effect on the response variable.

$$H_1: \text{at least one } \alpha_i \neq 0, \quad i = 1,2,\ldots,a$$

Hypothesis 2

Hypothesis regarding the impact of the second factor (factor B) in a mathematical form is:

$$H_0: \beta_1 = \beta_2 = \cdots = \beta_b = 0$$

Factor B does not affect the response variable (the response value is the same at all levels of factor B) as stated by the null hypothesis, while the complement of the null hypothesis is attached to the alternative hypothesis, which states that factor B has an effect on the response variable.

$$H_1: \text{at least one } \beta_j \neq 0, \quad j = 1,2,\ldots,b$$

Hypothesis 3

Hypothesis regarding the impact of interaction between the two factors A and B in a mathematical form is:

$$H_0: (\alpha\beta)_{ij} = 0 \text{ for all } i,j$$

Interaction does not exist between factor A and factor B (both factors are independent) as stated by the null hypothesis. Thus, the mathematical form of the alternative hypothesis is

$$H_1: \text{at least one } (\alpha\beta)_{ij} \neq 0$$

An interaction between factor A and factor B exists as stated by the alternative hypothesis.

Note:

- It is recommended that the main effect should not be explained if the interaction exists (significant). However, the main effect can be explained with some caution.

EXAMPLE 6.3 THE PERCENTAGE OF COD REDUCTION

A researcher wishes to run an experiment to investigate the impact of the shaking speed (rpm) and pH on the percentage of COD reduction. Three different shaking speeds (100, 120 and 140 rpm) and three pH values (4, 6 and 8) are chosen. Three replicates at each combination of the shaking speed and pH are run. Verify the hypothesis that there is no difference between the three shaking speeds, no difference between the three values of pH, and no interaction between the different shaking speeds and pH values on the percentage of COD reduction. Use a significance level of $\alpha = 0.05$ to carry out the analysis. Consider that the data for the percentage of COD reduction are normally distributed. The sample data are presented in Table 6.7.

TABLE 6.7 The Results of COD Reduction%

	Shaking speed								
pH		100			120			140	
4	75.26	72.47	69.85	53.33	55.24	50.78	41.60	41.60	34.79
6	48.35	43.75	51.44	61.08	60.82	67.01	75.47	79.58	76.85
8	55.67	55.15	51.80	58.56	57.45	55.45	56.44	54.64	76.80

We investigate the impact of two factors, namely, the shaking speed and the pH value, on the percentage of COD reduction as a response. The effect of interaction between the shaking speed and the pH value should be investigated as well.

Two-way analysis of variance was employed to investigate the effect of the two factors on the percentage of COD reduction. The commands and built-in functions in R employed to compute the entries of the two-way analysis of variance for the stored data `.CSV(Example6_3)` of the percentage of COD reduction are shown below.

```
pH = as.factor (Example6_3 $ pH)
Shaking = as.factor (Example6_3 $ Shaking)
COD = Example6_3 $ COD
aov (COD ~ pH * Shaking) # analysis of variance with interaction
summary (aov(COD ~ pH * Shaking))
```

The commands show that the independent variables should be defined as a factor before starting the analysis employing the function `as.factor ()`; the two variables are the shaking speed and pH. The response variable (percentage of COD reduction) was extracted and defined as COD. Moreover, an order was given to perform analysis of variance employing the function `aov ()` to produce general information regarding the sum of squares and degrees of freedom, the function `summary ()` was employed to display more information on the analysis.

```
> pH = as.factor (Example6_3 $ pH)
> Shaking = as.factor (Example6_3 $ Shaking)
> COD = Example6_3 $ COD
> aov (COD ~ pH*Shaking)          # analysis of variance with interaction
Call:
   aov(formula = COD ~ pH * Shaking)
Terms:
                          pH   Shaking pH:Shaking Residuals
Sum of Squares      272.1703   19.9557  3059.9828  435.5208
Deg. of Freedom            2         2          4        18
Residual standard error: 4.918902
Estimated effects may be unbalanced
> summary (aov(COD ~ pH*Shaking))
            Df  Sum Sq Mean Sq  F value    Pr(>F)
pH           2   272.2   136.1    5.624    0.0127 *
Shaking      2    20.0    10.0    0.412    0.6682
pH:Shaking   4  3060.0   765.0   31.617 6.42e-08 ***
Residuals   18   435.5    24.2
---
Signif. codes: 0 '***' 0.001 '**' 0.01 '*' 0.05 '.' 0.1 ' ' 1
```

The interaction between the shaking speed and the pH values is examined. We should explain the interaction (`pH:Shaking`) first. A significant interaction was exhibited between the shaking speed and the pH values because the p-value is very small (p-value < `6.42e-08`) as appeared in the analysis of variance table under `Pr(>F)`. This finding suggests that the collection of the shaking speed and the pH values influences the percentage of the COD reduction, and we should consider both factors in any future experiment regardless of the main effect (which may be significant or insignificant). The null hypothesis concerning the effect of the pH was rejected because the p-value < `0.0127`, which indicates that pH has an effect on the COD reduction, and insignificant effect was exhibited by the shaking speed (the null hypothesis is not rejected because the p-value < `0.6682`).

A plot for the interaction between shaking speed and pH can be generated to investigate the behavior of each factor in the presence of another factor and observing the percentage of COD reduction employing R commands and built-in functions. The function `interaction.plot ()` can be employed to produce the two-factor interaction. The interaction plot between the two factors, the shaking speed and pH, is given in Fig. 6.1.

```
interaction.plot (pH, Shaking, COD)
```

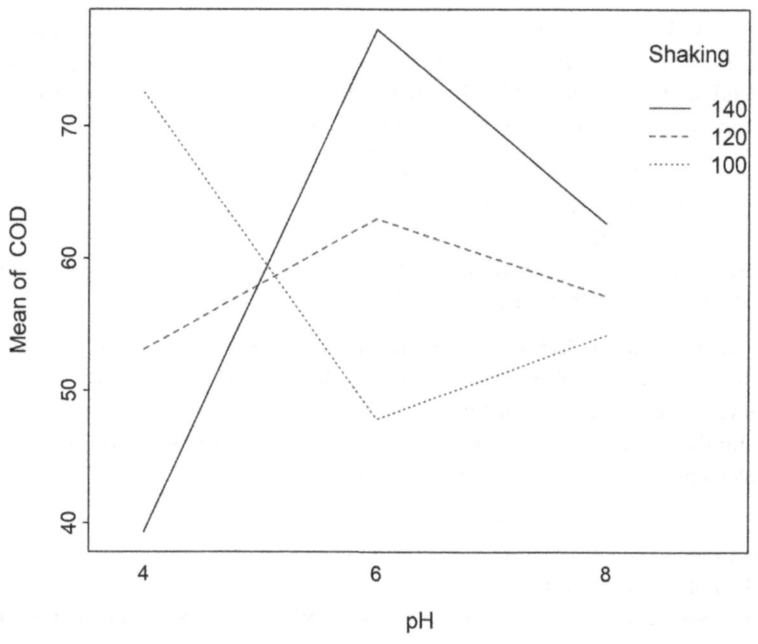

FIG. 6.1 Interaction plot between shaking speed and pH values.

EXAMPLE 6.4 THE PERCENTAGE OF TURBIDITY REDUCTION

An experiment was carried out to investigate the impact of the pH and pectin as a biopolymeric flocculant on the turbidity reduction as a response. Three pH values (3, 6, and 9) and three levels of the pectin dosage (mg/L) were chosen (1, 4.5, and 8 mg/L). Two replicates at each combination of the pH and pectin dosage were employed. The sample data obtained from the experiments are presented in Table 6.8.

TABLE 6.8 The Design With the Experimental Data Values for Turbidity Reduction

pH	Pectin	Turbidity	pH	Pectin	Turbidity
3	1	98.5	6	4.5	99.3
3	1	98.4	6	8	66.7
3	4.5	49.5	6	8	99.1
3	4.5	49.8	9	1	98.9
3	8	55.2	9	1	99.1
3	8	55.7	9	4.5	99.9
6	1	98.4	9	4.5	99.8
6	1	98.2	9	8	99.5
6	4.5	99.7	9	8	99.5

Consider that the data for percentage of turbidity reduction are normally distributed with a common variance, an analysis of variance should be carried out to find out whether the pH and pectin dosage as biopolymer flocculant affected the percentage of turbidity reduction at a significance level of $\alpha = 0.05$.

The pH and pectin dosage are the two variables of concern. A two-way analysis of variance should be employed to examine the effect of the pH, the pectin dosage, and the interaction between the pH and pectin dosage on the percentage of turbidity reduction. The commands and built-in functions in R will generate the necessary entries for a two-way analysis of variance table for the stored data `.CSV(Example6_4)` of the percentage of turbidity reduction.

```
pH = factor (Example6_4 $ pH, levels = c(3, 6, 9))
Pectin = factor (Example6_4 $ Pectin, levels = c(1, 4.5, 8))
Turbidity = Example6_4 $ Turbidity
aov (Turbidity ~ pH*Pectin)        # Analysis of variance with interaction
summary (aov (Turbidity ~ pH * Pectin))
```

The commands started with the function `factor ()` to define the two independent variables as a factor: the variables are the pH and `Pectin` dosage. The response variable (percentage of turbidity reduction) was extracted and defined as `Turbidity`. Furthermore, an analysis of variance was carried out employing the built-in function `aov ()` to generate the entries for ANOVA table, more details regarding the analysis can be produced by the built-in function `summary ()`. The entries of a two-way ANOVA Table for the percentage of turbidity reduction are presented below.

```
> pH = factor (Example6_4 $ pH, levels = c(3, 6, 9))
> Pectin = factor (Example6_4 $ Pectin,  levels = c(1, 4.5, 8))
> Turbidity = Example6_4 $ Turbidity
> aov (Turbidity ~ pH*Pectin)          # Analysis of variance with interaction
Call:
  aov(formula = Turbidity ~ pH * Pectin)
Terms:
                        pH   Pectin  pH:Pectin Residuals
Sum of Squares    3389.041 1258.288   1927.936   525.180
Deg. of Freedom          2        2          4         9
Residual standard error: 7.638935
Estimated effects may be unbalanced
> summary (aov (Turbidity ~ pH * Pectin))
            Df Sum Sq Mean Sq F value   Pr(>F)
pH           2   3389  1694.5   29.04 0.000119 ***
Pectin       2   1258   629.1   10.78 0.004080 **
pH:Pectin    4   1928   482.0    8.26 0.004409 **
Residuals    9    525    58.4
---
Signif. codes: 0 '***' 0.001 '**' 0.01 '*' 0.05 '.' 0.1 ' ' 1
```

To explain the outputs, we should examine the interaction between the pH and the pectin dosage first. The above results of the analysis of variance indicated that the interaction between the pH and the pectin dosage (`pH:Pectin`) was significant because the p-value is small and less than 0.05 as indicated under (`Pr(>F)`, p-value < `0.004409`). Thus, we should reject the null hypothesis about the interaction between the pH and pectin dosage, which indicates that the interaction between the pH and the pectin dosage exists.

A plot for the interaction between pH and pectin dosage can be generated to investigate the behavior of each factor in the presence of another factor and observing the percentage of turbidity reduction employing R commands and built-in functions. The function `interaction.plot ()` can be employed to produce the two-factor interaction between pH and the pectin dosage, as presented in Fig. 6.2.

```
interaction.plot (pH, Pectin, Turbidity, col = c ("darkgreen","blue", "black"),
  legend = T, ylab ="" , xlab = "", cex.axis = .9, tck = -0.01, mgp = c(3, .3, 0))
```

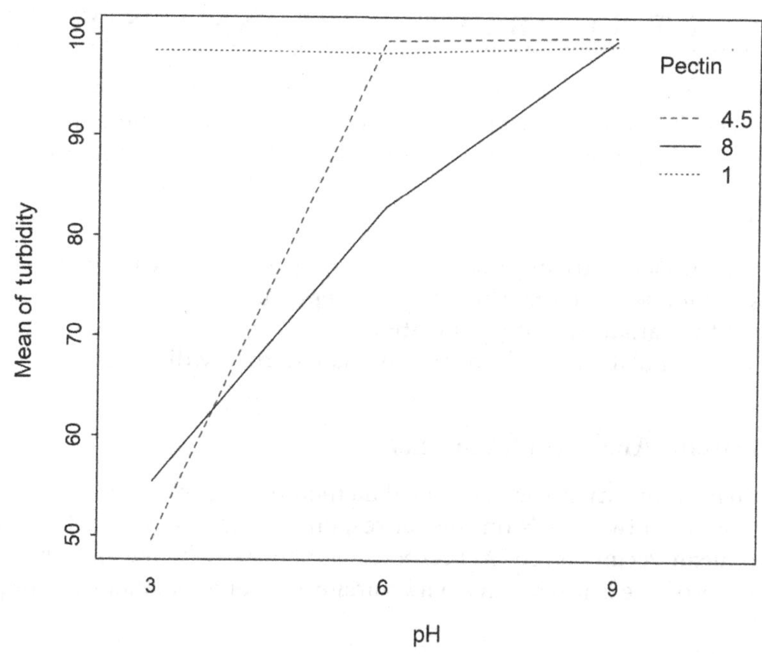

FIG. 6.2 Interaction plot between pH and pectin dosage.

The effect of the pH and the pectin dosage should be tested individually. The pH factor exhibited a significant effect, as the *p*-value is small (*p*-value < `0.000119`), while the pectin dosage exhibited a significant at *p*-value < `0.004080`.

In summary, we should consider the effect of the pH and pectin dosage in future work.

Note:

- An analysis of variance Table for *k* independent variables can be constructed using the same steps. For example, an analysis of variance Table for three independent variables is presented in Table 6.9.

TABLE 6.9 The Entries for a Three-Way ANOVA

Source of variation (S.O.V)	Degrees of freedom (d.f)	Sum of squares (S.S)	Mean sum of squares (M.S)	*F*-Value
A	$a-1$	SS_A	MS_A	$F_A=SS_A/SS_E$
B	$b-1$	SS_B	MS_B	$F_B=SS_B/SS_E$
C	$c-1$	SS_C	MS_C	$F_C=SS_C/SS_E$
AB	$(a-1)(b-1)$	SS_{AB}	MS_{AB}	$F_{AB}=SS_{AB}/SS_E$
AC	$(a-1)(c-1)$	SS_{AC}	MS_{AC}	$F_{AC}=SS_{AC}/SS_E$
BC	$(b-1)(c-1)$	SS_{BC}	MS_{BC}	$F_{BC}=SS_{BC}/SS_E$
ABC	$(a-1)(b-1)(c-1)$	SS_{ABC}	MS_{ABC}	$F_{ABC}=SS_{ABC}/SS_E$
Error	$abc(n-1)$	SS_E	MS_E	
Total	$abcn-1$	SS_T		

6.4 THE CONCEPT OF MULTIVARIATE ANALYSIS OF VARIANCE

The idea of analysis of variance (ANOVA) for examining the equality of more than two population means for a single response variable can be developed to include several response variables measured for each experimental unit. The method or procedure used to consider several response variables is **called** a multivariate analysis of variance (MANOVA).

Assumptions for MANOVA

- Normality: the populations under study must be normally or approximately normally distributed.
- Independence: samples should be randomly chosen and independent.
- Homogeneity: equality of the variances of the populations.
- Multivariate normal: Every population should be multivariate normally distributed.

6.4.1 One-Way Multivariate Analysis of Variance

One-way multivariate analysis of variance is a statistical method used to investigate the effect of only one independent variable (factor) at more than two levels on several response variables. Scientists are interested in testing the equality of more than two **mean vectors** on several response variables—whether the q means' vectors are the same or not. Consider that q samples of size n are chosen from k-variate normal populations with equal covariance matrices, as presented in Table 6.10.

6.4.1.1 Hypothesis Testing for One-Way MANOVA

In a one-way multivariate analysis of variance, the scientists are concerned in testing the equality of mean vectors; the null hypothesis in a mathematical form is given in (6.2).

$$H_0 : \boldsymbol{\mu}_1 = \boldsymbol{\mu}_2 = \cdots = \boldsymbol{\mu}_q$$

$$H_0 : \begin{bmatrix} \mu_{11} \\ \mu_{12} \\ \cdot \\ \cdot \\ \cdot \\ \mu_{1k} \end{bmatrix} = \begin{bmatrix} \mu_{21} \\ \mu_{22} \\ \cdot \\ \cdot \\ \cdot \\ \mu_{2k} \end{bmatrix} = \cdots = \begin{bmatrix} \mu_{q1} \\ \mu_{q2} \\ \cdot \\ \cdot \\ \cdot \\ \mu_{qk} \end{bmatrix} \tag{6.2}$$

The q population means are equal for each response variable as stated by the null hypothesis, and the alternative hypothesis states that at least two means are different.

$$H_1 : \text{at least two } \boldsymbol{\mu}'s \text{ are different}$$

where $\boldsymbol{\mu}$ is a vector.

If we reject the null hypothesis, we conclude that at least two means (groups) differ for at least one response variable, while if we fail to reject the null hypothesis, the q means are equal for each response variable.

The between sum of squares (B), the within sum of squares (W), and the total sum of squares (T) are presented in Table 6.11, including the formulas and associated degrees of freedom for each component.

TABLE 6.10 The Arrangement of Multivariate Data for a One-Way MANOVA

Sample 1	Y_{11}	Y_{12}	...	Y_{1n}
Sample 2	Y_{21}	Y_{22}	...	Y_{2n}
⋮	⋮	⋮	...	⋮
Sample q	Y_{q1}	Y_{q2}	...	Y_{qn}

We reject the null hypothesis if the likelihood ratio presented in (6.3) is less than or equal to the critical value.

$$\Lambda = \frac{|W|}{|W+B|}, \ 0 \le \Lambda \le 1 \tag{6.3}$$

TABLE 6.11 The Entries of a One-Way MANOVA Table

Source of variation (S.O.V)	Degrees of freedom (d.f)	Matrix of sum of squares and cross products (SSP)
Between	$q-1$	$B=n\sum_{i=1}^{q}(\overline{Y}_i-\overline{Y})(\overline{Y}_i-\overline{Y})'$
Error (residual)	$N-q$	$W=\sum_{i=1}^{q}\sum_{j=1}^{n}(Y_{ij}-\overline{Y}_i)(Y_{ij}-\overline{Y}_i)'$
Total	$N-1$	$T=\sum_{i=1}^{q}\sum_{j=1}^{n}(Y_{ij}-\overline{Y})(Y_{ij}-\overline{Y})'$

$$\text{Reject if } \Lambda \le \Lambda_{\alpha,k,g-1,N-g} \text{ (critical value)}$$

This test is called the Wilk's test.

Note:

- Small values of Λ leads to reject the null hypothesis.
- The null hypothesis for the multivariate case can be examined by employing other tests: Pillai, Lawley-Hotelling, and Roy's largest root. R supplies built-in functions and commands to carry out all multivariate test (MANOVA).

EXAMPLE 6.5 THE LEVEL OF THE PARTICULATE MATTER IN THE AIR OF A PALM OIL MILL

The impact of the location on the level of the particulate matter in the air of a palm oil mill was investigated. Four points (A, B, C, and D) and three responses were of concern: PM_{10}, $PM_{2.5}$, and PM_1 ($\mu g/m^3$). The four locations were generally located near the sterilizer, boiler, hydro cyclone, and nut/fiber separator. Three replicates at each location were employed. Verify whether the location affects the level of the particulate matter in the air of the palm oil mill. Use a significance level of $\alpha=0.05$ to carry out the analysis. Assume that the data for the particulate matter are normally distributed. The sample data are presented in Table 6.12. Conduct multivariate analysis of variance to examine whether the location of the chosen points influences the level of the particulate matter in the air of the palm oil mill.

TABLE 6.12 The Data for Particulate Matter in Air Quality of Palm Oil Mill ($\mu g/m^3$)

Point	PM_{10}	$PM_{2.5}$	PM_1
A	1182.74	219.86	130.07
A	1500.69	187.09	126.58
A	1659.14	272.10	188.51
B	223.20	88.55	36.88
B	480.17	127.63	61.03
B	82.94	28.24	11.72
C	257.83	111.32	36.86
C	650.96	134.09	64.43
C	1121.76	211.61	117.44
D	284.98	76.70	28.96
D	233.76	38.83	17.18
D	61.42	9.91	3.34

The location of sampling points is identified as an important factor that affects the level of the particulate matter in the air of a palm oil mill, namely PM_{10}, $PM_{2.5}$, and PM_1. We can perform a one-way multivariate analysis of variance to decide whether the four samples' mean vectors are the same or not. The function `manova ()` was employed to determine whether the behavior of the

PM_{10}, $PM_{2.5}$, and PM_1 is significantly affected by the location of the sampling point or not. We will clarify the use of `manova ()` function through this example to generate multivariate analysis of variance results employing R built-in functions and commands.

Moreover, the data for selected responses should be prepared for the function `manova ()` and placed in one file employing the function `cbind ()` function. More details on the analysis can be supplied employing the function `summary ()`. A multivariate analysis of variance was employed to analyze the data for the PM_{10}, $PM_{2.5}$, and PM_1 that had been stored as a `.CSV` (`Example6_5`) format.

```
PM10 = Example6_5 $ PM10     #Define and extract PM10 from the file
PM2.5 = Example6_5 $ PM2.5
PM1 = Example6_5 $ PM1
Location = as.factor (Example6_5 $ Point)
Responses <- cbind (PM10, PM2.5, PM1)  #To combine the responses in one place
MANOVA <- manova (Responses ~ Location)  #Perform multivariate analysis of variance
summary (MANOVA, test = "Wilks")
summary (MANOVA, test = "Hotelling-Lawley")
summary (MANOVA, test = "Roy")
summary (MANOVA, test = "Pillai")
```

One can observe that the responses (PM_{10}, $PM_{2.5}$, and PM_1) and the factor that is location were defined as presented by the first four rows of the R commands. The three responses (PM_{10}, $PM_{2.5}$, and PM_1) are combined in one file and the result saved in a new place called `Responses` employing the function `cbind ()`. Multivariate analysis of variance was performed for the air quality data employing the function `manova ()` and the output is saved in a new place called `MANOVA`. Furthermore, more details can be supplied by employing the function `summary ()`, the details are related to the four multivariate tests (Wilks, Hotelling-Lawley, Pillai, Roy) including the p-value and the value of test statistic for each test.

```
> PM10 = Example6_5 $ PM10  #Define and extract PM10 from the file
> PM2.5 = Example6_5 $ PM2.5
> PM1 = Example6_5 $ PM1
> Location = as.factor (Example6_5 $ Point)
> Responses <- cbind (PM10, PM2.5, PM1) #To combine the responses in one place
> MANOVA <- manova (Responses ~ Location) #Perform multivariate analysis of variance
> summary (MANOVA, test = "Wilks")
          Df    Wilks approx F num Df den Df  Pr(>F)
Location   3 0.064405   3.4195      9 14.753 0.01788 *
Residuals  8
---
Signif. codes: 0 '***' 0.001 '**' 0.01 '*' 0.05 '.' 0.1 ' ' 1
> summary (MANOVA, test = "Hotelling-Lawley")
          Df Hotelling-Lawley approx F num Df den Df   Pr(>F)
Location   3           6.3762   3.3062      9     14 0.02231 *
Residuals  8
---
Signif. codes: 0 '***' 0.001 '**' 0.01 '*' 0.05 '.' 0.1 ' ' 1
> summary (MANOVA, test = "Roy")
        Df    Roy approx F num Df den Df    Pr(>F)
Location 3 4.9985   13.329      3      8 0.001769 **
Residuals 8
---
Signif. codes: 0 '***' 0.001 '**' 0.01 '*' 0.05 '.' 0.1 ' ' 1
> summary (MANOVA, test = "Pillai")
          Df Pillai approx F num Df den Df  Pr(>F)
Location   3 1.5284   2.7696      9     24 0.02233 *
Residuals  8
---
Signif. codes: 0 '***' 0.001 '**' 0.01 '*' 0.05 '.' 0.1 ' ' 1
```

The multivariate analysis of variance clearly showed that the location of the sampling exhibited a significant effect on the PM_{10}, $PM_{2.5}$, and PM_1 because the p-value is small—less than 0.05—as appeared in the results under `pr (>F)` for all of the multivariate tests (`Wilks, p-value < 0.01788, Hotelling-Lawley, p-value < 0.02231, Roy, p-value < 0.001769, Pillai, p-value < 0.02233`), which indicates that the null hypothesis is rejected, indicating that the location influences the chosen responses PM_{10}, $PM_{2.5}$, and PM_1. This conclusion suggests that location has a significant impact on the PM_{10}, $PM_{2.5}$, and PM_1 and should be taken into account as an important factor for any future activity.

EXAMPLE 6.6 THE CONCENTRATION OF HEAVY METALS IN SEDIMENT

An experiment was conducted to measure the concentrations of five heavy metals in eight rivers with twenty-four sampling points across the Penang River Basin in the Penang State of Malaysia. Three sampling points were chosen from each location (Air Hitam River I, Air Hitam River II, Air Hitam River III, Dondang River, Air Puteh River, Air Terjun River, Jelutong River, and Pinang River). Assume that the data are normally distributed. Use a significance level of $\alpha = 0.05$ to carry out multivariate analysis of variance to examine whether the location of the river influences the heavy metals concentrations of the chosen rivers. The sample data are presented in Table 6.13.

TABLE 6.13 The Selected Heavy Metal Concentrations in Penang State (mg/kg)

Location	Cd	Pb	Zn	Cu	Fe
Air Hitam River I	0.15	3.45	76.39	9.40	1658.83
Air Hitam River I	0.70	11.65	101.29	17.25	1680.33
Air Hitam River I	0.01	9.30	120.09	19.75	1674.83
Air Hitam River II	0.35	6.50	75.84	20.45	1294.37
Air Hitam River II	0.01	0.10	52.04	8.40	1711.83
Air Hitam River II	0.02	10.40	72.34	46.25	1605.34
Air Hitam River III	0.02	20.05	58.14	43.90	1790.32
Air Hitam River III	0.01	22.85	67.09	19.35	1705.83
Air Hitam River III	0.01	23.95	120.54	15.00	1481.85
Dondang River	1.40	10.50	26.55	14.40	1713.33
Dondang River	1.10	0.10	68.14	40.70	1492.35
Dondang River	0.60	0.10	28.15	9.05	1674.83
Air Puteh River	0.35	0.01	34.45	4.50	1015.90
Air Puteh River	0.95	0.01	28.85	0.40	1558.84
Air Puteh River	0.50	0.02	48.25	1.10	1321.87
Air Terjun River	0.55	0.01	85.39	9.95	1648.84
Air Terjun River	1.95	1.90	59.14	8.15	1650.33
Air Terjun River	1.10	0.01	166.58	18.20	1589.84
Jelutong River	0.01	0.10	63.29	2.65	1585.84
Jelutong River	0.01	0.01	77.59	58.94	2069.29
Jelutong River	1.20	124.44	470.25	81.09	1947.81
Pinang River	3.90	83.79	259.67	59.49	2000.30
Pinang River	2.60	24.30	183.93	43.10	1930.81
Pinang River	2.45	36.05	298.67	53.29	1983.80

The scientist is concerned in measuring the concentration of five heavy metals (responses) to determine whether the concentration of the heavy metals in sediment is influenced by the location of the river (the factor), which would suggest that each river is affected by different surrounding areas, or the location of the river does not influence the concentration of chosen heavy metals, which would suggest that all of the chosen heavy metals have the same concentration. The effect of location of the river on the concentration of heavy metals was assessed employing multivariate analysis of variance. The multivariate analysis of variance was performed for the stored data (`.CSV (Example6_6)`) of heavy metals concentration in sediment.

```
location = as.factor (Example6_6 $ Location)
Responses <- cbind (Example6_6 $ Cd, Example6_6 $ Pb, Example6_6 $ Zn, Example6_6 $ Cu,
Example6_6 $ Fe)
MANOVA <- manova(Responses ~ location)
summary (MANOVA, "Wilks")
summary (MANOVA, "Hotelling-Lawley")
summary (MANOVA, "Pillai")
summary (MANOVA, "Roy")
```

The outputs of employing R commands and built-in functions to analyze the data for the chosen heavy metals concentrations measured from samples of sediment, including the four multivariate tests and the p-value, are given below.

The four test statistic values are shown in the outputs of employing R commands and built-in functions, the results include the value of the test statistic for the Wilk's, Hotelling's, Pillai, and Roy tests.

```
> location = as.factor (Example6_6 $ Location)
> Responses <- cbind (Example6_6 $ Cd, Example6_6 $ Pb, Example6_6 $ Zn, Example6_6 $ Cu,
Example6_6 $ Fe)
> MANOVA <- manova(Responses ~ location)
> summary (MANOVA, "Wilks")
          Df  Wilks approx F num Df den Df     Pr(>F)
location   7 0.0116   2.8493     35 52.909  0.0002861   ***
Residuals 16
---
Signif. codes: 0 '***' 0.001 '**' 0.01 '*' 0.05 '.' 0.1 ' ' 1
> summary (MANOVA, "Hotelling-Lawley")
          Df Hotelling-Lawley approx F num Df den Df    Pr(>F)
location   7           16.769   4.9827     35     52 1.071e-07 ***
Residuals 16
---
Signif. codes: 0 '***' 0.001 '**' 0.01 '*' 0.05 '.' 0.1 ' ' 1
> summary (MANOVA, "Pillai")
          Df Pillai approx F num Df den Df  Pr(>F)
location   7 2.1534   1.7292     35     80 0.02284 *
Residuals 16
---
Signif. codes: 0 '***' 0.001 '**' 0.01 '*' 0.05 '.' 0.1 ' ' 1
> summary (MANOVA, "Roy")
          Df    Roy approx F num Df den Df    Pr(>F)
location   7 13.974    31.94      7     16 3.016e-08   ***
Residuals 16
---
Signif. codes: 0 '***' 0.001 '**' 0.01 '*' 0.05 '.' 0.1 ' ' 1
```

The multivariate analysis of variance results showed that the location of the river exhibited a significant effect on the selected heavy metals concentrations in sediment because the p-values for all of the four multivariate tests are very small, as appeared in the results of employing R commands and built-in functions (`0.0002861, 1.071e-07, 0.02284`, and `3.016e-08`) for Wilks, Hotelling-Lawley, Pillai, and Roy test, respectively. This result could be due to the differences in the activities surrounding each river or other environmental reasons.

TABLE 6.14 The Entries of a Two-Way MANOVA

Source of variation (S.O.V)	Degrees of freedom (d.f)	Sum of squares and products matrices (SSP)
Factor (A)	$a-1$	$SS_A=\sum_{i=1}^{a} bn(\overline{Y}_{i.}-\overline{Y})(\overline{Y}_{i.}-\overline{Y})'$
Factor (B)	$b-1$	$SS_B=\sum_{j=1}^{b} kn(\overline{Y}_{.j}-\overline{Y})(\overline{Y}_{.j}-\overline{Y})'$
Interaction (AB)	$(a-1)(b-1)$	$SS_{AB}=\sum_{i=1}^{a}\sum_{j=1}^{b} n(\overline{Y}_{ij}-\overline{Y}_{i.}-\overline{Y}_{.j}+\overline{Y})(\overline{Y}_{ij}-\overline{Y}_{i.}-\overline{Y}_{.j}+\overline{Y})'$
Residual (Error)	$ab(n-1)$	$E=SS_{Error}=\sum_{i=1}^{a}\sum_{j=1}^{b}\sum_{k=1}^{n}(\overline{Y}_{ijk}-\overline{Y}_{ij})(\overline{Y}_{ijk}-\overline{Y}_{ij})'$
Total	$abn-1$	$Total=SS_T=\sum_{i=1}^{a}\sum_{j=1}^{b}\sum_{k=1}^{n}(\overline{Y}_{ijk}-\overline{Y})(\overline{Y}_{ijk}-\overline{Y})'$

6.4.2 Two-Way Multivariate Analysis of Variance

The concept of one-way multivariate analysis of variance can be used to include one more independent variable (factor) in the analysis. In a two-way multivariate analysis of variance (MANOVA), the scientists are concerned in testing three hypotheses; two hypotheses are for the effect of each factor and third hypothesis is for the interaction between the two factors on several response variables measured on the same research unit. Consider there are two independent variables (factors), the first variable is called A and has a levels while the second variable is called B and has b levels. The sum of squares for the two factors, the residual and the total sum of squares with the degrees of freedom associated with each component are presented in Table 6.14.

Wilk's test can be employed to examine the three hypotheses for the main effects and interaction, as presented in (6.4) and (6.5).

For the main effect,

$$\Lambda_A=\frac{|E|}{|E+SS_A|},\ \Lambda_B=\frac{|E|}{|E+SS_B|} \tag{6.4}$$

and for the interaction,

$$\Lambda_{AB}=\frac{|E|}{|E+SS_{AB}|} \tag{6.5}$$

EXAMPLE 6.7 EFFECT OF THE TIME AND LOCATION ON THE LEVEL OF PARTICULATE MATTER (μG/M³)

We wish to investigate the effect of two factors, namely, the location with four various places at a palm oil mill, A, B, C, and D, and the time of sampling with five various times (1, 2, 3, 4, and 5), on the level of the particulate matters in the air of a palm oil mill. Determine whether the various locations and various times have a significant impact on the chosen air quality parameters. Two replicates were employed for each combination of location and time. Assume that the data are normally distributed. Use a significance level of $\alpha=0.05$ to perform the analysis. The sample data are presented in Table 6.15.

TABLE 6.15 The Results of Air Quality Parameters in a Palm Oil Mill

Time	Location	PM_{10}	$PM_{2.5}$	PM_1	Time	Location	PM_{10}	$PM_{2.5}$	PM_1
1	A	611.22	190.25	76.54	3	A	1306.08	199.59	148.21
1	B	443.38	114.34	44.14	3	B	655.47	148.43	77.90
1	C	952.55	199.21	109.40	3	C	1663.73	262.92	179.83
1	D	327.34	93.19	39.35	3	D	429.52	58.00	29.10
1	A	1620.06	134.25	84.50	4	A	1500.69	187.09	126.58
1	B	173.46	31.00	9.54	4	B	480.17	127.63	61.03

Continued

TABLE 6.15 The Results of Air Quality Parameters in a Palm Oil Mill—cont'd

Time	Location	PM_{10}	$PM_{2.5}$	PM_1	Time	Location	PM_{10}	$PM_{2.5}$	PM_1
1	C	1464.18	202.91	155.91	4	C	650.96	134.09	64.43
1	D	164.69	40.85	14.06	4	D	233.76	38.83	17.18
2	A	461.14	89.08	53.91	4	A	2351.76	347.29	256.52
2	B	166.20	33.73	15.91	4	B	332.22	107.41	41.75
2	C	487.37	96.43	52.80	4	C	1261.46	186.51	132.50
2	D	179.20	4.55	1.21	4	D	58.34	10.10	3.80
2	A	1640.58	242.24	137.99	5	A	1500.69	187.09	126.58
2	B	644.33	163.92	87.10	5	B	480.17	127.63	61.03
2	C	322.75	143.19	56.73	5	C	650.96	134.09	64.43
2	D	226.58	52.75	22.27	5	D	233.76	38.83	17.18
3	A	1659.14	272.10	188.51	5	A	2351.76	347.29	256.52
3	B	82.94	28.24	11.72	5	B	332.22	107.41	41.75
3	C	1121.76	211.61	117.44	5	C	1261.46	186.51	132.50
3	D	61.42	9.91	3.34	5	D	58.34	10.10	3.80

The effects of the time and location on the level of particulate matters in the air of the palm oil mill were investigated and examined through the changes of three responses, namely, PM_1, $PM_{2.5}$, and PM_{10}.

The data for PM_1, $PM_{2.5}$, and PM_{10} in the air of the palm oil mill obtained from various locations and different times were analyzed employing a two-way multivariate analysis of variance (MANOVA) to verify the effect of various locations and various times on the level of the chosen particulate matters in the air of palm oil mill. Moreover, the effects of the interaction between the various locations and times on the level of the chosen particulate matters in the air of palm oil mill was studied and examined as well.

A two-way multivariate analysis of variance (MANOVA) was performed employing the function `manova ()` for the stored data `.CSV (Example6_7)` of the particulate matter in the air of palm oil mills. Moreover, the data for PM_1, $PM_{2.5}$, and PM_{10} were prepared for the function `manova ()` and placed in one file employing the function `cbind ()` function. More details on the analysis can be supplied employing the function `summary ()`. A multivariate analysis of variance was employed to analyze the data for the PM_{10}, $PM_{2.5}$, and PM_1. The built-in functions and commands in R to carry out a two-way multivariate analysis of variance for air quality of the palm oil mill are presented below.

```
PM10 = Example6_7 $ PM10
PM2.5 = Example6_7 $ PM2.5
PM1 = Example6_7 $ PM1
Responses <- cbind (PM10, PM2.5, PM1)
Time = as.factor (Example6_7 $ Time)
Location = as.factor (Example6_7 $ Location)
MANOVA <- manova(Responses ~ Time * Location)
summary (MANOVA, test = "Wilks")
summary (MANOVA, test = "Hotelling-Lawley")
summary (MANOVA, test = "Pillai")
summary (MANOVA, test = "Roy")
```

One can observe that the responses (PM_{10}, $PM_{2.5}$, and PM_1) and the two factors that are the location and time were defined as presented in the first three rows of the R commands. The three responses (PM_{10}, $PM_{2.5}$, and PM_1) are combined in one file and the result saved in a new place called `Responses` employing the function `cbind ()`. The output of employing the function `manova ()` to perform a two-way MANOVA for the particulate matters in the air of palm oil mills with the p-values for the four multivariate tests are presented below.

```
> PM10 = Example6_7 $ PM10
> PM2.5 = Example6_7 $ PM2.5
```

```
> PM1 = Example6_7 $ PM1
> Responses <- cbind (PM10, PM2.5, PM1)
> Time = as.factor (Example6_7 $ Time)
> Location = as.factor (Example6_7 $ Location)
> MANOVA <- manova(Responses ~ Time * Location)
> summary (MANOVA, test = "Wilks")
              Df    Wilks approx F num Df den Df    Pr(>F)
Time           4  0.65793   0.6846     12 47.915    0.7576
Location       3  0.15409   5.6482      9 43.958 3.586e-05   ***
Time:Location 12  0.32488   0.6934     36 53.911    0.8766
Residuals     20
---
Signif. codes: 0 '***' 0.001 '**' 0.01 '*' 0.05 '.' 0.1 ' ' 1
> summary (MANOVA, test = "Hotelling-Lawley")
             Df Hotelling-Lawley approx F num Df den Df    Pr(>F)
Time          4           0.4753   0.6602     12     50    0.7800
Location      3           4.0516   7.5030      9     50 7.581e-07   ***
Time:Location 12          1.4629   0.6773     36     50    0.8887
Residuals    20
---
Signif. codes: 0 '***' 0.001 '**' 0.01 '*' 0.05 '.' 0.1 ' ' 1
> summary (MANOVA, test = "Pillai")
             Df  Pillai approx F num Df den Df   Pr(>F)
Time          4 0.37147   0.7066     12     60 0.738896
Location      3 1.07294   3.7119      9     60 0.000944 ***
Time:Location 12 0.89384   0.7073     36     60 0.866794
Residuals    20
---
Signif. codes: 0 '***' 0.001 '**' 0.01 '*' 0.05 '.' 0.1 ' ' 1
> summary (MANOVA, test = "Roy")
              Df     Roy approx F num Df den Df    Pr(>F)
Time           4  0.3488   1.7440      4     20    0.1799
Location       3  3.6723  24.4822      3     20 6.697e-07 ***
Time:Location 12  0.9319   1.5532     12     20    0.1855
Residuals     20
---
Signif. codes: 0 '***' 0.001 '**' 0.01 '*' 0.05 '.' 0.1 ' ' 1
```

The multivariate analysis of variance clearly showed that the interaction between the location and time (`Time:Location`) did not exhibit a significant effect on the PM_{10}, $PM_{2.5}$, and PM_1 because the p-value is large and more than 0.05 as appeared in the results under `pr(>F)` for all of the multivariate tests (`Wilks, p-value < 0.8766, Hotelling-Lawley, p-value < 0.8887, Pillai, p-value < 0.866794, Roy, p-value < 0.1855`), which indicates that the null hypothesis for the interaction should not be rejected and that the interaction between the time and location does not exist. Moreover, the results revealed that various locations exhibited a significant effect on the behavior of at least one response because the p-value is small and less than < 0.05, while the second factor (time) did not exhibit a significant effect (p-value > 0.05), as appeared in the output and as a result, the location will affect the level of the chosen particulate matter.

EXAMPLE 6.8 EFFECT OF THE SEASONS AND LOCATIONS ON THE WATER QUALITY (mg/L)

A researcher wishes to study the impact of the seasons and locations on the physiochemical parameters of the surface water for the Beris Dam in the state of Kedah, Malaysia. The impact of two factors on the nine physiochemical parameters (responses), namely, the temperature, electrical conductivity (EC), pH, total dissolved solids (TDS), dissolved oxygen (DO), biochemical oxygen demand (BOD), chemical oxygen demand (COD), turbidity, and total suspended solids (TSS), are studied. The two factors are the location (before the dam, at the dam, and after the dam) and the seasons (the rainy season (R) and the dry season (D)). Twenty replicates at each combination of location and season were employed. Assume that the data are normally distributed. Use a significance level of $\alpha = 0.05$ to carry out the analysis. The sample data are presented in Table 6.16; they have been stored as a `.CSV (Example6_8)` format.

TABLE 6.16 The Data for Water Quality Concentration for the Beras Dam (mg/L)

Season	Location	Station	Temperature	Conductivity	pH	TDS	DO	BOD	COD	Turbidity	TSS
R	1	1	32.90	43.00	6.70	21.60	5.51	2.12	50.00	2.58	2.00
R	1	2	32.60	42.90	6.77	21.30	5.75	1.07	23.00	7.61	1.00
R	1	3	32.80	42.40	6.40	21.20	7.29	1.01	30.00	2.15	2.00
R	1	4	33.00	42.40	6.38	21.20	7.14	1.42	68.00	9.79	3.00
R	1	5	33.20	42.40	6.49	21.20	7.60	0.82	60.00	3.44	5.00
R	1	6	33.40	42.60	6.66	21.30	7.63	1.19	27.00	9.52	3.00
R	1	7	33.40	42.60	6.59	21.30	6.13	1.20	24.00	2.77	3.00
R	1	8	33.50	42.70	6.40	21.30	7.38	1.47	19.00	2.29	4.00
R	1	9	33.90	42.80	6.38	21.40	5.78	1.05	18.00	3.72	6.00
R	1	10	34.30	42.70	6.45	21.30	6.39	1.30	31.00	3.51	3.00
R	1	11	33.60	42.70	6.55	21.30	5.50	1.45	27.00	2.52	3.00
R	1	12	34.20	42.80	6.42	21.40	5.15	1.34	35.00	3.17	3.00
R	1	13	34.20	42.80	6.57	21.40	6.51	2.04	38.00	2.66	3.00
R	1	14	33.80	42.60	6.30	21.30	5.30	1.68	16.00	7.56	3.00
R	1	15	34.20	42.40	6.39	21.20	5.06	1.86	67.00	6.59	6.00
R	1	16	34.40	42.50	6.30	21.20	5.43	1.46	4.00	2.70	5.00
R	1	17	34.40	42.40	6.30	21.20	4.82	0.95	24.00	2.93	2.00
R	1	18	33.30	42.30	6.27	21.10	5.72	1.94	45.00	3.41	2.00
R	1	19	34.10	42.40	6.14	21.20	6.78	1.38	32.00	2.94	3.00
R	1	20	34.70	42.20	6.39	21.10	5.57	1.92	51.00	2.51	3.00
R	2	1	31.30	67.00	6.01	33.40	6.93	3.64	3.00	2.22	4.00
R	2	2	31.40	46.30	6.01	23.10	7.13	3.41	13.00	1.91	4.00
R	2	3	31.50	45.30	6.55	22.70	7.04	2.91	17.00	3.57	3.00
R	2	4	31.60	45.80	6.01	22.90	6.90	3.34	58.00	2.38	4.00
R	2	5	32.40	45.60	6.56	22.80	6.86	2.00	43.00	1.76	5.00
R	2	6	32.10	44.90	6.42	22.40	6.80	3.38	26.00	2.22	4.00
R	2	7	32.20	44.00	6.48	22.50	6.75	3.11	10.00	3.76	5.00
R	2	8	32.00	45.40	6.14	22.80	6.55	4.00	9.00	2.34	5.00
R	2	9	32.20	45.60	5.85	22.80	6.41	2.11	8.00	2.46	3.00
R	2	10	32.10	45.40	6.06	22.70	7.24	2.93	3.00	2.07	2.00
D	1	1	32.10	43.00	6.50	21.30	4.84	2.67	42.00	1.19	4.00
D	1	2	32.10	42.50	6.52	21.30	4.46	2.61	28.00	2.20	7.00
D	1	3	32.10	42.60	6.59	21.30	4.52	2.50	32.00	2.18	6.00
D	1	4	32.20	42.50	6.57	21.30	5.02	3.14	18.00	2.20	3.00

TABLE 6.16 The Data for Water Quality Concentration for the Beras Dam (mg/L)—cont'd

Season	Location	Station	Temperature	Conductivity	pH	TDS	DO	BOD	COD	Turbidity	TSS
D	1	5	32.20	42.50	6.49	21.20	5.11	1.18	24.00	2.15	5.00
D	1	6	32.10	42.90	6.76	21.20	4.71	1.54	20.00	1.89	4.00
D	1	7	32.30	42.80	6.31	21.20	5.04	2.18	35.00	2.11	4.00
D	1	8	32.30	42.80	6.42	21.50	6.13	1.89	15.00	3.01	3.00
D	1	9	32.30	42.80	6.42	21.70	6.45	2.54	24.00	3.11	3.00
D	1	10	32.30	42.30	6.53	21.70	6.21	3.21	28.00	2.70	2.00
D	1	11	32.20	42.30	6.70	21.30	6.47	2.78	32.00	3.17	4.00
D	1	12	32.20	42.30	6.45	21.30	6.30	3.21	27.00	3.30	7.00
D	1	13	31.80	42.30	6.40	21.30	7.10	2.87	18.00	2.49	8.00
D	1	14	31.70	42.30	6.49	21.80	7.11	2.57	20.00	2.52	9.00
D	1	15	31.70	42.20	6.51	21.80	6.89	2.18	40.00	2.79	10.00
D	1	16	32.10	42.20	6.73	22.30	5.49	1.98	35.00	4.67	7.00
D	1	17	32.10	42.30	6.78	22.30	5.31	2.45	21.00	5.41	8.00
D	1	18	32.20	42.30	6.50	22.30	5.40	2.87	19.00	3.20	10.00
D	1	19	32.00	42.20	6.49	21.90	5.45	2.65	27.00	2.70	6.00
D	1	20	32.00	42.30	6.71	22.30	5.40	1.88	38.00	2.90	7.00
D	2	1	32.00	43.10	6.15	23.50	6.21	3.15	40.00	3.14	8.00
D	2	2	32.00	43.10	6.32	23.50	5.78	3.05	32.00	2.17	4.00
D	2	3	32.10	43.10	6.04	23.20	6.01	2.78	28.00	2.89	4.00
D	2	4	32.20	43.80	6.15	23.20	6.01	4.01	40.00	3.33	4.00
D	2	5	32.20	44.00	6.06	23.20	5.56	3.11	32.00	2.21	5.00
D	2	6	32.10	44.00	6.31	23.40	5.75	3.89	24.00	2.19	3.00
D	2	7	32.10	43.90	6.44	23.20	5.95	2.18	36.00	3.18	4.00
D	2	8	32.20	43.90	6.40	23.50	5.43	3.05	17.00	2.01	7.00
D	2	9	32.20	43.80	6.32	23.10	5.82	3.24	25.00	2.89	3.00
D	2	10	32.20	44.00	6.11	23.10	4.89	2.14	30.00	3.45	2.00
R	2	11	32.30	45.30	6.43	22.70	6.66	3.37	34.00	3.48	3.00
R	2	12	32.10	45.10	6.36	22.60	5.98	0.59	24.00	2.38	4.00
R	2	13	32.20	45.40	6.57	22.70	6.86	3.23	3.00	2.47	3.00
R	2	14	32.30	45.60	6.52	22.80	6.03	3.96	16.00	3.19	4.00
R	2	15	32.30	45.40	6.69	22.70	5.77	3.02	21.00	2.53	3.00
R	2	16	32.30	47.80	6.37	47.80	7.19	2.04	15.00	2.35	5.00
R	2	17	32.10	43.60	6.31	43.60	6.93	2.67	21.00	3.17	6.00
R	2	18	32.10	42.70	6.43	42.70	2.51	1.73	13.00	3.25	7.00
R	2	19	32.20	42.90	6.32	42.90	2.08	0.98	6.00	2.86	4.00
R	2	20	32.20	43.10	6.40	42.50	2.35	1.25	10.00	2.97	6.00
R	3	1	32.20	47.80	6.37	23.90	7.19	2.04	15.00	17.03	15.00
R	3	2	32.60	43.60	6.31	31.80	6.93	2.67	12.00	24.33	22.00

Continued

TABLE 6.16 The Data for Water Quality Concentration for the Beras Dam (mg/L)—cont'd

Season	Location	Station	Temperature	Conductivity	pH	TDS	DO	BOD	COD	Turbidity	TSS
R	3	3	32.60	42.70	6.43	21.30	2.51	1.73	18.00	13.79	11.00
R	3	4	32.60	42.90	6.32	21.40	2.08	0.98	15.00	15.32	18.00
R	3	5	32.60	256.00	3.88	128.20	1.65	3.61	35.00	32.51	10.00
R	3	6	33.80	144.90	5.19	72.40	1.83	0.79	50.00	45.88	16.00
R	3	7	34.10	137.50	5.24	68.70	1.38	1.97	53.00	28.90	13.00
R	3	8	35.10	141.60	5.24	70.70	1.93	1.39	35.00	16.89	12.00
R	3	9	32.20	246.00	5.63	123.30	1.76	1.41	32.00	35.21	16.00
R	3	10	30.90	256.00	5.87	127.80	2.53	3.46	26.00	24.62	12.00
R	3	11	31.40	255.00	3.91	127.60	1.65	1.99	38.00	52.36	25.00
R	3	12	31.00	263.00	5.59	131.60	1.68	1.26	24.00	55.30	32.00
R	3	13	30.60	264.00	5.62	132.00	1.59	1.15	26.00	79.50	48.00
R	3	14	31.00	262.00	5.55	131.10	2.71	0.96	34.00	106.00	38.00
R	3	15	32.20	250.00	3.94	125.20	7.44	2.38	49.00	34.31	15.00
R	3	16	32.20	253.00	3.98	126.40	4.18	1.61	53.00	32.59	28.00
R	3	17	30.80	260.00	5.57	130.20	7.28	2.45	56.00	34.71	16.00
R	3	18	31.20	251.00	3.94	125.40	6.84	2.29	29.00	45.30	19.00
R	3	19	31.30	251.00	3.96	125.60	3.94	2.58	37.00	45.16	15.00
R	3	20	31.00	150.00	3.95	125.20	3.48	2.48	33.00	36.10	20.00
D	2	11	32.20	44.40	6.30	23.10	4.76	2.18	19.00	3.41	4.00
D	2	12	32.40	44.20	6.33	23.20	5.11	2.70	12.00	3.20	2.00
D	2	13	32.40	44.20	6.31	23.30	5.01	3.19	22.00	2.79	3.00
D	2	14	32.40	40.90	6.20	23.20	4.79	3.22	15.00	2.87	3.00
D	2	15	32.20	40.90	6.52	23.40	4.91	3.05	18.00	3.50	5.00
D	2	16	32.20	40.90	6.46	23.30	5.20	3.11	27.00	3.67	4.00
D	2	17	32.10	42.60	6.52	23.10	4.71	2.76	29.00	3.20	5.00
D	2	18	32.10	42.60	6.30	23.10	4.01	2.87	32.00	2.90	3.00
D	2	19	32.10	42.60	6.70	23.50	4.13	3.01	10.00	3.17	3.00
D	2	20	32.20	42.70	6.54	23.50	4.13	3.02	12.00	3.04	2.00
D	3	1	33.00	149.00	5.30	70.40	1.89	3.18	32.00	13.00	35.00
D	3	2	33.00	151.00	5.52	45.60	1.92	1.17	38.00	17.40	30.00
D	3	3	33.10	200.40	6.01	68.50	2.11	2.32	42.00	13.20	32.00
D	3	4	33.20	243.00	5.39	68.50	2.51	3.15	17.00	27.30	28.00
D	3	5	32.80	215.00	5.85	58.40	2.08	1.89	25.00	45.10	50.00
D	3	6	32.80	180.00	6.07	87.20	1.75	1.15	22.00	60.70	52.00
D	3	7	33.20	178.00	5.70	70.70	3.01	3.20	28.00	64.50	60.00
D	3	8	33.40	180.20	5.75	86.00	2.52	3.07	31.00	86.00	51.00
D	3	9	33.40	246.00	6.11	104.00	2.53	2.17	36.00	110.00	48.00
D	3	10	33.40	230.00	6.00	132.00	3.20	3.01	17.00	115.00	43.00

TABLE 6.16 The Data for Water Quality Concentration for the Beras Dam (mg/L)—cont'd

Season	Location	Station	Temperature	Conductivity	pH	TDS	DO	BOD	COD	Turbidity	TSS
D	3	11	33.60	245.00	5.43	124.00	2.19	2.87	16.00	132.00	70.00
D	3	12	33.30	250.00	3.98	126.00	1.63	2.11	42.00	82.00	67.00
D	3	13	33.10	186.00	4.94	110.40	2.53	2.08	44.00	108.00	50.00
D	3	14	33.20	190.00	5.40	115.20	3.01	2.92	38.00	96.00	50.00
D	3	15	33.30	190.00	5.10	114.30	3.51	3.11	29.00	113.00	42.00
D	3	16	33.00	201.00	5.00	132.20	3.33	2.82	30.00	48.00	46.00
D	3	17	33.20	201.00	4.89	132.20	4.00	3.05	30.00	60.00	37.00
D	3	18	33.20	232.00	4.31	133.00	4.18	2.18	31.00	64.00	37.00
D	3	19	33.30	232.00	5.20	133.30	4.05	2.27	31.00	64.30	42.00
D	3	20	33.40	232.00	5.20	132.80	4.11	2.29	38.00	60.00	45.00

The effect of the locations and the seasons on water quality were investigated and tested through the changes of nine responses (physiochemical parameters): the temperature, conductivity, pH, TDS, DO, COD, BOD, turbidity, and TSS. The data for the physiochemical parameters obtained from Beris Dam were analyzed employing a two-way multivariate analysis of variance (MANOVA) to examine the effect of various locations and various seasons on physiochemical parameters of surface water. Moreover, the effect of the interaction between the various locations and seasons on the physiochemical parameters of surface water was studied and examined.

The outputs of employing R commands and built-in functions to analyze the impact of the location and the season on the physiochemical parameters of water are given below.

```
> Temperature = Example6_8 $ Temperature
> Conductivity = Example6_8 $ Conductivity
> pH = Example6_8 $ pH
> TDS = Example6_8 $ TDS
> DO = Example6_8 $ DO
> BOD = Example6_8 $ BOD
> COD = Example6_8 $ COD
> Turbidity = Example6_8 $ Turbidity
> TSS = Example6_8 $ TSS
> Responses <- cbind(Temperature, Conductivity, pH, TDS, DO, BOD, COD,Turbidity,TSS)
> Season = as.factor(Example6_8 $ Season)
> Location = as.factor(Example6_8 $ Location)
> MANOVA <- manova(Responses ~ Season * Location)
> summary (MANOVA, test="Wilks")
                 Df   Wilks approx F num Df den Df    Pr(>F)
Season            1 0.43619   15.224      9    106 1.219e-15 ***
Location          2 0.04911   41.367     18    212 < 2.2e-16 ***
Season:Location   2 0.19914   14.615     18    212 < 2.2e-16 ***
Residuals       114
---
Signif. codes: 0 '***' 0.001 '**' 0.01 '*' 0.05 '.' 0.1 ' ' 1
> summary (MANOVA, test="Pillai")
                 Df  Pillai approx F num Df den Df    Pr(>F)
Season            1 0.56381   15.224      9    106 1.219e-15 ***
Location          2 1.39885   27.665     18    214 < 2.2e-16 ***
Season:Location   2 1.00358   11.974     18    214 < 2.2e-16 ***
Residuals       114
---
Signif. codes: 0 '***' 0.001 '**' 0.01 '*' 0.05 '.' 0.1 ' ' 1
```

```
> summary (MANOVA, test="Hotelling-Lawley")
                 Df Hotelling-Lawley approx F num Df den Df    Pr(>F)
Season            1           1.2926   15.224      9    106 1.219e-15 ***
Location          2          10.2402   59.734     18    210 < 2.2e-16 ***
Season:Location  2           3.0036   17.521     18    210 < 2.2e-16 ***
Residuals       114
---
Signif. codes: 0 '***' 0.001 '**' 0.01 '*' 0.05 '.' 0.1 ' ' 1
> summary (MANOVA, test="Roy")
                 Df    Roy approx F num Df den Df    Pr(>F)
Season            1 1.2926   15.224      9    106 1.219e-15 ***
Location          2 9.2546  110.027      9    107 < 2.2e-16 ***
Season:Location   2 2.6142   31.080      9    107 < 2.2e-16 ***
Residuals       114
---
Signif. codes: 0 '***' 0.001 '**' 0.01 '*' 0.05 '.' 0.1 ' ' 1
```

One can observe that the results exhibited a significant interaction between the seasons and the locations (`Season: Location`) because the p-value is very small and less than < 0.05 as presented in the `Pr(>F)` column for all multivariate tests ((`Wilks, p-value < 2.2e-16, Pillai, p-value < 2.2e-16, Hotelling-Lawley, p-value < 2.2e-16, Roy, p-value < 2.2e-16`), this indicates that the hypothesis regarding the interaction between seasons and locations is rejected and indicates that the two factors work together to affect at least one of the chosen responses of the temperature, conductivity, pH, TDS, DO, COD, BOD, turbidity, and TSS. Moreover, both the seasons and the location had a significant effect on at least one of the chosen responses. Thus, the seasons and the location should be taken into account as influential factors for any further study.

Further Reading

Alkarkhi, A.F.M., Alqaraghuli, W.A.A., 2019. Easy Statistics For Food Science With R, first ed. Academic Press.

Alkarkhi, A.F.M., Low, H.C., 2012. Elementary Statistics For Technologist. Universiti Sains Malaysia Press, Pulau Pinang.

Bryan, F.J.M., 1991. Multivariate Statistical Methods: A Primer. Chapman & Hall, Great Britain.

Daniel & Hocking, 2013. Blog Archives, High Resolution Figures In R . [Online]. R-Bloggers. Available: https://www.r-bloggers.com/Author/Daniel-Hocking/. (Accessed 8 December 2018).

Farah, N.M.S., Nik, N.N.A.R., Alkarkhi, A.F.M., 2012. Assessment of heavy metal pollution in sediments of Pinang River, Malay. In: International Conference On Environmental Research And Technology (ICERT 2012). Universiti Sains Malaysia, Penang, Malaysia, pp. 470–475.

Ho, Y.C., Norli, I., Alkarkhi, A.F., Morad, N., 2010. Characterization of biopolymeric flocculant (Pectin) and organic synthetic flocculant (Pam): a comparative study on treatment and optimization in Kaolin suspension. Biores. Technol. 101, 1166–1174.

Johnson, R.A., Wichern, D.W., 2002. Applied Multivariate Statistical Analysis. Prentice Hall, New Jersey.

Kabacoff, R.I., 2012. Quick-R, Anova . [Online]. Data Camp. Available: https://www.statmethods.net/stats/anova.html. (Accessed 6 June 2018).

Mark, G., 2018. Using R For Statistical Analyses—Introduction . [Online]. Available: http://www.gardenersown.co.uk/Education/Lectures/R/. (Accessed 28 May 2018).

Md. Azmal, H., Alkarkhi, A.F.M., Nik, N.N.A., 2008. Statistical and trend analysis of water quality data for the Baris Dam of Darul Aman River In Kedah, Malaysia. In: International Conference on Environmental Research And Rechnologt (ICERT 200s). Universiti Sains Malaysia, Penang, Malaysia, pp. 568–572.

Ngu, C.C., Nik, N.N.A.R., Alkarkhi, A.F.M., Anees, A., Omar, M., K, A., 2010. Assessment of Major Solid Wastes Generated in Palm Oil Mills. Int. J. Environ. Technol. Manage. 13, 245–252.

Nik, N.N.A.R., Ngu, C.C., Al-Karkhi, A.F.M., Mohd., R., Mohd. Omar, A.K., 2017. Analysis of particulate matters in air of palm oil mills—a statistical assessment. Environ. Eng. Manage. J. 16, 2537–2543.

Rencher, A.C., 2002. Methods of Multivariate Analysis. J. Wiley, New York.

Sarah, S., 2012. Instant R-Creating An Interaction Plot In R . [Online]. Wordpress. Available: http://www.instantr.com/2012/12/13/creating-an-interaction-plot-in-r/. (Accessed 6 June 2018).

CHAPTER

7

Regression Analysis

LEARNING OBJECTIVES

After careful consideration of this chapter, you should be able:

- *To describe the concept of linear regression.*
- *To explain linear regression analysis (simple and multiple).*
- *To explain multiple multivariate regression analysis.*
- *To explain the procedure of employing analysis of variance (ANOVA) in the regression analysis.*
- *To apply statistical hypothesis testing in simple, multiple, and multiple multivariate linear regression.*
- *To describe the procedure of testing the overall significance of the regression model.*
- *To describe the procedure of testing the individual regression coefficients.*
- *To describe the procedure of testing a subset of the regression coefficients.*
- *To describe the style of explaining regression models.*
- *To know how to use and interpret the coefficient of determination.*
- *To describe how to employ R built-in functions and commands to carry out linear regression models.*
- *To draw smart conclusions and recommendations.*

7.1 THE CONCEPT OF REGRESSION ANALYSIS

Regression analysis is a powerful statistical method employed to examine the relationship between one or more independent variables (predictor, explanatory) that influence one or more response variables (dependent). Regression analysis can also be employed to forecast the results of one or more dependent variables with a regression line that tests the relationship between the one or more independent variables and one response variable. The model that describes the relationship between the two types of variables is called a regression model.

Three types of regression analysis can be recognized based on the number of variables included in the analysis; the three types of regression models are presented below.

1. The relationship between only one response variable Y and one predictor (independent variable) X can be described by a model called simple linear regression. For example, an environmentalist wishes to investigate the impact of distance as a predictor (independent) variable on the traffic noise level as a response (dependent) variable.
2. The relationship between only one response variable Y and two or more predictors (k, independent variables) X_1, $X_2, \ldots, X_k$ can be described by a model called multiple linear regression. For example, we study the behavior of the indoor air components with respect to different polyurethane factories. Two independent variables were of interest, relative humidity (Rh %) and dry bulb temperature (Td, C0). In addition, there was one dependent variable toluene diisocyanate (TDI).
3. The relationship between several response variables and several predictors (independent variables) can be described by a model called multivariate multiple linear regression: For example, Environmental monitoring: stations yield measurements on ozone, NO, CO, and $PM_{2.5}$.

Applied Statistics for Environmental Science with R
https://doi.org/10.1016/B978-0-12-818622-0.00007-1

7.2 REGRESSION MODELS IN R

Several built-in functions are provided by R statistical software to carry out regression analysis. These functions are usually employed to fit a regression model, carry out statistical hypothesis testing for the regression equation, compute estimated values and errors (residuals), and predict new values. We can carry out regression analysis in R using the built-in functions and commands given below.

1. Two built-in functions `lm ()` and `Coef (lm())` are provided by R to estimate the intercept and the slope of the regression model (a regression coefficients).

```
lm (y ~ x, data = data frame)
or
coef (lm(y ~ x, data = data frame))
```

   where `y` refers to the response variable and `x` refers to the predictor.

2. More details can be provided by employing the built-in function `summary ()`, including the estimated coefficients, *t*-test, residuals, *p*-value, estimated error, and coefficient of determination (R^2).

```
Summary (data frame)
```

3. An analysis of variance table can be generated by employing the built-in function `anova ()` in R, showing the sum of squares (`Sum sq`), degrees of freedom (`Df`), mean sum of squares (`Mean sq`), `F value`, and *p*-value (`pr(>F)`).

```
anova (y ~ x, data = data frame)
or
anova (lm (y ~ x, data = data frame))
```

   where `y` refers to the response variable and `x` refers to the predictor.

4. A multiple linear regression model can be fitted employing the two built-in functions `lm ()` and `coef (lm ())`.

```
lm (y ~ x1 + x2 + ... + Xk, data = data frame)
or
coef (lm(y ~ x1 + x2 + ... + Xk, data = data frame))
```

   where `x1, x2, ...,` xk represent the *k* predictors and `y` refers to the response variable (dependent).

5. A prediction can be made by employing the built-in function `predict ()` to produce the lower and upper confidence limits with the predicted value. We can predict any value of the response variable using three built-in functions; the functions are `lm (), data.frame ()` and `Predict ()` as given below.

```
Fitting = lm (y ~ x, data = data frame)
Nv = data.frame (x = given value)
predict (Fitting, Nv, interval = "predict")
```

6. The estimated values can be computed employing the built-in function `fitted ()` function.

```
fitted (data frame)
```

7. The residuals of the model can be produced by employing the built-in function `resid ()`.

```
resid (data frame)
```

7.3 THE CONCEPT OF SIMPLE LINEAR REGRESSION MODEL

The relationship between one independent variable X and one dependent variable Y can be described and tested employing a simple linear regression model. The mathematical formula for this relationship is presented in (7.1).

$$Y = \beta_0 + \beta_1 X + \varepsilon \tag{7.1}$$

where β_0 and β_1 refer to the intercept and the slope of the regression model, respectively, and ε is the error term.

The sample data are usually employed to estimate the parameters β_0 and β_1; the estimated values are b_0 and b_1, respectively. The estimated regression model (also called predicted or fitted model) is presented in (7.2).

$$\hat{Y} = b_0 + b_1 X \tag{7.2}$$

where $\hat{Y}$ refers to the estimated response.

The least squares method is employed to compute the values of b_0 and b_1 as presented in (7.3) and (7.4).

$$b_1 = \frac{n\sum_{i=1}^{n} X_i Y_i - \sum_{i=1}^{n} X_i \sum_{i=1}^{n} Y_i}{n\sum_{i=1}^{n} X_i^2 - \left[\sum_{i=1}^{n} X_i\right]^2} \tag{7.3}$$

$$b_0 = \overline{Y} - b_1 \overline{X} \tag{7.4}$$

Note:

- Researchers can plot the regression line to get the best-fitting line.
- The data must be normally distributed.
- The direction of the relationship between the response variable and the independent variable can be identified by employing the sign of b_1, whether it is negative (decrease) or positive (increase).

7.3.1 Hypothesis Testing for Regression Models

We can examine the impact of the chosen independent variable on the dependent variable employing hypothesis testing related to the simple linear regression, deciding whether or not the chosen independent variable takes part in interpreting the total variation in the response values. Testing the contribution of an independent variable can be performed by examining the slope β_1 of the regression model. The two hypotheses regarding the slope of the regression model in a mathematical form are presented in (7.5).

$$\begin{aligned} H_0 &: \beta_1 = \beta_{1,0} \\ H_1 &: \beta_1 \neq \beta_{1,0} \end{aligned} \tag{7.5}$$

The test statistic employed for testing the hypothesis regarding the slope is given in (7.6).

$$t = \frac{b_1 - \beta_{1,0}}{se(b_1)} \tag{7.6}$$

with $n-2$ degrees of freedom.

where $\beta_{1,\,0}$ is a constant (given value), b_1 is the least squares estimate of β_1, and $se(b_1)$ is the standard error.

The value of $se(b_1)$ can be computed employing the formula given in (7.7).

$$se(\hat{\beta}_1) = \sqrt{\frac{\sum_{i=1}^{n} e_i^2/(n-2)}{\sum_{i=1}^{n} (X_i - \overline{X})^2}} \tag{7.7}$$

There is no impact of the chosen independent variable on the dependent variable, as stated by the null hypothesis (H_0); while the opposite direction of the null hypothesis states that the chosen predictor (independent variable) has an impact on the dependent and should be retained in the regression model, as stated by the alternative hypothesis (H_1). The relationship between the chosen independent variable and the dependent variable exists if the null hypothesis is rejected; this indicates that the chosen independent variable contributes in explaining the variation in the measured values of the response variable. The chosen independent variable does not help in interpreting the total variation in the dependent variable if the null hypothesis is not rejected, which indicates that the relationship between the chosen independent variable and the dependent variable does not exist.

The hypothesis on the intercept β_0 can be examined employing similar procedure used for β_1. The two hypotheses regarding β_0 in a mathematical form are presented in (7.8):

$$H_0 : \beta_0 = \beta_{0,0}$$
$$H_1 : \beta_0 \neq \beta_{0,0} \tag{7.8}$$

We can make a decision either to keep the intercept β_0 or not in the model based on the test statistic result. Moreover, the *t*-test statistic employed to test the hypothesis concerning β_0 is presented in (7.9).

$$t = \frac{b_0 - \beta_{0,0}}{se(b_0)} \tag{7.9}$$

where b_0 is the least squares estimate of β_0 and $se(b_0)$ is the standard error of b_0.

The value of $se(b_0)$ can be computed employing the formula presented in (7.10).

$$se(b_0) = \sqrt{\frac{\sum_{i=1}^{n} e_i^2}{n-2}\left[\frac{1}{n} + \frac{\overline{X}^2}{\sum_{i=1}^{n}(X_i - \overline{X})^2}\right]} \tag{7.10}$$

Note:

- Two approaches can be employed to test the significance of the slope and intercept of a regression model; the two approaches are the *t*-test and an analysis of variance (ANOVA).

7.3.2 Explanation of Regression Model

The regression model has been found ready to be employed for other purposes. Thus, the next step in the regression analysis is to interpret the regression model and identify the direction of the linear relationship between the chosen independent variable and the response. We can employ the coefficient associated with the independent variable to show the impact of the chosen independent variable on the dependent variable, and the sign of the coefficient can help in identifying the direction, positive or negative. If the sign of the coefficient in the regression model is positive, this indicates that the independent variable can affect the response values positively (increase), while if the sign of the coefficient is negative, this indicates that the independent variable can affect the response values negatively (decrease).

EXAMPLE 7.1 THE CONCENTRATION OF CADMIUM AND COPPER IN THE WATER

An environmentalist wishes to analyze the impact of the concentration of copper (Cu) on the concentration of cadmium (Cd) in the water. Fit a regression model for the sample data for the concentration of Cu and Cd in the water. The sample data for the Cu and the Cd in the water are presented in Table 7.1.

TABLE 7.1 The Concentration of Cu and Cd in the Water

X = Cu	Y = Cd	X = Cu	Y = Cd
0.077	0.134	0.014	0.194
0.048	0.127	0.001	0.193
0.065	0.129	0.005	0.21
0.055	0.135	0.007	0.181
0.048	0.127	0.007	0.177
0.045	0.138	0.016	0.211
0.082	0.132	0.021	0.275
0.076	0.117	0.028	0.263
0.032	0.157	0.061	0.252
0.016	0.168	0.062	0.129

We can employ the formulas (7.3) and (7.4) to compute the intercept β_0 and the slope β_1 of the regression model that describes the linear relationship between the concentration of cadmium (Cd) as a dependent variable and the concentration of copper (Cu) as an independent variable(predictor). The commands and built-in functions provided by R can be employed to compute the intercept β_0 and the slope β_1 of the regression model.

Two equivalent built-in functions are offered by R to calculate the coefficient of the regression model; the two built-in functions are `lm ()` and `coef (lm (y ~ x))`. Thus, one can use either function to compute the regression coefficients for the stored data a. `CSV (Example7_1)` of the concentration of cadmium and copper in the water. The relationship between cadmium and copper can be described by the linear regression that connects these two variables. The fitted model that connects Cd (y) as a dependent variable and Cu (x) as an independent variable was calculated employing the built-in function `lm ()` and the results produced by the function `lm ()` were stored in a new place called `Fitting`. The two results of employing `lm ()` function are the intercept and the slope.

```
> Fitting = lm (y ~ x, data = Example7_1) #Fit the regression model
> Fitting
Call:
lm(formula = y ~ x, data = Example7_1)
Coefficients:
(Intercept)              x
     0.2112        -1.0124
```

One can observe that the results generated by employing the built-in function (`lm (y ~ x, data=Example7_1)`) to fit a linear regression model for the stored data are two values for the b_0 and b_1. The estimated coefficients b_0 and the b_1 for the linear regression model are `0.2112` and `-1.0124`, respectively. These two values can be placed in a mathematical form to represent the regression model.

$$\hat{Y} = 0.2112 - 1.0124X.$$

It can be observed that the sign of b_1 in the estimated regression model is negative. This sign marks that the concentration of copper (X) influences negatively the concentration of cadmium (Y), which shows that if the copper increases by one unit, the Cadmium will decrease by as much as −1.0124.

We should examine the significance of the estimated model to verify whether the copper is an influential variable and should be kept in the model or not. Thus, we can examine the model through an analysis of variance (ANOVA) table and make a decision about the significance of the regression model for the copper in the water (based on the p-value associated with the estimated coefficient). An ANOVA table for testing the significance of the regression model can be generated by employing the built-in function `anova ()` in R.

```
> anova (lm (y ~ x, data = Example7_1))
Analysis of Variance Table
Response: y
          Df   Sum Sq   Mean Sq F value   Pr(>F)
x          1 0.013959 0.0139593  7.8753 0.01168 *
Residuals 18 0.031906 0.0017725
—
Signif. codes: 0 '***' 0.001 '**' 0.01 '*' 0.05 '.' 0.1 ' ' 1
```

The outputs of the analysis of variance exhibited that the regression model is significant, because the p-value is small (p-value `<0.01168`), which indicates that the copper takes part in interpreting the differences in the cadmium. In other words, the concentration of cadmium relies on the concentration of copper. This conclusion is a general one, and we should examine the coefficients individually.

Scientists require detailed information regarding the analysis to gain an overall picture of the research of interest. More information on the model can be provided by employing the built-in function `summary ()`, containing the estimated coefficients of the regression model (`Estimate`), t-test to examine each coefficient (`t value`), standard error for each coefficient (`Std. Error`), p-value to guide us making a decision regarding the importance of each coefficient (`pr(>|t|)`, `residual standard error`, coefficient of determination `R-squared` (R^2), and `adjusted R-squared`.

```
> summary (lm(y ~ x, data = Example7_1))
Call:
lm(formula = y ~ x, data = Example7_1)
```

```
Residuals:
      Min        1Q    Median        3Q       Max
-0.035630 -0.024109 -0.017248 0.003802 0.102531

Coefficients:
            Estimate Std. Error t value Pr(>|t|)
(Intercept) 0.21122    0.01672  12.634  2.19e-10 ***
x           -1.01237   0.36075  -2.806    0.0117 *
—
Signif. codes: 0 '***' 0.001 '**' 0.01 '*' 0.05 '.' 0.1 ' ' 1
Residual standard error: 0.0421 on 18 degrees of freedom
Multiple R-squared: 0.3044, Adjusted R-squared: 0.2657
F-statistic: 7.875 on 1 and 18 DF, p-value: 0.01168
```

One can observe a significant impact was exhibited by the intercept term b_0, as the p-value for the t-test under the column `Pr(>|t|)` is small (`2.19e-10`). However, the impact of the independent variable (Cu) (X) was also significant on the response variable (Cd) Y (p-value < `0.0117`). Thus, an intercept will be kept in the fitted model and the slope as well. The linear relationship between the concentration of Cd and the concentration of Cu can be expressed mathematically as a linear regression model.

$$Cd = 0.21112 - 1.01237Cu$$

Moreover, the contribution of the chosen independent variable (Cu) in explaining the differences in the dependent variable (Cd) can be measured through the value of R^2 (the coefficient of determination), which is usually displayed at the bottom of the results as `Multiple R-squared: 0.3044`. Only 30.44% of the total differences (fluctuations) in the data are interpreted by one predictor (independent variable): Cu (X).

EXAMPLE 7.2 THE CONCENTRATION OF CADMIUM AND COPPER IN THE SEDIMENT

A researcher wishes to investigate the relationship between the concentration of cadmium (Y = Cd) as a dependent variable and the concentration of copper (X = Cu) as an independent variable (mg/L). The sample data for the concentration of the two heavy metals in sediment are presented in Table 7.2.

TABLE 7.2 The Concentration of Cadmium (Cd) and Copper (Cu) in the Sediment

$X = Cu$	$Y = Cd$	$X = Cu$	$Y = Cd$
0.63	1.95	0.36	1.95
0.73	1.99	0.63	1.98
0.35	1.94	0.52	1.93
0.76	1.98	0.55	1.97
0.6	1.94	0.47	1.92

We can investigate the relationship between the concentration of cadmium and concentration of copper by employing the formulas (7.3) and (7.4) to compute the coefficients of the regression equation (fitted model) that represents the form of the connection between the two variables, the concentration of cadmium and copper in the sediment. The regression coefficients (b_0 and b_1) for the model can easily be computed by employing the built-in function `lm (Cd ~ Cu)` in R for the stored data `.CSV` (`Example7_2`) of the cadmium and copper concentrations in the sediment. The results of employing `lm ()` function were stored in a new place called `Fitting`.

```
> Fitting = lm (Cd ~ Cu, data = Example7_2)
> Fitting

Call:
lm(formula = Cd ~ Cu, data = Example7_2)
```

```
Coefficients:
(Intercept)          Cu
     1.8910      0.1142
```

The built-in function `lm (Cd ~ Cu)` produced two values as presented in the last row of the output, which represents the coefficients of the regression model b_0 and b_1. The two values are $b_0 = $ `1.8910` (intercept) and $b_1 = $ `0.1142` (the slope). The fitted regression model with both b_0 and b_1 is given below.

$$\hat{Y} = 1.890 + 0.1142X$$

The concentration of cadmium ($Y = $ Cd) and the concentration of copper ($X = $ Cu) are positively correlated as indicated by the sign of b_1 in the fitted model, this result indicates that if copper (Cu) increases by of one unit, the concentration of cadmium (Cd) will increase by as much as 0.1142.

The two built-in functions `anova (lm (Cd ~ Cu))` and `summary (lm (Cd ~ Cu))` provide detailed information on the fitted model as presented below.

```
> anova (lm (Cd ~ Cu, data = Example7_2))
Analysis of Variance Table

Response: Cd
           Df     Sum Sq    Mean Sq F value    Pr(>F)
Cu          1  0.0022733 0.00227331  6.5497  0.03369 *
Residuals   8  0.0027767 0.00034709
- - -
Signif. codes: 0 '***' 0.001 '**' 0.01 '*' 0.05 '.' 0.1 ' ' 1
> summary (lm (Cd ~ Cu, data = Example7_2))

Call:
lm(formula = Cd ~ Cu, data = Example7_2)

Residuals:
      Min         1Q     Median        3Q        Max
 -0.024719  -0.017926  0.005571  0.016002  0.017847

Coefficients:
            Estimate Std. Error  t value   Pr(>|t|)
(Intercept)  1.89103    0.02568   73.634   1.29e-12   ***
Cu           0.11424    0.04464    2.559     0.0337   *
- - -
Signif. codes: 0 '***' 0.001 '**' 0.01 '*' 0.05 '.' 0.1 ' ' 1

Residual standard error: 0.01863 on 8 degrees of freedom
Multiple R-squared: 0.4502, Adjusted R-squared: 0.3814
F-statistic: 6.55 on 1 and 8 DF, p-value: 0.03369
```

Detailed information was produced by the two functions `anova ()` and `summary ()` regarding the significance of the regression model. The model is significant, as exhibited by the p-value for the Cu (`pr(>F) =0.03369`). This conclusion indicates that the relationship between the concentration of cadmium and the concentration of copper in the sediment exists. Furthermore, the concentration of each variable depends on whether the other variable is low or high in the sediment.

Moreover, the effect of the intercept term (b_0) on the model was significant, as indicated by the result of t-test (`t value`) and because the p-value is small (`1.29e-12`). However, the concentration of copper has a significant effect on the concentration of cadmium at a p-value $<$`0.0337`. The value of R^2(`Multiple R-squared`) is low (`0.4502`) which indicates that most of the differences in the concentration of cadmium are not captured by the computed regression model.

7.3.3 Prediction Using a Regression Equation

Predicting the response variable at a particular value of the independent variable (X) is a significant objective of the regression analysis. The predicted Y value can easily be computed by substituting the value of the independent variable (X) into the regression equation. We can predict the response value if the relationship between the response variable and predictor (independent variable) exists, this indicates that the two variables are significantly correlated, and the value of the correlation differs from 0 (linear correlation exists). The mean of the data values of Y is considered as the best prediction of Y when the correlation between the variables is insignificant.

Note:

- Use the available data values to make prediction, it is better not to predict beyond the range of data collected.

EXAMPLE 7.3 PREDICTION OF CADMIUM (CD) CONCENTRATION IN THE WATER

Predict the concentration of cadmium (Cd) in the water when the concentration of copper (Cu) is 0.03 by employing the regression model computed in Example 7.1. The relationship between cadmium as a response and copper as a predictor can be expressed by the regression model as given below.

$$\hat{Y} = 0.2112 - 1.0124X$$

The prediction of the concentration of cadmium in the water when the concentration of copper (Cu) is 0.03 requires three steps:

Step 1: Fitting the regression model for the available data by employing the built-in function `lm ()` in R.

Step 2: We should employ the built-in function `data.frame ()` to define the specified (particular) value of X.

Step 3: The built-in function `predict ()` should be employed to produce the predicted value with the lower and upper confidence level.

R commands and built-in functions can be employed to compute the predicted value of the cadmium concentration when the copper concentration is equal to 0.03 using the three steps for calculating the predicted value.

```
> Fitting = (lm(y ~ x, data = Example7_1)) #Step 1: Fitting the regression model
> newdata = data.frame (x = 0.03)            #Step 2: Define the required value
> predict (Fitting, newdata, interval = "predict") # Predict
        fit        lwr       upr
1 0.1808527 0.08999833 0.271707
```

The function `predict ()` produced three values including the fitted value (`fit`), the lower confidence level (`lwr=0.08999833`), and the upper confidence level (`upr=0.271707`). If the concentration of the Cu is 0.03, then the predicted value of the concentration of the Cd in the water is `0.1808527`. The regression equation for the linear relationship between cadmium (Y) and copper (X) is:

$$\hat{Y} = 0.2112 - 1.0124X$$

and the predicted value of the concentration of the Cd in the water is:

$$\hat{Y} = 0.2112 - 1.0124\,[0.03] = 0.180828.$$

Note:

- The built-in function `fitted ()` can be employed to calculate the fitted values.

```
fitted ()
```

7.3.4 Extreme Values and Influential Observations

An outlier can be defined as a data point that falls far away from the overall points. An outlier value could be any value that has a significant effect on the coefficient of the regression model. We should conduct an investigation to decide whether this data point is an outlier or not. A scatter diagram can be used to confirm whether this value is an effective point or not. A decision must be made regarding each outlier in the data set—whether to keep or discard it in final analysis. The decision depends on the importance of the data point, if it is real and there is a scientific reason behind it, then we include it in the final analysis. Otherwise we should exclude it from the analysis. The impact of an outlier data point is to drag the regression line towards the data point itself.

EXAMPLE 7.4 OUTLIER OBSERVATION

The data for concentration of cadmium and the concentration of copper in the sediment presented in Example 7.2 will be used after changing the last data value to Cu = 1.5 and Cd = 4. The sample data for the chosen variables are presented in Table 7.3.

TABLE 7.3 The Concentration of Cadmium and Copper in the Sediment After Changing the Two Values (Highlighted)

Before		After	
Cu	**Cd**	**Cu**	**Cd**
0.63	1.95	0.63	1.95
0.73	1.99	0.73	1.99
0.35	1.94	0.35	1.94
0.76	1.98	0.76	1.98
0.6	1.94	0.6	1.94
0.36	1.95	0.36	1.95
0.63	1.98	0.63	1.98
0.52	1.93	0.52	1.93
0.55	1.97	0.55	1.97
0.47	1.92	**1.5**	**4**

The regression model and the scatter diagram should be found for both cases before and after modifying the value of the concentration of copper to clarify the impact of the outlier value. The two fitted regression equations (before and after modifying the value) were computed employing the built-in function `lm (Cd ~ Cu)` in R for the stored data. CSV (`Example7_4B`) and (`Example7_4A`) for the two situations before and after we modified the value, respectively. The results of employing the built-in function `lm ()` in R for fitting the two regression models are stored in new places `Fitting_B` and `Fitting_A` for before and after as given below.

```
> Cu_B = Example7_4B $ Cu
> Cd_B = Example7_4B $ Cd
> Fitting_B = lm (Cd ~ Cu, data = Example7_4B) #Fitting Before changing
> Fitting_B

Call:
lm(formula = Cd ~ Cu, data = Example7_4B)

Coefficients:
(Intercept)          Cu
     1.8910      0.1142

> Cu_A = Example7_4A $ Cu
> Cd_A = Example7_4A $ Cd
> Fitting_A = lm (Cd1 ~ Cu1, data = Example7_4A) # Fitting After changing
> Fitting_A

Call:
lm(formula = Cd1 ~ Cu1, data = Example7_4A)

Coefficients:
(Intercept)         Cu1
     0.9511      1.8279
```

The two regression models produced by the built-in function `lm ()` are presented below.

$$\hat{Y} = 1.8910 + 0.1142\,X \quad \text{(Before)}$$

$$\hat{Y} = 0.9511 + 1.8279\,X \quad \text{(After)}$$

Two scatter diagrams with a regression line were generated for before and after we modified the data values to display the impacts of the outlier values graphically, Employing the built-in functions and commands in R to the data values before and after modifying the data value will produce Fig. 7.1A and B, respectively.

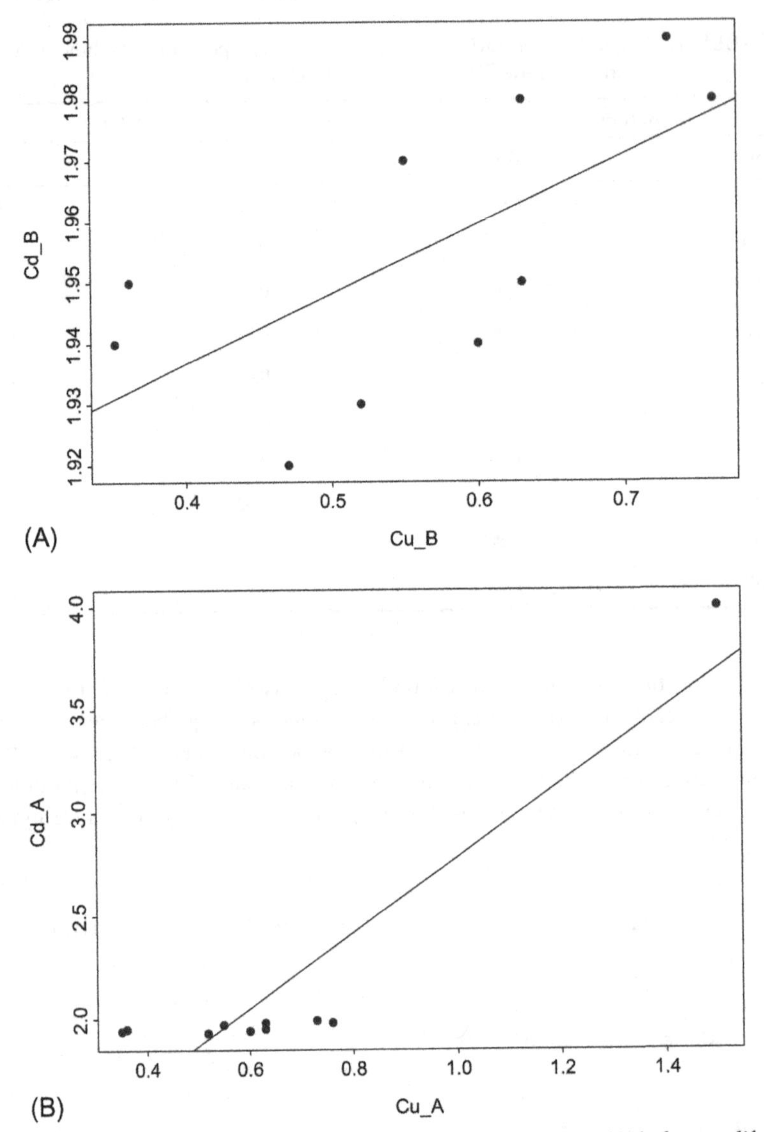

FIG. 7.1 Illustrating the impact of an outlier on regression equation through a Scatter diagram: (A) before modifying the data value and (B) after modifying the data value.

One can observe Fig. 7.1A and B clearly for before and after the change to understand the impact of the data point ($X = Cu = 1.5$, $Y = Cd = 4$) and how the regression line is dragged towards this point. The relationship between the independent variable and dependent variable will be inaccurate because of this data value and produce incorrect inferences regarding the nature of the relationship between the chosen variables. Thus, scientists should decide to include or exclude this data value based on scientific reasoning. A strong significant relationship was exhibited between the chosen variables after changing the data value, as clearly appeared in the results of analysis of variance with a very small p-value under the column `Pr(>F) 0.0001919`.

```
> anova(lm(Cd~Cu, data=Example7_4A))
Analysis of Variance Table
```

```
Response: Cd
           Df   Sum Sq  Mean Sq  F value      Pr(>F)
Cu          1  3.15275  3.15275   42.005  0.0001919 ***
Residuals   8  0.60046  0.07506
- - -
Signif. codes: 0 '***' 0.001 '**' 0.01 '*' 0.05 '.' 0.1 ' ' 1
> summary(lm(Cd~Cu, data=Example7_4A))

Call:
lm(formula = Cd ~ Cu, data = Example7_4A)

Residuals:
     Min       1Q    Median       3Q      Max
-0.36030 -0.14518  -0.04715  0.23739  0.34913

Coefficients:
            Estimate Std. Error t value  Pr(>|t|)
(Intercept)   0.9511     0.2061   4.615 0.001721 **
Cu            1.8279     0.2820   6.481 0.000192 ***
- - -
Signif. codes: 0 '***' 0.001 '**' 0.01 '*' 0.05 '.' 0.1 ' ' 1

Residual standard error: 0.274 on 8 degrees of freedom
Multiple R-squared: 0.84, Adjusted R-squared: 0.82
F-statistic: 42 on 1 and 8 DF, p-value: 0.0001919
```

7.3.5 Residuals

The difference between the actual data values of the dependent variable (Y) and the fitted data values $\hat{Y}$ for a given value of the independent variable (X) is called residual. We can define residuals as the vertical distance between the fitted regression line and the actual data value of the dependent variable (Y), as illustrated in Fig. 7.2.

Note:

- The sum of all of the residuals is always equal to 0.
- The smallest possible sum of squares is the sum of squares of the residuals.
- The regression line is also called the least squares line.

7.3.6 Explained and Unexplained Variation

Identifying the source of changes that influence the dependent variable is considered as the main aim of all studies. Furthermore, scientists wish to identify the origin of differences in the output (dependent variable) Y. The differences in the dependent variable could be due to known or unknown sources, identifying the source of differences will help researchers comprehend the behavior of the variable under investigation.

The sum of squares of the differences between the actual data values Y from the grand mean $\overline{Y}$ is called the total variation and is represented by $\sum(Y-\overline{Y})^2$. In general, we can divide the total variation into two terms:

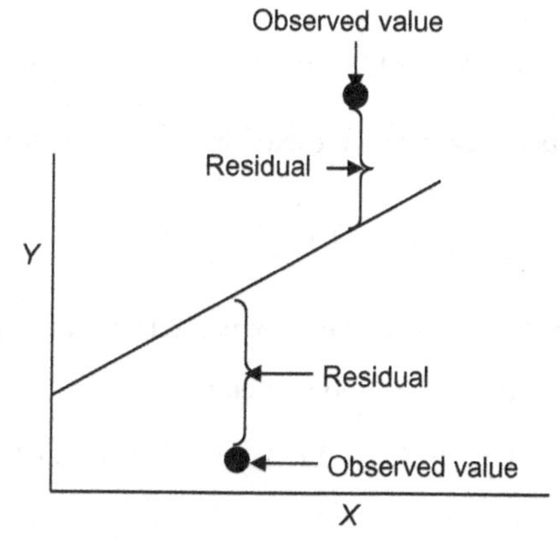

FIG. 7.2 Illustrating the residual as the difference (distance) between the observed and fitted values.

- *Explained variation (residual)*: The amount of variance captured by the regression model which is due to the effect of the independent variable X is called the explained variation and represented by $\sum\left(\hat{Y}-\overline{Y}\right)^2$.
- *Unexplained variation*: The amount of variation that is not captured by the regression model which is due to an unknown source is called unexplained variation and represented by $\sum\left(Y-\hat{Y}\right)^2$. Unexplained variation is also called the residual.

We can express the total variation in a mathematical equation as the sum of explained and unexplained variations, as presented in (7.11).

$$\sum(Y-\overline{Y})^2=\sum\left(\hat{Y}-\overline{Y}\right)^2+\sum\left(Y-\hat{Y}\right)^2 \tag{7.11}$$

EXAMPLE 7.5 EXPLAINED AND UNEXPLAINED VARIATION

Employ the regression model produced in Example 7.1 to compute:

1. The residual.
2. Explained variation.
3. Unexplained variation.
4. Total variation.

We computed the fitted regression model in Example 7.1 as presented in (7.12).

$$\hat{Y}=0.2112-1.0124X \tag{7.12}$$

(1) *The residual*:

The residual of the model can be computed employing the formula $Y-\hat{Y}$. We can compute the value of $\hat{Y}$ employing the fitted model for the data, substitute the values of the independent variable (X) into the fitted Eq. (7.12) as shown below.

The estimated value $\hat{Y}$ for the first value of the independent variable equals 0.133 as shown below,

$$\hat{Y}=0.2112-1.0124X=0.2112-1.0124[0.077]=0.133245=0.133$$

repeating the same process to substitute other values of the independent variable X to compute the fitted values.

(2) *Total variation*:

The formula $\sum(Y-\overline{Y})^2$ is usually employed to calculate the total variation. The total variation for the first value employing the formula $\sum(Y-\overline{Y})^2$ is 0.0014.

$$(Y-\overline{Y})=(0.134-0.172)^2=0.00144=0.0014$$

Repeat the same operation for the other values to calculate the total variation for all data values and then compute the sum of all values.

$$\sum(Y-\overline{Y})^2=0.045869$$

(3) *Explained variation*:

The formula $\sum\left(\hat{Y}-\overline{Y}\right)^2$ is usually employed to calculate the explained variation. The explained variation for the first data value employing the formula $\sum\left(\hat{Y}-\overline{Y}\right)^2$ is 0.002.

$$\left(\hat{Y}-\overline{Y}\right)=(0.133-0172)^2=0.0015=0.002$$

Repeat the same operation for the other values to calculate the explained variation for all data values and then compute the sum of all values.

$$\sum\left(\hat{Y}-\overline{Y}\right)^2=0.01401085$$

(4) *Unexplained variation*:

The formula $\sum\left(Y-\hat{Y}\right)^2$ can be employed to calculate the unexplained variation. The unexplained variation for the first data value employing the formula $\sum\left(Y-\hat{Y}\right)^2$ is 0.00001 as shown below

$$\left(Y-\hat{Y}\right)=(0.134-0.133)^2=(0.001)^2=000001$$

Repeat the same operation for the other values to calculate the unexplained variation for all data values and then compute the sum of all values.

$$\sum\left(Y-\hat{Y}\right)^2 = 0.032015$$

(5) *Total variation*:

One can observe that the sum of the two components (explained and unexplained variations) will represent the total variation.

$$\text{Total variation} = \text{Explained variation} + \text{Unexplained variation}$$
$$0.04586495 = 0.01401085 + 0.032015$$

The built-in function `predict ()` in R can easily generate the fitted values. Moreover, the residuals for the regression model can be generated by employing the built-in function `resid ()` in R. One can use the facilities provided by R to compute the explained, unexplained, total variations, predicted values, and residual for the model presented in Example 7.1; Below are the outputs of employing R built-in functions and commands.

```
> Fitting = lm (y ~ x, data = Example7_5)
> Fitting

Call:
lm(formula = y ~ x, data = Example7_5)

Coefficients:
(Intercept)            x
     0.2112      -1.0124

> anova (lm (y ~ x, data = Example7_5))
Analysis of Variance Table

Response: y
          Df    Sum Sq   Mean Sq F value   Pr(>F)
x          1  0.013959 0.0139593  7.8753  0.01168 *
Residuals 18  0.031906 0.0017725
---
Signif. codes: 0 '***' 0.001 '**' 0.01 '*' 0.05 '.' 0.1 ' ' 1
> summary (lm(y ~ x, data = Example7_5))

Call:
lm(formula = y ~ x, data = Example7_5)

Residuals:
      Min        1Q     Median        3Q       Max
-0.035630 -0.024109  -0.017248  0.003802  0.102531

Coefficients:
            Estimate Std. Error t value Pr(>|t|)
(Intercept)  0.21122    0.01672  12.634 2.19e-10  ***
x           -1.01237    0.36075  -2.806 0.0117    *
---
Signif. codes: 0 '***' 0.001 '**' 0.01 '*' 0.05 '.' 0.1 ' ' 1

Residual standard error: 0.0421 on 18 degrees of freedom
Multiple R-squared: 0.3044, Adjusted R-squared: 0.2657
F-statistic: 7.875 on 1 and 18 DF, p-value: 0.01168

> predicted = round (predict (Fitting), digits = 3)
> predicted
    1     2     3     4     5     6     7     8     9    10    11    12    13
0.133 0.163 0.145 0.156 0.163 0.166 0.128 0.134 0.179 0.195 0.197 0.210 0.206
   14    15    16    17    18    19    20
0.204 0.204 0.195 0.190 0.183 0.149 0.148
> #or fitted
```

```
> round (fitted (Fitting), digits = 3)
    1     2     3     4     5     6     7     8     9    10    11    12    13
0.133 0.163 0.145 0.156 0.163 0.166 0.128 0.134 0.179 0.195 0.197 0.210 0.206
   14    15    16    17    18    19    20
0.204 0.204 0.195 0.190 0.183 0.149 0.148
> round (resid (Fitting), digits = 3) #residual
    1      2      3      4      5      6     7      8      9     10     11
0.001 -0.036 -0.016 -0.021 -0.036 -0.028 0.004 -0.017 -0.022 -0.027 -0.003
    12     13     14     15     16     17     18     19     20
-0.017  0.004 -0.023 -0.027  0.016  0.085  0.080  0.103 -0.019
> Mean = round (mean (Example7_5 $ y), digits = 3)
> Mean
[1] 0.172
> (Example7_5 $ y - Mean)^2
 [1] 0.001444 0.002025 0.001849 0.001369 0.002025 0.001156 0.001600 0.003025
 [9] 0.000225 0.000016 0.000484 0.000441 0.001444 0.000081 0.000025 0.001521
[17] 0.010609 0.008281 0.006400 0.001849
> sum((Example7_5 $ y - Mean)^2)
[1] 0.045869
> sum(round((Example7_5 $ y - Mean)^2, digits = 3)) # total variance
[1] 0.044
> ((predicted - Mean)^2)
       1        2        3        4        5        6        7        8        9
0.001521 0.000081 0.000729 0.000256 0.000081 0.000036 0.001936 0.001444 0.000049
      10       11       12       13       14       15       16       17       18
0.000529 0.000625 0.001444 0.001156 0.001024 0.001024 0.000529 0.000324 0.000121
      19       20
0.000529 0.000576
> sum ((predicted - Mean)^2) #Explained variation
[1] 0.014014
> Example7_5 $ y - predicted #Error
    1      2      3      4      5      6     7      8      9     10     11
0.001 -0.036 -0.016 -0.021 -0.036 -0.028 0.004 -0.017 -0.022 -0.027 -0.003
    12    13     14     15    16    17    18    19     20
-0.017 0.004 -0.023 -0.027 0.016 0.085 0.080 0.103 -0.019
> round ((Example7_5 $ y - predicted)^2, digits = 3)
    1     2     3     4     5     6     7     8     9    10    11    12    13
0.000 0.001 0.000 0.000 0.001 0.001 0.000 0.000 0.000 0.001 0.000 0.000 0.000
   14    15    16    17    18    19    20
0.001 0.001 0.000 0.007 0.006 0.011 0.000
> sum ((Example7_5 $ y - predicted)^2) #Unexplained variation
[1] 0.032015
```

The built-in function `lm ()` was employed to fit a regression equation, and then an analysis of variance was performed to examine significance of the relationship between the variables in the model employing the built-in function `anova ()`. The two built-in functions `predict (Fitting)` and `resid (Fitting)` are used to compute the fitted values and residual values, respectively. R provided one more built-in function to compute the fitted values as an alternative function, the built-in function is `fitted ()`. Moreover, the commands and built-in functions were employed to compute the mean for the response variable, the total variation, the explained variation, and the last result was the sum of the unexplained variation. The results were rounded employing the built-in function `round ()`.

7.3.7 Coefficient of Determination

The variation in the dependent variable that is due to the effect of the selected independent variable in the model can be measured by the coefficient of determination (R^2). Thus, the percentage of the variation (differences) in the data values captured by the regression equation is called the coefficient of determination and can be computed as defined in (7.13).

$$R^2 = \frac{Explained\ Variance}{Total\ Variation} \quad (7.13)$$

Note:

- A percentage is commonly used to represent the result of R^2.
- The range of the values of R^2 varies between 0 and 1; $0 \le R^2 \le 1$.
- R^2 is considered as a measure of the goodness of fit of the regression model.
- The correlation coefficient can be produced by taking the square root of the coefficient of determination.

The coefficient of determination is commonly used to show the importance of the selected independent variables in the model, the large value of R^2 indicates that the chosen independent variables contribute significantly in explaining the differences in the dependent variable values, and the small value of R^2 indicates that the independent variable (variables) in the model do not contribute significantly in explaining the differences in the data values. The value of R^2 represents the percentage of the variation that is captured by the model and $1-R^2$ refers to the percentage of variation in the data values that is unexplained by the model (residual, unknown source).

EXAMPLE 7.6 COEFFICIENT OF DETERMINATION

Employ the formula in (7.13) to compute the coefficient of determination (R^2) for the sample data given in Example 7.5.

The values of the explained and total variations were computed in Example 7.5 to be 0.0101085 and 0.04586495, respectively; thus, the value of the coefficient of variation employing the formula given in (7.9) is 0.3044.

$$R^2 = \frac{Explained\ Variance}{Total\ Variation} = \frac{0.01401085}{0.04586495} = 0.3044$$

One can observe that only 30.44% of the total variation in the data values for the concentration of Cd is due to the concentration of copper (Cu) (independent variable) and 69.56% is unexplained, which is due to unknown sources.

The value of the coefficient of determination (R^2) can be produced by employing the built-in function `summary ()` in R. The outputs of employing the `summary ()` function appear at the bottom of the results with the *F*-statistics and *p*-value, which showed that the R^2 value is 0.3044.

7.4 MULTIPLE LINEAR REGRESSION MODEL

Simple linear regression model covered the case where there is only one independent variable and one dependent variable. The relationship between the two variables can be represented in a mathematical equation called a simple linear regression model.

$$\hat{Y} = b_0 + b_1 X$$

We can develop the concept of simple linear regression to call several variables that affect the only response to be in the model. Consider that there are k independent variables $X_1, X_2, \ldots, X_k$ that influence the result of the experiment represented as the dependent variable Y. The regression model that describes the relationship between several independent variable and the response is called a multiple regression model. The relationship between k independent variables and the response variable Y can be expressed by a linear model as presented in (7.14).

The general form of a multiple regression model can be represented mathematically as presented in (7.14).

$$Y_i = \beta_0 + \beta_1 X_1 + \beta_2 X_2 + \ldots + \beta_k X_k + \varepsilon_i \quad (7.14)$$

where Y represents the dependent variable, $X_1, X_2, \ldots, X_k$ are the independent variables, and $\beta_1, \beta_2, \ldots, \beta_k$ are parameters to be estimated.

Furthermore, the estimated regression equation is presented in (7.15):

$$\hat{Y} = b_0 + b_1 X_1 + b_2 X_2 + \ldots + b_k X_k \quad (7.15)$$

where $\hat{Y}$ refers to the fitted (predicted, estimated) value of the response variable and $b_0, b_1, \ldots, b_k$ are estimates of $\beta_0, \beta_1, \beta_2, \ldots, \beta_k$.

We usually use the least squares method to obtain the estimates of $\beta_0, \beta_1, \beta_2, \dots, \beta_k$ employing the formula presented in (7.16).

$$b = \left(X'X\right)^{-1} X'Y \tag{7.16}$$

where $(X'X)$ is a nonsingular matrix, which means that $n > k+1$, the sample size (n, number of observation) must be greater than the number of selected independent variables in the model at least by 1.

Note:

- Knowledge of advanced mathematics related to matrices, such as matrix multiplication, inverse matrix, and others is needed to carry out the calculations for a multiple regression model. Fortunately, this knowledge is not needed if we use statistical packages for analyzing the data and fit a multiple linear regression with any number of the independent variables. Readers should focus on the explanation of the output and the regression model to draw smart conclusions.

7.5 HYPOTHESIS TESTING FOR MULTIPLE LINEAR REGRESSION

Hypothesis testing for a multiple linear regression model includes three types of tests regarding the regression coefficients β's, as summarized below:

Type 1: Analysis of variance: This type of test will examine the overall regression, whether the regression model is significant or not.

Type 2: The *t*-test: This type of test will examine the effect of each independent variable in the model, whether an independent variable influences the dependent variable or not.

Type 3: Partial *F* test: This type of test will examine the effect of a subset of the independent variables on the dependent variable.

7.5.1 Hypothesis Testing for Overall Significance of the Regression Model

Hypothesis testing regarding the overall significance provides general information about the significance of the model without detailed information on the significance of each independent variable. The general information can be generated employing an analysis of variance (ANOVA) to test whether the linear relationship between k independent variables and the dependent variable exists or not.

The null and alternative hypotheses for a multiple linear regression model with k independent variables are presented in (7.17).

$$\begin{aligned} H_0 &: \beta_1 = \beta_2 = \cdots = \beta_k = 0 \\ H_1 &: \beta_i \neq 0 \text{ for at least one } i \end{aligned} \tag{7.17}$$

None of the chosen independent variables influence the dependent variable, as stated by the null hypothesis, i.e., the linear relationship between the response variable and the k independent variables does not exist; while the complement of the null hypothesis states that the relationship exist (at least one predictor influences the response), which represent the alternative hypothesis. Failing to reject the null hypothesis indicates that the independent variables do not influence the response (the effect of all independent variables is 0) and the regression model cannot be used to describe the linear relationship between the two sets of variables. Rejecting the null hypothesis indicates that the produced regression model can be used to describe the relationship between the response variable and the k independent variables, it can be concluded that at least one independent variable influences the value of the response variable.

7.5.2 Hypothesis Testing on Individual Regression Coefficients

Hypothesis testing regarding each independent variable can be performed if the overall test was significant. The suitable test for individual independent variable is the *t*-test to examine the effect of each independent variable on the response variable. This type of testing usually uses the regression coefficient associated with the independent variable and makes a decision regarding the importance of the independent variable. Significant variables will be kept in the regression model and we exclude insignificant variables. The null and alternative hypotheses regarding individual regression coefficient are expressed in a mathematical form as presented in (7.18).

$$H_0 : \beta_i \neq 0$$
$$H_1 : \beta_i \neq 0 \tag{7.18}$$

The t-test statistic value can be calculated employing the formula given below:

$$t = \frac{b_i}{se(b_i)}, \text{ with } n-2 \text{ degrees of freedom}$$

Rejecting the null hypothesis indicates that the independent variable is important and contributes in explaining the variation in the response variable; failing to reject indicates that the independent variable does not contribute in explaining the variation in the response variable and as a result this variable should be excluded.

7.5.3 Hypothesis Testing on a Subset of the Regression Coefficients

Hypothesis testing regarding a subset of regression coefficients is about the contribution of the subset of the independent variables in explaining the variation of the dependent variable. In other words; this test is employed to examine whether adding (deleting) independent variables to the regression model will contribute in explaining the differences in the data values or not.

Consider that the regression coefficient $\beta's$ are placed into two vectors; one vector takes $k-r+1$ coefficients, and r $\beta's$ are in the second vector. The two vectors can be placed in one vector as shown in (7.19).

$$\boldsymbol{\beta} = \begin{bmatrix} \boldsymbol{\beta}_1 \\ \boldsymbol{\beta}_2 \end{bmatrix} \tag{7.19}$$

where $\boldsymbol{\beta}_1 = [\beta_0, \beta_1, \ldots, \beta_{k-r}]$ and $\boldsymbol{\beta}_2 = [\beta_{k-r+1}, \beta_{k-r+2}, \ldots, \beta_{k}.]$

The null and alternative hypotheses regarding hypothesis testing on a subset of the regression coefficients are expressed in a mathematical form as presented in (7.20).

$$H_0 : \boldsymbol{\beta}_2 = 0$$
$$H_1 : \boldsymbol{\beta}_2 \neq 0 \tag{7.20}$$

The test statistic employed to examine the contribution of the variables in $\boldsymbol{\beta}_2$ is called the F test statistic, which is usually employed to examine the null hypothesis in (7.20); for instance testing the significance of the independent variables in the vector $\boldsymbol{\beta}_2$ can be performed employing the F test statistic presented in (7.21).

$$F = \frac{SS_R(\boldsymbol{\beta}_2 | \boldsymbol{\beta}_1)/r}{MSE} \tag{7.21}$$

where $SS_R = (\boldsymbol{\beta}_2 | \boldsymbol{\beta}_1)$ refers to the effect of the independent variables in the vector $\boldsymbol{\beta}_2$.

In other words, it refers to the extra amount of the sum of squares in the model when the independent variables in the vector $\boldsymbol{\beta}_2$ are added to a regression equation already including the independent variables in the vector $\boldsymbol{\beta}_1$. Rejecting the null hypothesis indicates that at least one of the independent variables in the vector $\boldsymbol{\beta}_2$ influences significantly the dependent variable.

7.6 ADJUSTED COEFFICIENT OF DETERMINATION

We have discussed the concept of the coefficient of determination R^2 and the purpose of using it in regression analysis. However, the value of R^2 increases (or could remain the same) as more independent variables are added in the analysis. Thus, to avoid this drawback of R^2, it is preferable to employ an adjusted R^2, considering the number of variables in the analysis and the sample size as well. Adjusted R^2 can be computed employing the formula presented in (7.22).

$$\text{Adjusted } R^2 = 1 - \frac{(n-1)}{[n-(k+1)]}(1-R^2) \tag{7.22}$$

where k refers to the number of independent variables included in the model, and n refers to the sample size.

One can observe in the R output for Example 7.5 that the adjusted R^2 value is 0.2657 (`Adjusted R-squared: 0.2657`).

$$\text{Adjusted } R^2 = 1 - \frac{n-1}{[n-(k+1)]}(1-R^2) = 1 - \frac{7-1}{[7-(1+1)]}(1-0.3044) = 0.2657$$

EXAMPLE 7.7 INDOOR AIR POLLUTION

The relationship between multi-factor scores and metabolite of toluene (TDI) concentration as a response (Y) was studied to understand the behavior of indoor air components with respect to different polyurethane factories. The two independent variables of interest are relative humidity (Rh %) and dry bulb temperature (Td, C^0), and one dependent variable, toluene diisocyanate (TDI) ($\mu g/m^3$). The sample data for the chosen variables are presented in Table 7.4.

TABLE 7.4 The Data for Metabolite of Toluene (TDI) Concentration, Relative Humidity (Rh %) and Dry Bulb Temperature (Td, CO)

y	x1	x2
81	50	35
79	50	35
78	51	33
76	51	33
75	53	33
59	40	30
58	40	28
57	40	28
55	41	27
53	43	27

The built-in function `lm ()` in R can be employed to fit a multiple linear regression model for the stored data. CSV (`Example7_7`) of the indoor air pollution. R built-in functions and commands were used to fit the regression model with two independent variables regarding the indoor air pollution data. The data produced by the function `lm ()` were stored in a new place called `Fitting`.

```
> Fitting <- lm( y ~ x1 + x2, data = Example7_7)
> Fitting

Call:
lm(formula = y ~ x1 + x2, data = Example7_7)

Coefficients:

(Intercept)          x1          x2
   -40.2601      0.5842      2.6066
```

The outputs of employing the built-in function `lm ()` are the coefficients of the regression equation, including the model intercept ($b_0 =$ `-40.2601`), the slope of the first predictor X_1($b_1=$ `0.5842`), and the slope of the second predictor X_2 ($b_2=$ `2.6066`). A mathematical equation can be employed to describe the linear relationship between the TDI (Y) and the two predictors for the relative humidity (Rh) (X_1), and the dry bulb temperature (Td) (X_2), as given below.

$$\hat{Y} = -40.3 + 0.58X_1 + 2.61X_2$$

The importance of the regression model can be tested by employing the analysis of variance table, which shows detailed information regarding the significance of the model including the F-value, p-value, and the degrees of freedom, sum of squares, and the mean sum of squares. Furthermore, a t-test for each independent variable is provided to decide about the impact of each selected independent variable on the dependent variable (TDI).

```
> anova (lm (y ~ x1 + x2, data = Example7_7)) # anova table
Analysis of Variance Table

Response: y
          Df  Sum Sq Mean Sq F value      Pr(>F)
```

```
x1          1 1006.53 1006.53 262.183   8.338e-07 ***
x2          1  157.50  157.50  41.026   0.0003655 ***
Residuals 7   26.87    3.84
- - -
Signif. codes: 0 '***' 0.001 '**' 0.01 '*' 0.05 '.' 0.1 ' ' 1
> summary (lm (y ~ x1 + x2, data = Example7_7)) # show results

Call:
lm(formula = y ~ x1 + x2, data = Example7_7)

Residuals:
    Min      1Q Median      3Q     Max
-2.3071 -1.5870 0.6321 0.9229  2.4466

Coefficients:
            Estimate Std. Error  t value Pr(>|t|)
(Intercept) -40.2601     6.2463   -6.445 0.000352 ***
x1            0.5842     0.2400    2.435 0.045119 *
x2            2.6066     0.4069    6.405 0.000365 ***
- - -
Signif. codes: 0 '***' 0.001 '**' 0.01 '*' 0.05 '.' 0.1 ' ' 1

Residual standard error: 1.959 on 7 degrees of freedom
Multiple R-squared: 0.9774, Adjusted R-squared: 0.971
F-statistic: 151.6 on 2 and 7 DF, p-value: 1.726e-06
```

The regression model is significant as shown by the outputs of the analysis of variance because the p-value is very small (`8.338e-07`). This conclusion indicates that one or both of the chosen independent variables (X_1 and X_2) influence the dependent variable (Y), as the p-values (`Pr(>F)`) are small: the p-value $<$`0.045119` for X_1 and the p-value $<$`0.000365` for X_2. Moreover, a significant impact was shown by the intercept (b_0), as indicated by the `t-value` test (the p-value is small $<$`0.000352`).

The value of R^2 is 0.9774 (`Multiple R-squared: 0.9774`), which is high and indicates that most of the variance in the response (TDI) is captured by the fitted regression model.

We conclude that the model helps us describe the relationship between the toluene diisocyanate (TDI) and the two independent variables (relative humidity (Rh %) and dry bulb temperature (Td).

7.7 MULTIVARIATE MULTIPLE LINEAR REGRESSION MODEL

The concept of multivariate multiple regression is similar to the concept of multiple regression and the only difference is that the multivariate multiple regression includes more than one dependent variable. The meaning of a multivariate multiple regression model; multivariate indicates that the number of dependent variables is more than one dependent variable, and multiple indicates that the number of independent variables is more than or equal two. Consider that there are p dependent variables and k independent variables; the model that describes the relation between the p dependent variables and k independent variables is called multivariate multiple linear regression model. There will be one regression model for each dependent variable; the fitted regression models for the p response are presented below.

$$\hat{Y}_1 = b_{01} + b_{11}X_1 + \ldots + b_{k1}X_k$$

$$\hat{Y}_2 = b_{02} + b_{12}X_1 + \ldots + b_{k2}X_k$$

$$\hat{Y}_p = b_{0p} + b_{1p}X_1 + \ldots + b_{kp}X_k$$

The least squares method is usually employed to find the coefficients of the regression models using the formula presented in (7.23).

$$b = \left(\acute{X}X\right)^{-1}\acute{X}Y \tag{7.23}$$

where:

$$\mathbf{Y} = \begin{bmatrix} Y_{11} & Y_{12}\ldots Y_{1p} \\ Y_{21} & Y_{22}\ldots Y_{2p} \\ . & . & . \\ . & . & . \\ . & . & . \\ Y_{n1} & Y_{n2}\ldots Y_{np} \end{bmatrix} = \begin{bmatrix} \acute{Y}_1 \\ \acute{Y}_2 \\ . \\ . \\ . \\ \acute{Y}_n \end{bmatrix}$$

$$\mathbf{X} = \begin{bmatrix} 1 & X_{11} & X_{12}\ldots X_{1k} \\ 1 & X_{21} & X_{22}\ldots X_{2k} \\ . & . & \ldots & . \\ . & . & \ldots & . \\ . & . & \ldots & . \\ 1 & X_{n1} & X_{n2} & X_{nk} \end{bmatrix}$$

$$\boldsymbol{\beta} = \begin{bmatrix} \beta_{01} & \beta_{02}\cdots\beta_{0p} \\ \beta_{11} & \beta_{12}\cdots\beta_{1p} \\ . & . \ldots & . \\ . & . \ldots & . \\ . & . \ldots & . \\ \beta_{k1} & \beta_{k2} & \beta_{kp} \end{bmatrix}$$

Note:

- Each multiple linear regression model should be fitted individually by considering each response with the selected independent variables and then combine all models in one form to represent a multivariate multiple regression model.

Further Readings

Abbas, F.M., Alkarkhi, N.I., Ahmed, A., Easa, A.M., 2009. Analysis of heavy metal concentrations in sediments of selected estuaries of Malaysia—a statistical assessment. Environ. Monit. Assess. 153, 179–185.

Alkarkhi, A.F.M., Alqaraghuli, W.A.A., 2019. Easy Statistics for Food Science with R, first ed. Academic Press.

Alkarkhi, A.F.M., Anees, A., Azhar, M.E., 2009. Assessment of surface water quality of selected estuaries of Malaysia-multivariate statistical techniques. Environmentalist 29, 255–262.

Alkarkhi, A.F.M., Low, H.C., 2012. Elementary Statistics for Technologist. Universiti Sains Malaysia Press, Pulau Pinang.

Chi, Y. 2009. R Tutorial, An Introduction To Statistics—Estimated Simple Regression Equation [Online]. Available: http://www.r-tutor.com/elementary-statistics/simple-linear-regression/estimated-simple-regression-equation [Accessed 10 January 2019].

Daniel & Hocking. 2013. Blog archives, high resolution figures in R [Online]. R-Bloggers. Available: https://www.r-bloggers.com/Author/Daniel-Hocking/ [Accessed 26 February 2019].

Draper, N.R., Smith, H., 1988. Applied Regression Analysis. Wiley, New York.

Johnson, R.A., Wichern, D.W., 2002. Applied Multivariate Statistical Analysis. Prentice Hall, New Jersey.

Kabacoff, R. I. 2017. Quick_R, multiple (linear) regression. [Online]. Data Camp, Available: https://www.statmethods.net/stats/regression.html [Accessed 18 January 2019].

Mirtagi, M., M. Hakimi, I., Aness, A., M., A. F., Norizam, E. M., Omar, A. K., Mohammadyan, M. & Mirashrafi, S. B. 2009, Indoor air pollution study on toluene Diiscocyanate (Tdi) and biological assessment of toluene Diamine (Tda) in the polyurethane industries. World Appl. Sci. J., 6, 242–247.

Rencher, A.C., 2002. Methods of Multivariate Analysis. J. Wiley, New York.

C H A P T E R

8

Principal Components

LEARNING OBJECTIVES

After careful consideration of this chapter, you should be able:

- *To explain principal components analysis (PCA) as a data reduction.*
- *To know how to choose the number of components.*
- *To understand how to employ principal components analysis for environmental data.*
- *To understand how to interpret the components.*
- *To describe built-in functions and commands in R for principal components analysis.*
- *To comprehend the results of R output for PCA and how to employ it in the analysis.*
- *To draw smart conclusions and write useful reports.*

8.1 INTRODUCTION

Principal components analysis (PCA) is a multivariate method employed to minimize the number of dimensions to interpret the total variance in the available data with a few equations constructed from the original variables, called components, that are uncorrelated. If the selected variables are strongly correlated positively or negatively then principal components analysis will provide superior outcomes. For example, consider there are 16 variables that are strongly correlated; it is highly likely that only two or three PCAs can explain most of the variance in the available data.

8.2 PRINCIPAL COMPONENTS ANALYSIS IN R

We should install `GPArotation` and `psych` packages from the R library before performing principal components analysis. R offers a set of built-in functions from various packages to perform principal components analysis. The required built-in functions and commands for performing principal analysis with their corresponding packages are provided in this chapter. Furthermore, the structure of each built-in function and command with detailed explanation is provided.

1. The eigenvalues of the variance–covariance matrix (or correlation matrix) can be computed employing the built-in function `princomp ()`. This function is installed by default with the package `stats`.

```
princomp (data frame, cor = TRUE)
```

The eigenvalues are computed from the correlation matrix if the option `cor=` is chosen to be TRUE (`cor=TRUE`), whereas the eigenvalues are computed from the variance–covariance matrix if the option `cor=` is chosen to be FALSE (`cor=FALSE`). We will employ this function to analyze the available data in this chapter. One can employ the other built-in functions provided by R to perform principal components analysis. The necessary built-in functions associated with employing the function `princomp ()` to perform PCA and extracting the needful information are provided.

Applied Statistics for Environmental Science with R
https://doi.org/10.1016/B978-0-12-818622-0.00008-3

2. The proportion of the variance explained by each component and the cumulative proportion of the explained variance can be produced by employing the built-in function `summary ()`.

```
summary (data frame)
```

3. The loadings for all of the components can be generated by employing the built-in function `$loadings`.

```
Name of the principal component $ loadings
```

4. The values of the principal components are called scores, and can be generated by R for all observations employing the built-in function `$scores`.

```
Name of the principal component $ scores
```

5. The built-in function `screeplot ()` can be employed to generate the Scree plot for the extracted eigenvalues.

```
screeplot (data frame, npcs = number of components)
```

6. The `$ sdev` function can be employed to locate the eigenvalues.

```
Name of principal component $ sdev
```

More built-in functions are supplied by R to carry out principal components analysis. Some of these functions are presented below.

7. The package `Stats` offers the built-in function `prcomp ()` to compute the eigenvalues associated with each component, this function sets up by default.

```
prcomp (data frame, scale = TRUE )
```

The eigenvalues are computed from the correlation matrix if the option `scale=` is chosen to be TRUE (`scale=TRUE`), whereas the eigenvalues are computed from the variance–covariance matrix if the option `scale=` is chosen to be FALSE (`cor=FALSE`).

Other built-in functions can be employed to extract the results, e.g., the built-in function `$x` can be employed to produce the scores. The loadings can be produced by employing the built-in function `Name of principal component $ rotation`.

8. The package (`FactoMineR`) offers the built-in function `PCA ()`.

```
PCA (data frame, graph = TRUE)
```

The function `PCA` offers option to produce the results with a Scree plot, if the option `graph=` is chosen to be TRUE (`graph=TRUE`), and without a Scree plot if the option `graph=` is sets FALSE (`graph=FALSE`).

The percentage of the variance and the percentage of cumulative variance can be produced by employing the built-in function `$ eig`, and the scores can be generated by employing the built-in function `$ svd`.

9. The package `ade4` offers a new built-in function to perform principal components, the built-in function is `dudi.pca ()`.

```
dudi.pca (data frame, scale = FALSE, scannf = FALSE, nf = No of components)
```

The function `dudi.pca` offers option to produce the results with a Scree plot, if the option `scannf=` is chosen to be TRUE (`scannf=TRUE`), and without a Scree plot if the option `scannf ==` is sets FALSE (`scannf=FALSE`). The option `nf=No of components` allows the user to set the number of components to be produced. The loadings can be generated by employing the built-in function `$ c1`, and the scores can be produced by employing the built-in function `$eig`.

10. The package `amap` offers the built-in function `acp ()` to carry out principal analysis

```
acp (data frame)
```

Note:

- The results of employing various built-in functions to carry out principal components will be the same, including the number of eigenvalues and their values, proportions of variance explained by each component, cumulative proportion of the explained variances, loadings, scores, and related information.
- More built-in functions are offered by different packages to generate the eigenvalues, loadings, and scores based on the commands employed to perform principal components analysis.

8.3 DESCRIBING PRINCIPAL COMPONENTS

The configuration of the data for a principal components analysis will be described before discussing how to compute the PCA. Consider n samples are selected, and k response variables (Y) are measured from each sample, then the configuration of the data for principal components analysis is presented in Table 8.1.

The steps for describing principal components that capture most of the differences (variance) in the data can be summarized as follows:

Step 1: Employ the k measured response variables $Y_1, Y_2, \ldots, Y_k$ to generate k components $Z_1, Z_2, \ldots, Z_k$ that are uncorrelated; the generated components represent the linear combination of the original variables.

Step 2: Arrange the k extracted components in order, so that the maximum difference in the data is captured by the first principal component Z_1, Z_1 is constructed from the original response variables in the form of linear combination, the second largest amount of variance was captured by the second principal component Z_2, which is less than the amount captured by Z_1, and so on for other components. We can arrange the extracted components in order based on the amount of variance captured by each component, var.(Z_1) $\geq$var.(Z_2) $\geq \ldots \geq$var.(Z_k).

Step 3: The steps for constructing the principal components can be expressed in mathematical equations to represent each principal component as a linear combination of all selected response variables $Y_1, Y_2, \ldots, Y_k$, as presented below.

$$Z_1 = a_{11}Y_1 + a_{12}Y_2 + a_{13}Y_3 + \ldots + a_{1k}Y_k$$
$$Z_2 = a_{21}Y_1 + a_{22}Y_2 + a_{23}Y_3 + \ldots + a_{2k}Y_k$$
$$Z_k = a_{k1}Y_1 + a_{k2}Y_2 + a_{k3}Y_3 + \ldots + a_{kk}Y_k$$

where the a's are the coefficients of the principal component.

The first equation (Z_1) represents the first principal component that maximizes var(Z_1) (var(Z_1) as large as possible) and is subject to Eq. (8.1).

$$a_{11}^2 + a_{12}^2 + \ldots + a_{1k}^2 = 1 \tag{8.1}$$

The variance of the first principal component Z_1 can be increased by increasing the value of any coefficient (a_{1i}) in the Eq. (8.1). The second equation Z_2 represents the second principal component which maximizes var. (Z_2) (var(Z_2) is as large as possible) subject to:

$$a_{21}^2 + a_{22}^2 + \ldots + a_{2k}^2 = 1$$

Z_1 and Z_2 are uncorrelated, the covariance between the two components is 0 (cov (Z_1, Z_2)=0).

We can define other principal components in a similar way until we obtain k uncorrelated components.

Note:

- The assumption of normality is not required for principal components analysis.
- Scientists usually adapt the first few components that capture the maximum amount of the variance and discard other components that capture minimal amount of the variation.

TABLE 8.1 The General Configuration of Data for Principal Components Analysis

	Measured variable			
Sample No.	Y_1	Y_2	...	Y_k
1	Y_{11}	Y_{12}	...	Y_{1k}
2	Y_{21}	Y_{22}	...	Y_{2k}
⋮	⋮	⋮	...	⋮
n	Y_{n1}	Y_{n2}	...	Y_{nk}

8.4 COMMON PROCEDURE FOR COMPUTING PRINCIPAL COMPONENTS

Statistical software is usually employed to perform principal components analysis and produce the required and necessary results with detailed information regarding the analysis. Thus, it is preferable to deliver a common idea of the steps for computing the principal components rather than discussing how the equations (principal components) are mathematically derived. We can summarize the steps for computing the principal components from the covariance matrix employing the following steps:

Step 1: Prepare the variance–covariance matrix for k selected response variables under study, the variance is represented by the values allocated on the diagonal of the matrix, while the covariance of the variables is represented by the off-diagonal values.

Step 2: The eigenvalues λ_i of the covariance matrix should be computed; k eigenvalues can be computed, which is equal to the number of selected response variables. The eigenvalues can be ordered based on the size of each eigenvalue; $\lambda_1 \geq \lambda_2 \geq \ldots \geq \lambda_k \geq 0$. Each eigenvalue represents the variance of the principal component Z_i it is associated with.

Step 3: The eigenvectors corresponding to each eigenvalue should be computed. The coefficients of the principal components $a_1, a_2, a_3, \ldots, a_k$ are represented by the elements of its eigenvector. The values of each eigenvector are called the loadings. The k eigenvectors are orthogonal, if the eigenvalues are distinct.

Step 4: The last step is to choose a few components that capture most of the variation in the gathered data.

Note:

- The sum of the diagonal values of the covariance matrix is equal to the sum of all the eigenvalues. The variances of the original variables are represented by sum of all eigenvalues

$$\sum_{i=1}^{k} S_{ii} = \lambda_1 + \lambda_2 + \cdots + \lambda_k$$

  where S_{ii} refers to the variance of the selected variables.
- The importance of each variable to the principal component can be measured by the loadings a_i.
- Each principal component explains various aspects in the data set.

8.5 EXTRACT PRINCIPAL COMPONENTS FROM CORRELATION MATRIX

Researchers can face situations where the variances are unequal or the measurement units are not commensurate (different), thus, a correlation matrix should be employed to extract the principal components rather than a covariance matrix. The first reason to employ the correlation matrix instead of a covariance matrix is to stop giving the opportunity to the variables with large variances to contribute more than the variables with small variances. The second reason is that changing the scales of the selected variables will not influence the components; this means that correlation matrix is an invariant scale. Moreover, the components are more interpretable when we employ a correlation matrix. The same procedure is used for computing the principal components from the covariance matrix to compute the components from the correlation matrix.

8.6 STANDARDIZATION

Sometimes the variables under study could be measured on various scales that cause unequal variances of the chosen variables. Thus, in order to avoid this case, it is important to standardize the selected variables and get invariant scales. We can perform standardization employing the z score given in (8.2).

$$Z = \frac{y_i - \overline{y}}{s_i} \tag{8.2}$$

The produced matrix for standardized values will be similar to the correlation matrix.

8.7 SELECTING THE NUMBER OF COMPONENTS

Principal components analysis is considered as a data reduction technique. Thus, the aim of employing PCA is to express the gathered data by a few equations that describe most of the differences in the data. This stage is an important stage because the analysis will depend on the number of components needing to be kept. We can provide some tips on how to select the most influential components. Suitable selection will consider the rules given below:

Rule 1: If a correlation matrix is used to extract the components, keep the components whose eigenvalues are >1. For a covariance matrix, keep the components whose eigenvalues are greater than the average.

Rule 2: Employ a Scree plot (the plot of λ_i versus i) and locate the break between the large and small eigenvalues.

EXAMPLE 8.1 PARTICULATE MATTER IN THE AIR OF CHOSEN PALM OIL MILLS

An environmentalist wishes to study the particulate matter (such as the total particulate (T.P), PM_{10}, $PM_{2.5}$, and PM_1) in five palm oil mills distributed in the Penang and Kedah States, Malaysia. Twelve samples are selected from each mill and tested for the selected particulate matter. The sample data for the particulate matter in the air of the selected palm oil mills are presented in Table 8.2.

TABLE 8.2 The Results of Particulate Matter for the Chosen Palm Oil Mills

Mill	T.P	PM_{10}	$PM_{2.5}$	PM_1	Mill	T.P	PM_{10}	$PM_{2.5}$	PM_1
1.00	587.42	487.19	124.11	58.19	3.00	448.13	226.63	28.02	10.65
1.00	858.93	716.40	121.91	67.31	3.00	616.24	340.34	61.91	21.97
1.00	1310.82	1164.88	168.29	120.28	3.00	207.95	105.00	26.53	11.38
1.00	728.29	583.62	149.25	67.36	3.00	286.46	173.42	173.42	15.65
1.00	819.91	731.32	130.47	80.25	3.00	619.89	434.62	85.68	35.83
1.00	1117.88	1013.70	167.24	108.76	3.00	287.84	168.49	24.13	8.74
1.00	735.79	853.55	148.43	74.08	4.00	796.47	709.26	124.27	76.92
1.00	418.12	302.52	95.11	40.61	4.00	742.62	622.17	97.07	65.12
1.00	418.19	323.48	55.95	30.96	4.00	686.95	514.38	83.34	47.11
1.00	1111.88	1000.95	162.83	108.64	4.00	620.18	500.10	82.56	46.71
1.00	986.81	855.60	102.25	66.00	4.00	534.26	444.23	85.42	50.32
1.00	865.56	708.56	150.53	76.02	4.00	624.16	396.64	13.02	4.29
2.00	362.05	231.92	57.75	21.07	4.00	1167.05	1080.82	177.42	114.60
2.00	968.91	593.17	119.83	53.66	4.00	1139.09	1049.77	192.12	117.61
2.00	310.93	173.71	36.54	11.46	4.00	1115.66	958.86	174.21	101.80
2.00	280.03	178.46	33.41	13.33	4.00	404.94	190.93	15.25	3.85
2.00	723.32	564.43	94.88	50.76	4.00	246.15	217.64	88.68	36.82
2.00	441.19	208.30	25.70	8.69	4.00	984.15	818.53	135.02	77.47
2.00	241.77	135.39	26.57	8.31	5.00	375.01	169.85	27.05	10.51
2.00	150.39	76.85	17.40	6.76	5.00	512.83	346.06	46.41	15.39
2.00	236.16	122.25	23.84	10.87	5.00	256.75	165.88	19.89	7.97
2.00	237.69	118.79	23.72	11.37	5.00	213.19	123.86	18.19	7.14
2.00	273.03	141.48	40.98	22.10	5.00	670.62	466.57	55.82	18.94

Continued

TABLE 8.2 The Results of Particulate Matter for the Chosen Palm Oil Mills—cont'd

Mill	T.P	PM_{10}	$PM_{2.5}$	PM_1	Mill	T.P	PM_{10}	$PM_{2.5}$	PM_1
2.00	314.26	163.28	29.26	13.22	5.00	492.37	344.92	34.82	15.25
3.00	847.05	712.03	105.57	59.86	5.00	274.87	200.81	53.40	13.94
3.00	491.06	281.55	61.12	25.52	5.00	447.94	284.14	21.86	6.97
3.00	296.51	172.97	36.97	14.50	5.00	395.93	287.49	30.76	10.01
3.00	353.66	175.70	31.56	11.28	5.00	204.58	101.90	8.21	1.95
3.00	329.20	166.51	21.63	7.96	5.00	187.17	79.84	14.03	3.94
3.00	507.64	318.96	38.45	15.25	5.00	175.42	93.54	12.62	3.78

We should examine the variances of chosen random variables of the particulate matter to have a general idea regarding the variation of the data set for each variable. One can compute the variances of the 4 chosen variables under investigation by employing the built-in function `sapply ()` in R. The built-in function `round ()` was employed to round the outputs of the variances generated by R built-in functions to two decimals.

```
> round (sapply (Example8_1[2:5], var), digits = 2)
     T.P      PM10    PM2.5      PM1
93069.07 92110.34 2982.22 1191.66
```

The outputs of employing the function `sapply ()` for the stored data. CSV (`Example8_1`) of the particulate matter in the air of chosen palm oil mills exhibited that the variances of the total particulate and PM_{10} are **93,069.07** and **92,110.34**, respectively. These two variances are considered very large values compared to the other variances, which indicates that the principal components will be dominated by these two variables. Thus, the correlation matrix should be used to compute the eigenvalues rather than the covariance matrix. The built-in function `corr.test ()` was employed to compute the correlation matrix for the particulate matter. We employed the function `corr.test ()` instead of `cor ()` because the significance of the correlation is not provided by the built-in function `cor ()`.

```
corr.test (Example8_1[,2:5])
> corr.test(Example8_1[,2:5])
Call:corr.test(x = Example8_1[, 2:5])
Correlation matrix
       T.P PM10 PM2.5  PM1
T.P   1.00 0.98  0.84 0.92
PM10  0.98 1.00  0.88 0.96
PM2.5 0.84 0.88  1.00 0.93
PM1   0.92 0.96  0.93 1.00
Sample Size
[1] 60
Probability values (Entries above the diagonal are adjusted for multiple tests.)
      T.P PM10 PM2.5 PM1
T.P     0    0     0   0
PM10    0    0     0   0
PM2.5   0    0     0   0
PM1     0    0     0   0

 To see confidence intervals of the correlations, print with the short=FALSE option
```

The correlation matrix for the particulate matters in the air of chosen palm oil mills was checked. It was noticed that all chosen variables showed a strong positive relationship with other variables. One can see that the total particulate (T.P) exhibited a strong positive correlation with PM_{10}, $PM_{2.5}$, and PM_1. This feedback suggests that a strong association of these variables exits with the other chosen variables and shares an original source.

Principal components analysis was performed on the particulate matter in the air data set, including four parameters to recognize the source of differences between various mills. The two built-in functions `princomp (data frame, cor=TRUE)` and `summary ()` in R were employed to carry out the principal components analysis. The command `cor=TRUE` was employed to indicate that the principal components are computed from the correlation matrix rather than covariance matrix. The results produced by the function `princomp ()` were stored in a place called `Components`.

```
Components <- princomp (Example8_1[2:5], cor = TRUE)
summary (Components)
```

The square root of the eigenvalue, which represents the standard deviation associated with each principal component, was generated by employing the built-in function `princomp ()` in R. Moreover, the proportion of the variance in the data set interpreted by each component and the cumulative proportion of the interpreted variance are produced by employing the built-in function `summary (Components)`. One can see that employing the built-in functions and commands in R to carry out principal components analysis for the particulate matters in the air of chosen palm oil mills produced four principal components, including the `standard deviation`, the `proportion of the variance`, and the `cumulative proportion` that is explained by the components.

```
> Components <- princomp (Example8_1[2:5], cor = TRUE)
> summary (Components)
Importance of components:
                          Comp.1     Comp.2     Comp.3
Standard deviation     1.9358320 0.43605357 0.22429450
Proportion of Variance 0.9368613 0.04753568 0.01257701
Cumulative Proportion  0.9368613 0.98439702 0.99697403
                           Comp.4
Standard deviation     0.11001764
Proportion of Variance 0.00302597
Cumulative Proportion  1.00000000
```

One can see that >93% of the total variance in the data was captured by only one component (`Comp.1`), which represents the first component. The first component interprets `0.9368`% of the total variance $[((1.9358320)^2/4)=(1.936/4)=93.686\%]$, which is the highest contribution, and the second-highest contribution was shown by the second principal component, which interprets `4.753568`% $[((0.436)^2/4)=(3.858/13)=4.753\%]$ of the total variance. The other components can be interpreted similarly.

We should make a decision on how many components are needed to interpret the variation in the data. One can follow the procedure given in Section 8.7 to choose the number of components to be retained; the summary outputs showed that only one component with eigenvalue >1 can be kept. The significance of the component is measured by eigenvalues; the component with the highest eigenvalue is the most important and is accountable for interpreting the largest amount of variation in the data. The function `screeplot ()` in R was employed to generate the Scree plot to get a clear view of the various eigenvalues. The Scree plot for the particulate matter in the air of chosen palm oil mills employing R built-in functions is presented in Fig. 8.1.

`Screeplot (Components, npcs=4, type="barplot", xaxt="n", yaxt="n").`

The Scree plot will be generated by this function as a bar chart. A Scree plot can be generated with connected lines employing R built-in functions and commands; for instance, employing the function `type="lines"` will produce a plot with connected lines. Furthermore, the number of components to be plotted can be fixed by employing the function `npcs=`.

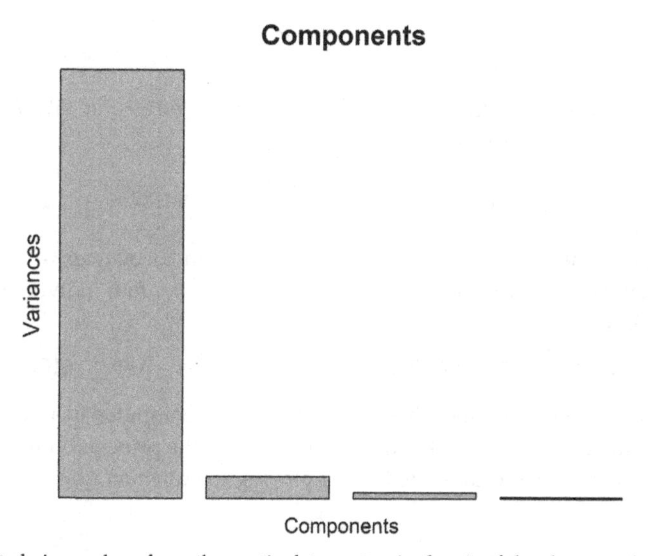

FIG. 8.1 Scree plot for the computed eigenvalues from the particulate matter in the air of the chosen palm oil mills data.

One can see clearly from Fig. 8.1 that most of the variance in the particulate matter in the air of the chosen palm oil mills data was captured by only one component.

We can employ the built-in function `$loadings` to compute the loadings for the extracted component. This function will not show excessively small coefficients; thus, one can employ the built-in function `unclass ()` to display all of the coefficients. The principal components are stored in a place called (`Components`); thus, the structure of the built-in function `unclass ()` is:

```
Unclass (Components $ loadings)
```

and the results can be rounded employing the function `round ()`.

```
round (unclass (Components $ loadings), digits = 3)
```

Calling the function `Components $ loadings` in R will generate the loadings for all extracted components regarding the particulate matters in the air of the chosen palm oil mills data, and the function `unclass` was employed to show all of the values. The outputs of employing R built-in functions and commands to compute the loadings for the four extracted components with their coefficients (loadings) are given below and rounded to three decimals.

```
> round (unclass (Components $ loadings), digits = 3) #To get the loadings
      Comp.1 Comp.2 Comp.3 Comp.4
T.P    0.498  0.545  0.473  0.481
PM10   0.509  0.327 -0.111 -0.788
PM2.5  0.485 -0.760  0.429 -0.063
PM1    0.508 -0.137 -0.762  0.378
```

The purpose of principal components analysis is to express the data by a few linear combinations, because the principal components method is a data reduction technique. Thus, for this case one can see clearly that only one component can interpret >93% of the total variance (its eigenvalue is >1). The other principal components captured very little of the variance in the data, as presented by the results of employing the built-in function `summary ()`.

The principal component in a mathematical form is presented in Eq. (8.3) where X_i in the same series appeared with R outputs. For instance, X_1 refers to the total particulate (T.P), X_2 refers to PM_{10}, and so on for the other parameters.

$$Z_1 = 0.496\,X_1 + 0.509\,X_2 + 0.485\,X_3 + 0.508\,X_4 \tag{8.3}$$

The loadings (coefficients) associated with each variable in Eq. (8.3) can be employed to measure the contribution of the selected variables to the first principal component. One can arrange the contribution of all variables in ascending order based on the size of each coefficient: PM_{10} (X_2) > PM_1 (X_4) > T.P (X_1) > $PM_{1.5}$ (X_3). Moreover, whereas a positive sign associated with the coefficients indicates that the variable has the power to increase the value of the principal component. The component in Eq. (8.3) refers to the average of the effects of all chosen variables.

The values of the principal component for each of the n observations are called the scores of the principal component. The values can be employed for displaying the differences or other purposes, such as to sketch a scatter plot or other plots that help for further analysis. The standardized values are usually employed in calculating the scores (the values of the principal components). One can try to show that the value of the first individual of the first principal component is 0.94 as shown below:

First, we should standardize the chosen variables employing the formula in Eq. (8.2):

$$Z = \frac{y_i - \bar{y}}{s_i}$$

The average value and standard deviation for the total particulate (T.P) are 550.99 and 350.07, respectively. We apply the formula to standardize the value of the total particulate as follows:

$$\text{T.P} = X_1 = \frac{587.42 - 550.99}{305.07} = 0.1194187$$

The other standardized values for other variables can be computed in a similar manner.

Second, we substitute the standardized values in Eq. (8.3). The value of the first individual (observation) of the principal component is computed to be 0.94.

$$Z_1 = 0.498(0.1194187) + 0.509(0.23817875) + 0.485(0.9259409) + 0.508(0.61043056) = 0.939895$$

The other values of the principal component for the other observations were computed in a similar manner. R built-in functions and commands can easily compute the scores for all of the observations of the principal component. The scores for the particulate matters in the air of palm oil mills were computed by employing the function `$scores`.

```
round (Components $ scores, digits = 3)
```

The scores for all of the observations of the principal component were extracted from the file `Components` employing the built-in function and command `Components $ scores`. The results of the scores for all of the observations of the particulate matters in the air of palm oil mills are rounded to three decimals as presented below.

```
> round (Components $ scores, digits = 3)
       Comp.1 Comp.2 Comp.3 Comp.4
 [1,]  0.947 -0.650 -0.038  0.043
 [2,]  1.898  0.083  0.081 -0.022
 [3,]  4.601  0.522 -0.190  0.054
 [4,]  1.704 -0.681  0.141  0.087
 [5,]  2.127 -0.143 -0.206  0.010
 [6,]  3.848  0.071 -0.188  0.017
 [7,]  2.265 -0.389 -0.103 -0.533
 [8,] -0.164 -0.678 -0.073  0.097
 [9,] -0.622 -0.067 -0.176 -0.020
[10,]  3.775  0.108 -0.225  0.045
[11,]  2.148  0.746  0.103 -0.175
[12,]  2.281 -0.350  0.127  0.072
[13,] -1.000 -0.253  0.004  0.020
[14,]  1.649  0.231  0.585  0.327
[15,] -1.515 -0.073 -0.008  0.009
[16,] -1.558 -0.087 -0.124 -0.028
[17,]  0.930  0.119  0.079  0.009
[18,] -1.380  0.363  0.159  0.108
[19,] -1.830 -0.086 -0.111 -0.024
[20,] -2.184 -0.180 -0.270 -0.022
[21,] -1.848 -0.083 -0.193  0.033
[22,] -1.845 -0.084 -0.202  0.050
[23,] -1.434 -0.281 -0.257  0.146
[24,] -1.566  0.017 -0.096  0.070
[25,]  1.614  0.316  0.100 -0.093
[26,] -0.607 -0.032  0.115  0.140
[27,] -1.491 -0.118 -0.094  0.021
[28,] -1.489  0.077  0.023  0.076
[29,] -1.683  0.176 -0.017  0.036
[30,] -0.872  0.398  0.177 -0.019
[31,] -1.288  0.355  0.137  0.090
[32,] -0.348  0.260  0.374  0.145
[33,] -1.892 -0.192 -0.221  0.036
[34,] -0.268 -2.054  0.944 -0.141
[35,]  0.236 -0.020  0.225  0.030
[36,] -1.713  0.065 -0.080 -0.030
[37,]  1.946 -0.109 -0.209  0.001
[38,]  1.292  0.128 -0.214  0.044
[39,]  0.628  0.176  0.031  0.055
[40,]  0.481  0.052 -0.065 -0.017
[41,]  0.324 -0.218 -0.237  0.029
[42,] -0.939  1.093  0.372 -0.128
[43,]  4.220  0.066 -0.185 -0.028
[44,]  4.297 -0.237 -0.168  0.025
[45,]  3.710 -0.063  0.039  0.072
[46,] -1.635  0.445  0.133  0.054
[47,] -0.704 -0.975 -0.277  0.011
```

```
[48,]  2.545  0.195  0.116  0.007
[49,] -1.515  0.175  0.040  0.122
[50,] -0.744  0.324  0.235 -0.089
[51,] -1.818  0.069 -0.144 -0.076
[52,] -1.989 -0.028 -0.191 -0.042
[53,] -0.144  0.593  0.432 -0.125
[54,] -0.886  0.449  0.114 -0.107
[55,] -1.341 -0.355  0.004 -0.111
[56,] -1.301  0.518  0.149 -0.095
[57,] -1.256  0.291  0.069 -0.163
[58,] -2.206  0.093 -0.160 -0.044
[59,] -2.191 -0.052 -0.178  0.001
[60,] -2.202 -0.038 -0.209 -0.054
```

A bar chart can be built employing the values of the principal component to illustrate the behavior of various observations in a pictorial form. The built-in function `barplot ()` in R can be employed to generate a bar chart for the values of the first principal component for the particulate matter in the air of chosen palm oil mills, as given in Fig. 8.2.

```
barplot (A [,1], xaxt ="n", yaxt ="n", ylim = c(-5, 5), col = c(rep('gray',12),rep('blue',12),
rep('darkkhaki',12), rep('black',12), rep('orange',12)))
```

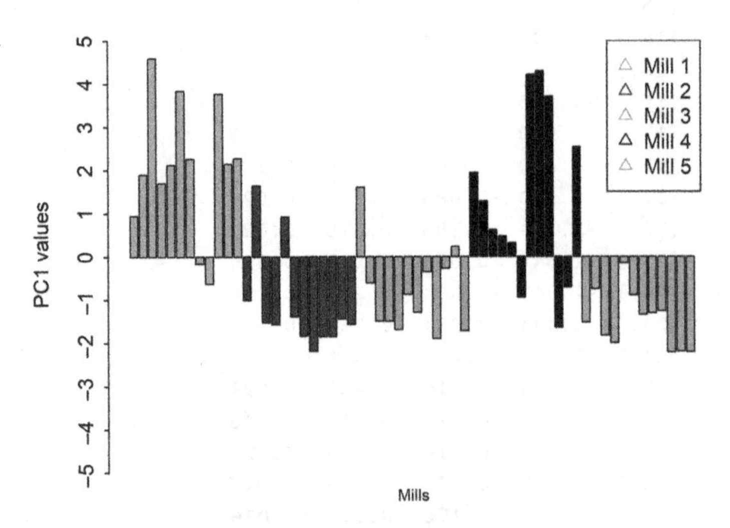

FIG. 8.2 The values of the first principal component for the particulate matter in the air of the selected palm oil mills.

The behavior of various particulate matter observations in connection to the first principal component can be investigated in Fig. 8.2. One can see that special behavior was presented by various observations and based on the mill location, which means that the location of the mill would affect the contribution of each observation. For instance, the observations of the first mill (mill 1), represented by the first 12 bars of Fig.8.2, displayed a positive effect except for two observations; the positive effect is mainly due to high values of all chosen particulate matters in the air (T.P, PM_{10}, $PM_{2.5}$, PM_1), whereas the negative effect is mainly due to low values of all chosen particulate matters (T.P, PM_{10}, $PM_{2.5}$, PM_1). However, in the next 24 observations, which belong to mill 2 and mill 3, respectively, all of the observations exhibited a negative effect to the first principal component except for two observations each. The fourth 12 bars belong to mill 4, which showed a positive effect except for three samples. The last 12 samples belong to mill 5; all samples presented a negative effect to the first principal component.

In summary, one can conclude that the differences in the contribution support the idea that the palm oil mills had different characteristics. Positive contribution was mainly due to high level of all selected parameters whilst the negative contribution was mainly due to low level of all parameters.

EXAMPLE 8.2 ASSESSMENT OF SURFACE WATER

An assessment of the water quality in the Juru and Jejawi Rivers in the Penang state of Malaysia was conducted by monitoring ten physiochemical parameters; namely, temperature, pH, electrical conductivity (EC), dissolved oxygen (DO), biochemical oxygen demand (BOD), chemical oxygen demand (COD), turbidity, total suspended solids (TSS), phosphate, and nitrate. The APHA Standard Methods for the Examination of Water and Wastewater were applied to analyze the concentration of the above-mentioned parameters. The sample data obtained from 10 different sites each are presented in Table 1.2.

One can employ the function `sapply ()` in R to compute the variances for the stored data`.CSV` (`Example8_2`) of the physicochemical parameters (variables) for the water quality in the Juru and Jejawi Rivers involved in this research.

```
> round (sapply (Example8_2[2:11], var), digits = 2)
Temperature            PH          DO         BOD         COD
       0.29          0.06        0.41      243.82   218828.31
        TSS  Conductivity   Turbidity   Phosphate     Nitrate
   30475.02        126.61      154.42   256059.27       35.26
```

The variances of the COD (**218,828.31**), TSS (**30,475.02**) and phosphate (**256,059.27**) are very large compared to the other variances. Thus, employing the correlation matrix is the suitable and right decision to extract the components. The built-in function `corr.test ()` in R was employed to compute the correlation matrix for the physicochemical parameters of the surface water in the Juru and Jejawi Rivers. The result of employing the `corr.test ()` built-in function is the bivariate correlation matrix for all possible combination of two variables of the selected physicochemical parameters of the surface water, including the correlation coefficients and the significance level for the various variables.

```
> corr.test(Example8_2[2:11])
Call:corr.test(x = Example8_2[2:11])
Correlation matrix
              Temperature    PH    DO   BOD   COD   TSS
  Temperature        1.00 -0.08 -0.15 -0.81  0.50 -0.35
  PH                -0.08  1.00  0.86 -0.37  0.65 -0.07
  DO                -0.15  0.86  1.00 -0.34  0.62 -0.03
  BOD               -0.81 -0.37 -0.34  1.00 -0.84  0.46
  COD                0.50  0.65  0.62 -0.84  1.00 -0.37
  TSS               -0.35 -0.07 -0.03  0.46 -0.37  1.00
  Conductivity       0.39  0.78  0.71 -0.71  0.84 -0.02
  Turbidity          0.91  0.07  0.03 -0.91  0.70 -0.42
  Phosphate         -0.75 -0.41 -0.41  0.95 -0.86  0.46
  Nitrate            0.58  0.15  0.29 -0.77  0.54 -0.28
             Conductivity Turbidity Phosphate Nitrate
Temperature          0.39      0.91     -0.75    0.58
PH                   0.78      0.07     -0.41    0.15
DO                   0.71      0.03     -0.41    0.29
BOD                 -0.71     -0.91      0.95   -0.77
COD                  0.84      0.70     -0.86    0.54
TSS                 -0.02     -0.42      0.46   -0.28
Conductivity         1.00      0.52     -0.71    0.57
Turbidity            0.52      1.00     -0.86    0.72
Phosphate           -0.71     -0.86      1.00   -0.70
Nitrate              0.57      0.72     -0.70    1.00
Sample Size
[1] 20
Probability values (Entries above the diagonal are adjusted for multiple tests.)
            Temperature   PH   DO  BOD  COD  TSS
Temperature        0.00 1.00 1.00 0.00 0.50 1.00
PH                 0.72 0.00 0.00 1.00 0.05 1.00
DO                 0.52 0.00 0.00 1.00 0.09 1.00
BOD                0.00 0.10 0.14 0.00 0.00 0.79
```

```
COD               0.02 0.00 0.00 0.00 0.00 1.00
TSS               0.13 0.77 0.90 0.04 0.11 0.00
Conductivity      0.09 0.00 0.00 0.00 0.00 0.92
Turbidity         0.00 0.77 0.90 0.00 0.00 0.07
Phosphate         0.00 0.07 0.07 0.00 0.00 0.04
Nitrate           0.01 0.53 0.21 0.00 0.01 0.24
             Conductivity Turbidity Phosphate Nitrate
Temperature           1.00      0.00      0.01    0.17
PH                    0.00      1.00      1.00    1.00
DO                    0.01      1.00      1.00    1.00
BOD                   0.01      0.00      0.00    0.00
COD                   0.00      0.02      0.00    0.33
TSS                   1.00      1.00      0.79    1.00
Conductivity          0.00      0.39      0.01    0.21
Turbidity             0.02      0.00      0.00    0.01
Phosphate             0.00      0.00      0.00    0.02
Nitrate               0.01      0.00      0.00    0.00
 To see confidence intervals of the correlations, print with the short=FALSE option
```

It was noticed that some parameters (variables) showed a strong relationship (positive or negative) with other variables, as exhibited by the correlation matrix for the physicochemical parameters of the surface water. For instance, one can observe that a strong positive correlation was exhibited between temperature and turbidity; BOD and phosphate showed a strong negative correlation. Strong correlations reflect the behavior of the various variables in the study with a high possibility of sharing the original source.

Principal components analysis was performed on the physiochemical parameters data set of the surface water including 10 parameters (variables) to identify the root of the differences in the sample data. Two built-in functions, `princomp (data frame, cor=TRUE)` and `summary ()`, were employed to carry out the principal components analysis for the stored data. CSV`(Example8_2)` of the physicochemical parameters of the surface water. The principal components are computed from the correlation matrix of the chosen variables. The outputs of employing the function `princomp ()` were stored in a place called `Components`.

```
Components <- princomp (Example8_2[2:11], cor = TRUE)
Summary (Components)
```

We employed the function `princomp ()` to compute the standard deviation associated with each principal component. More details were provided by the built-in function `summary (Components)` including the proportion and the cumulative proportion of the variance in the data captured by each component. Ten principal components were generated by employing the R built-in functions and commands for the physicochemical parameters of the surface water, including the `standard deviation`, the `proportion of the variance` captured by each component and the `cumulative proportion`.

```
> Components <- princomp (Example8_2[2:11], cor = TRUE)
> summary(Components)
Importance of components:
                          Comp.1    Comp.2     Comp.3     Comp.4
Standard deviation     2.4349118 1.5175967 0.93177727 0.68621973
Proportion of Variance 0.5928795 0.2303100 0.08682089 0.04708975
Cumulative Proportion  0.5928795 0.8231895 0.91001039 0.95710015
                           Comp.5    Comp.6      Comp.7      Comp.8
Standard deviation     0.39767663 0.3468689 0.257666926 0.216362532
Proportion of Variance 0.01581467 0.0120318 0.006639224 0.004681275
Cumulative Proportion  0.97291482 0.9849466 0.991585842 0.996267117
                            Comp.9     Comp.10
Standard deviation     0.164099882 0.101980700
Proportion of Variance 0.002692877 0.001040006
Cumulative Proportion  0.998959994 1.000000000
```

One can observe that >82% of the total variance in the data of the physicochemical parameters of the surface water was captured by the first two components. The first component captured `0.5928795`of the total variance [$((2.4349118)^2/10)=59.2879547\%$], which is the highest contribution, and the second-highest contribution was exhibited by the second principal component, which captured `23.031`% [$((1.5175967)^2/10)=23.031\%$] variance. The other components can be interpreted similarly.

How many principal components should be kept to interpret the dispersion in the data of the physicochemical parameters of surface water? Based on the procedure given in Section 8.7, one can decide to keep only two components with eigenvalues >1. The Scree plot for the eigenvalues can be generated by employing the function `screeplot ()` in R to gain an obvious picture of the various eigenvalues. The extracted eigenvalues for the physicochemical parameters of the surface water are plotted employing R built-in functions and commands to produce a Scree plot, as given in Fig. 8.3.

```
screeplot (Components, xaxt ="n", yaxt ="n")
```

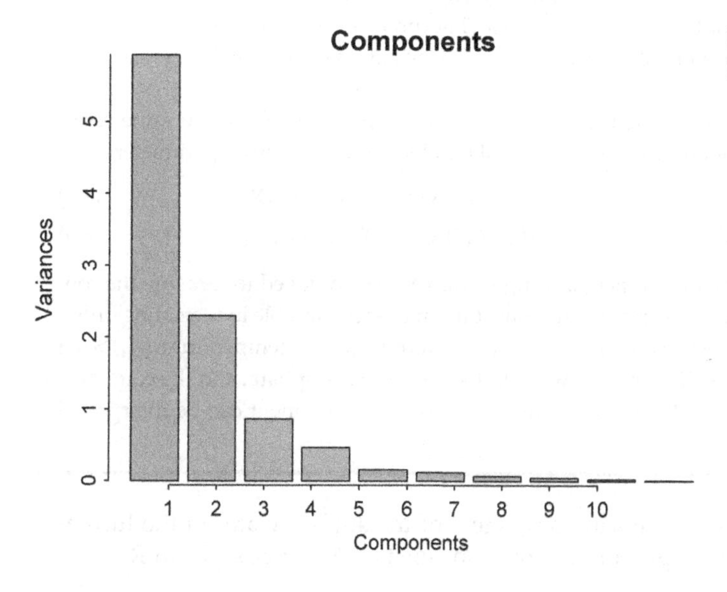

FIG. 8.3 The Scree plot for the computed eigenvalues from the physicochemical parameters of the surface water.

It is clear from Fig. 8.3 that most of the variance in the data was captured by only two components.

We can employ the built-in function `$loadings` to compute the loadings for each component. One can employ the built-in function `unclass ()` to display all of the coefficients. The extracted principal components are stored in a place called `(Components)`, thus, the structure of the command for extracting the loadings are:

```
round (unclass (Components $ loadings), digits = 2)
```

Calling the function `Components $ loadings` in R will generate the loadings for all components regarding the physicochemical parameters of the surface water data, and the function `unclass` was employed to show all of the values. The outputs of employing R built-in functions and commands to compute the loadings produced ten components with their coefficients (loadings). The results of loadings were rounded to two decimals.

```
> round (unclass (Components $ loadings), digits = 2)
             Comp.1 Comp.2 Comp.3 Comp.4 Comp.5 Comp.6 Comp.7
Temperature   0.30   0.40   0.18   0.30   0.23   0.46   0.49
PH            0.21  -0.53  -0.14   0.19   0.49   0.09  -0.32
DO            0.21  -0.53  -0.09  -0.27  -0.39   0.42   0.43
BOD          -0.40  -0.12  -0.01  -0.02   0.01  -0.09   0.14
COD           0.38  -0.15  -0.09   0.24  -0.44  -0.56   0.18
TSS          -0.18  -0.21   0.89   0.14  -0.23   0.06  -0.15
Conductivity  0.34  -0.29   0.26   0.10   0.41  -0.25   0.19
Turbidity     0.35   0.31   0.12   0.11  -0.10  -0.24  -0.05
Phosphate    -0.39  -0.07   0.05  -0.09   0.31  -0.38   0.60
Nitrate       0.31   0.13   0.23  -0.83   0.16  -0.12  -0.06
```

```
             Comp.8 Comp.9 Comp.10
Temperature    0.06   0.10    0.34
PH            -0.35   0.00    0.38
DO            -0.15  -0.20   -0.12
BOD            0.39  -0.59    0.55
COD            0.09   0.29    0.36
TSS           -0.13   0.09    0.09
Conductivity   0.49  -0.17   -0.43
Turbidity     -0.47  -0.68   -0.09
Phosphate     -0.46   0.13   -0.05
Nitrate        0.02   0.09    0.30
```

The objective of principal components analysis is to represent the available the data by a few linear combinations of the original variables. Thus, for this example one can clearly see that only two components can interpret >82% of the total variance (their eigenvalues are >1). The other principal components captured very little of the variance in the data, as shown in the `summary ()` outputs.

The two principal components in a mathematical form are presented in Eqs. (8.4) and (8.5), where X_i in the same series appeared with R outputs. For instance, X_1 refers to the temperature, X_2 refers to pH and so on for the other parameters.

$$Z_1 = 0.30X_1 + 0.21X_2 + 0.21X_3 - 0.40X_4 + 0.38X_5 - 0.18X_6 + 0.34X_7 + 0.35X_8 - 0.39X_9 + 0.31X_{10} \quad (8.4)$$

$$Z_2 = 0.4X_1 - 0.53X_2 - 0.53X_3 - 0.12X_4 - 0.15X_5 - 0.21X_6 - 0.29X_7 + 0.31X_8 - 0.07X_9 + 0.13X_{10} \quad (8.5)$$

The loadings (coefficient) associated with each variable in the first principal components can be employed to measure the contribution of the variables to the first principal component. One can arrange the contribution of each variable in ascending order: BOD (X_4) > phosphate (X_9) > COD (X_5) > turbidity (X_8) > electrical conductivity (X_7) > nitrate (X_{10}) > temperature (X_1) > pH (X_2) = DO (X_3) > TSS (X_6). The first component represents the difference between BOD, TSS, and phosphate, and the combined effect of all other variables. This difference interprets >59% of the total variance. The second component can be interpreted similarly.

The values of the principal components for the physicochemical parameters of the surface water in the Juru and Jejawi Rivers were computed for all of the observations employing the built-in function `$ scores ()` in R.

```
round (Components $ scores, digits = 3)
```

```
> round (Components $ scores, digits = 3)
      Comp.1 Comp.2 Comp.3 Comp.4 Comp.5 Comp.6
 [1,]  2.689 -2.263  0.512  0.355 -0.280  0.044
 [2,]  2.721 -2.060  0.394 -0.049  0.221 -0.186
 [3,]  2.469  0.525  0.528 -0.103 -1.104  0.656
 [4,]  2.676 -1.318  0.226  0.694  0.106  0.225
 [5,]  2.406 -1.662  0.082  0.462  0.109 -0.132
 [6,]  2.936 -1.366  0.074 -0.576  0.668 -0.209
 [7,]  2.548  1.475  0.057 -1.673  0.124  0.030
 [8,]  1.919  1.520 -1.032 -0.037 -0.157 -0.044
 [9,]  1.793  3.481 -0.338  1.297  0.262  0.459
[10,]  1.951  3.025 -0.561  0.124  0.061 -0.852
[11,] -2.253 -0.329 -1.518 -0.008  0.571  0.601
[12,] -2.290 -1.415 -1.129  0.261 -0.347 -0.303
[13,] -2.483 -0.950 -1.314  0.214  0.284  0.121
[14,] -2.752 -0.334 -1.013  0.440 -0.236 -0.165
[15,] -1.757 -0.282 -1.007 -1.426 -0.581 -0.056
[16,] -2.124  0.730  1.195 -0.732  0.250  0.007
[17,] -2.341  0.562  1.413  0.457 -0.009 -0.271
[18,] -2.567  0.298  1.454  0.388 -0.117 -0.249
[19,] -2.762 -0.039  0.622  0.464 -0.246 -0.158
[20,] -2.779  0.403  1.355 -0.551  0.421  0.481
```

```
       Comp.7 Comp.8 Comp.9 Comp.10
 [1,]  0.208 -0.137  0.113  0.006
 [2,]  0.014 -0.034 -0.021  0.009
 [3,]  0.213  0.058  0.095 -0.095
 [4,]  0.095 -0.071 -0.409  0.179
 [5,] -0.129 -0.075  0.369 -0.012
 [6,] -0.235  0.146 -0.008 -0.070
 [7,] -0.137  0.124 -0.155 -0.072
 [8,] -0.395  0.335 -0.024 -0.044
 [9,] -0.209 -0.071 -0.023  0.101
[10,]  0.618 -0.157  0.088 -0.028
[11,] -0.085 -0.322  0.166 -0.128
[12,]  0.012  0.350 -0.084  0.027
[13,]  0.448 -0.008 -0.204 -0.066
[14,] -0.130  0.275  0.193  0.130
[15,] -0.147 -0.458 -0.070  0.119
[16,]  0.006  0.022  0.185  0.239
[17,] -0.287 -0.342 -0.089 -0.081
[18,] -0.181  0.015 -0.067 -0.052
[19,] -0.109  0.046 -0.109 -0.156
[20,]  0.429  0.304  0.053 -0.007
```

The behavior of the physiochemical parameters of the surface water for each river regarding each principal component can be illustrated in a pictorial form employing the values of each component for all sampling points to build a bar chart. The bar chart for the values of the first principal component for various sampling points of the rivers is presented in Fig. 8.4 to investigate the relationship between the component values and the samples from various sampling points (sites) of the rivers. Whereas the Juru River affected positively to the first principal component, the Jejawi River affected negatively to the first component. Positive contribution was due to the concentration of COD, turbidity, nitrate, temperature, pH, and DO, whereas negative contribution was due to the concentration of BOD, phosphate, and TSS. This difference in the contribution level indicates that the Juru River and Jejawi River are different in terms of first principal component. The high BOD and phosphate concentrations indicate relatively high waste dumping activity in the Jejawi river.

The values for the second principal component are shown in Fig. 8.5. The behavior of the physiochemical parameters of the surface water for each river regarding the second principal component can be interpreted in a similar manner.

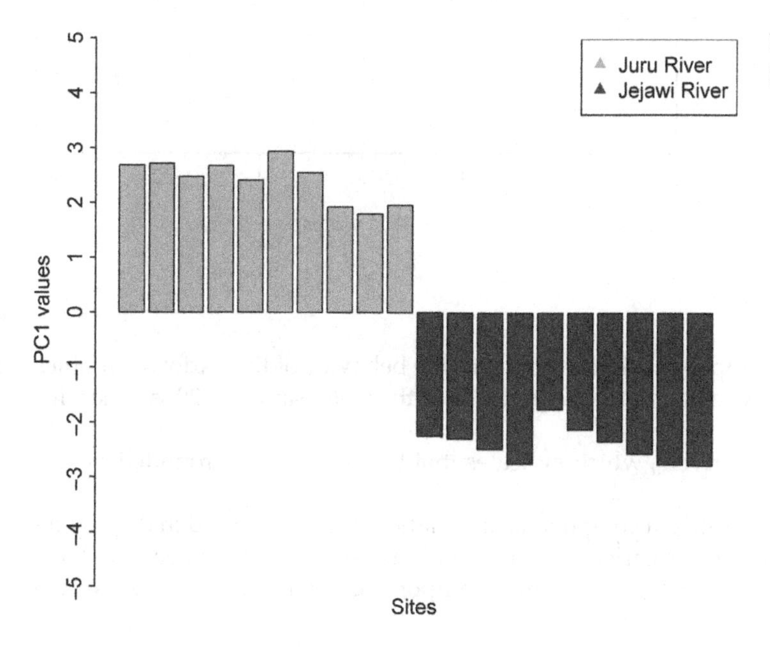

FIG. 8.4 The values of the first principal component for the physicochemical parameters of the surface water.

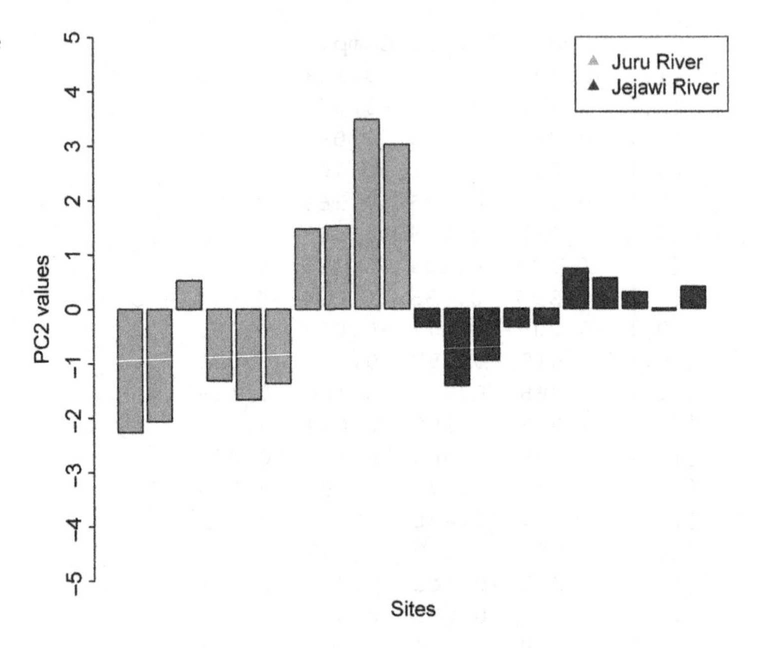

FIG. 8.5 The values of the second principal component for the physicochemical parameters of the surface water.

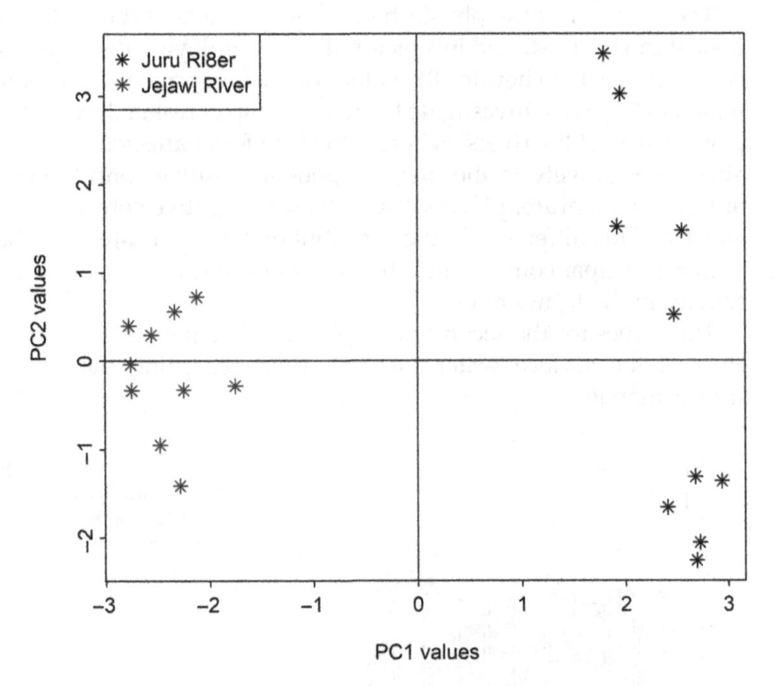

FIG. 8.6 Plot of the scores of the first and second principal components.

It is helpful to graph the values of the first two components to comprehend the behavior of the various variables about the first two principal components. Fig. 8.6 shows a plot of the values for all of the water samples (20 samples) for the first two principal components.

One can clearly observe that there are two different rivers, which indicates that the rivers are surrounded by different conditions.

In summary, we have observed the importance of principal components as a data reduction method in displaying the hidden information regarding the behavior of the various variables that are included in the study. Indeed, we have represented 10 physicochemical variables by only two linear combinations (components) that captured >82% of the total variation in the data.

Further Reading

Alkarkhi, A.F.M., Alqaraghuli, W.A.A., 2019. Easy Statistics for Food Science with R, first ed. Academic Press.

Alkarkhi, A.F.M., Anees, A., Azhar, M.E., 2009. Assessment of surface water quality of selected estuaries of Malaysia-multivariate statistical techniques. Environmentalist 29, 255–262.

Bryan, F.J.M., 1991. Multivariate Statistical Methods: A Primer. Chapman & Hall, Great Britain.

Daniel & Hocking. 2013. Blog archives, High resolution figures in R [Online]. R-Bloggers. Available: https://www.r-bloggers.com/Author/Daniel-Hocking/ [Accessed 26 February 2019].

Johnson, R.A., Wichern, D.W., 2002. Applied Multivariate Statistical Analysis. Prentice Hall, New Jersey.

Kabacoff, R.I., 2017. Quick R Accessing The Power Of R, Principal Components And Factor Analysis. [Online] Available: https://www.statmethods.net/advstats/factor.html. [(Accessed 6 December 2018)].

Nik, N.N.A.R., Ngu, C.C., Al-Karkhi, A.F.M., Mohd., R., Mohd. Omar, A.K., 2017. Analysis of particulate matters in air of palm oil mills—a statistical assessment. Environ. Eng. Manage. J. 16, 2537–2543.

Rencher, A.C., 2002. Methods of Multivariate Analysis. J. Wiley, New York.

Sanchez, G., 2012. 5 Functions To Do Principal Components Analysis In R [Online]. Available: http://www.gastonsanchez.com/visually-enforced/how-to/2012/06/17/PCA-in-R/. [(Accessed 16 January 2019)].

Sarah, S., 2012. Instant R, Performing A Principal Component Analysis In R. [Online]. Wordpress | Nest Theme by Ychong. Available: http://www.instantr.com/2012/12/18/Performing-A-Principal-Component-Analysis-In-R/. [(Accessed 9 January 2019)].

CHAPTER

9

Factor Analysis

LEARNING OBJECTIVES

After careful consideration of this chapter, you should be able:

- *To explain factor analysis (FA) as a data reduction.*
- *To know how to choose the number of factors.*
- *To understand how to employ factor analysis to environmental data.*
- *To understand how to interpret the extracted factor.*
- *To describe built-in functions and commands in R for factor analysis.*
- *To comprehend the results of R output for factor analysis and how to employed in the analysis.*
- *To draw smart conclusions and write useful reports.*

9.1 INTRODUCTION

Environmental studies generate huge data with many measured variables; thus, studying the relationship between various variables to identify the source of variation in the data is impossible with the raw data. Factor analysis (FA) is a multivariate technique and can be used to express huge data and measured variables by a few equations called factors to represent the entire data and to identify the responsibility of each group of variables associated with each factor in explaining the total variation in the data. Factor analysis produces a wonderful result if the original measured variables are highly correlated. The variables associated with each factor are highly correlated among themselves and have a very weak relationship with other factors.

9.2 FACTOR ANALYSIS IN R

R statistical software provides a variety of built-in functions to perform factor analysis and generate the required information. We should install and load `psych` and `GPArotation` packages from R library and then start performing factor analysis.

Step 1: The package `psych` in R library offers the built-in function `principal ()` to carry out factor analysis. The structure of this call is:

```
principal (data frame, nfactors = Number of factor, rotate = "varimax", covar = FALSE)
```

`covar = FALSE` indicates that the analysis is carried out employing a correlation matrix and `covar = TRUE` indicates that the analysis is carried out employing covariance matrix, R usually employ a correlation matrix as a default. There are a few options for rotation including `quatimax`, `varimax`, `promax`, `oblimin`, `simplimax`, `cluster`, and `none`.

Applied Statistics for Environmental Science with R
https://doi.org/10.1016/B978-0-12-818622-0.00009-5

Step 2: One more built-in function is offered by the psych package to carry out factor analysis; the function is factor.pa (). The structure of this call is:

```
factor.pa (data, nfactors = Number of factor, rotation = "varimax")
```

A raw data matrix or a covariance matrix can be used for this built-in function. "varimax" or "promax" can be employed as the option for rotations.

Step 3: The package stats in R library offers another built-in function factanal () to carry out factor analysis. The structure of this call is:

```
factanal (data frame, factors = Number of factor)
```

There are a few options for rotation including quatimax, varimax, promax, oblimin, simplimax, cluster, and none. Varimax is the default with this function regardless of whether rotation= is included or not.

```
factanal (data frame, factors = ?, rotation = "varimax")
```

Step 4: The loadings for the factors can be generated by employing the built-in function $loadings. The structure for this function is:

```
Name of the stored FACTOR $ loadings
```

Step 5: The scores for all of the observations can be generated by employing the built-in function $scores.

```
Name of the stored FACTOR $ scores
```

Step 6: Readers can explore more built-in functions in the packe nFactors to help one select the number of factors.

Step 7: More built-in functions for exploratory factor analysis are provided by the package FactoMineR, such as the use of both qualitative and quantitative variables.

9.3 GENERAL MODEL FOR FACTOR ANALYSIS

It is important to know the configuration of the data for any multivariate technique. Thus, the first step is to give the configuration of the data for factor analysis, the data for factor analysis should be arranged in the form as presented in Table 9.1.

TABLE 9.1 The General Configuration of Data for Factor Analysis

	Measured variable			
Sample No.	Y_1	Y_2	...	Y_k
1	Y_{11}	Y_{12}	...	Y_{1k}
2	Y_{21}	Y_{22}	...	Y_{2k}
⋮	⋮	⋮	...	⋮
n	Y_{n1}	Y_{n2}	...	Y_{nk}

Consider a random sample is selected and k variables are measured from each sample, the mean and covariance matrix are $\boldsymbol{\mu}$ and $\sum$, respectively. A linear combination of the hidden (unobservable) variables m $(m<k)$ can be employed to express the original variables; the hidden variables $(f_1, f_2, \ldots, f_m)$ are called common factors, and the error, which is a source for unknown source of variation $\varepsilon_1, \varepsilon_2, \ldots, \varepsilon_k$, is called the specific factors (error). Suppose there are k observable variables and m common factors, then the model that represents the factor analysis model is presented below.

$$
\begin{aligned}
Y_1 &= d_{11}f_1 + d_{12}f_2 + \cdots + d_{1m}f_m + \varepsilon_1 \\
Y_2 &= d_{21}f_1 + d_{22}f_2 + \cdots + d_{2m}f_m + \varepsilon_2 \\
&\vdots \\
Y_k &= d_{k1}f_1 + d_{k2}f_2 + \cdots + d_{km}f_m + \varepsilon_k
\end{aligned}
$$

The k linear combinations can be expressed by a general model:

$$Y_i = d_{i1}f_1 + d_{i2}f_2 + \cdots + d_{im}f_m + \varepsilon_i$$

Because factor analysis is a data reduction method, in order to meet this aim the number of extracted factors f_i from the original variables should be less than the original variables Y_i.

The contribution of each factor to the original variable can be measured directly by the coefficients associated with each factor d_{ij}, these coefficients are called loadings (how much the original variable depends on the extracted factors ($f's$)). The mean and variance of the extracted factors are zero and unit variance respectively, the extracted factors (common factors) $f_1, f_2, \ldots, f_m$ are uncorrelated. Moreover, the specific factors are uncorrelated with extracted factors (common factors) and have zero mean.

The variance of the original variable (Y_i) is given in (9.1)

$$\begin{aligned} Var(Y_i) &= d_{i1}^2 var(f_1) + d_{i2}^2 var(f_i) + \cdots + d_{im}^2 var(f_m) + var(\varepsilon_i) \\ Var(Y_i) &= d_{i1}{}^2 + d_{i2}{}^2 + \cdots + d_{im}{}^2 + var(\varepsilon_i) \end{aligned} \tag{9.1}$$

One can see that two components presented in Eq. (9.1) represent the variance of Y_i; the first component is due to the extracted factors called communality (common variance), as shown in (9.2).

$$d_{i1}^2 + d_{i2}^2 + \cdots + d_{im}^2 \tag{9.2}$$

The second component is called the specificity of Y_i (specific variance or residual) var.(ε_i). An idea on how much of the variance in each of the original variables is explained by the extracted factors can be gained and measured directly through examining the communalities.

Note:

- One can compute the correlation between Y_i and Y_j employing the formula given in (9.3).

$$r_{ij} = d_{i1}d_{j1} + d_{i2}d_{j2} + \cdots + d_{im}d_{jm} \tag{9.3}$$

9.4 COMMON STEPS FOR FACTOR ANALYSIS

It is very significant to understand the common steps for extracting the factors, rotation, loadings, and computing the scores (the value of the factor for all observations). The common steps for performing factor analysis are presented below.

Step 1: There are several techniques to estimate the loadings and communalities when performing factor analysis, such as maximum likelihood, principal component method, Alpha, principal factor, image factoring, and others. Scientists usually prefer the popular technique used for extracting the factors called the principal component method.

Step 2: Rotation will show clearly the difference in the coefficient value d_{ij} associated with each factor; highly correlated variables with a factor will have a value of d_{ij} that is far from zero, and weak correlation, or no correlation with a factor, will be reflected by a small value of d_{ij} (close to zero). The rotation can be either orthogonal or oblique, depending on the technique employed for rotation. Several techniques are available to rotate the factors such as oblique rotation, varimax or orthogonal rotation, and quartimax.

Step 3: The values of the factors for each of n samples should be computed; these values are called factor scores to be employed for further analysis, such as to draw a scatter plot or bar chart, or other purposes, such as to study the behavior of the samples and identify the positive and negative contribution with each factor.

The type of extraction, type of rotation, and the number of factors to be kept for the analysis are usually depend on the researcher's opinion, since all the options are provided by most of the statistical packages.

9.4.1 Principal Component Method

There are several methods to extract factors from a correlation matrix or covariance matrix between various variables. The most popular method is called the principal component method, which consists of six common steps to extract the factors as shown below.

Step 1: The sample variance–covariance matrix or correlation matrix for the k chosen variables should be computed first, and then the eigenvalues and the corresponding eigenvectors for each eigenvalue will be computed. The principal components can be expressed mathematically as a linear combination of the original variables. These factors are unrotated factors as presented below.

$$\begin{aligned}
Z_1 &= a_{11}Y_1 + a_{12}Y_2 + a_{13}Y_3 + \cdots + a_{1k}Y_k \\
Z_2 &= a_{21}Y_1 + a_{22}Y_2 + a_{23}Y_3 + \cdots + a_{2k}Y_k \\
&\vdots \\
Y_k &= a_{1k}Z_1 + a_{2k}Z_2 + a_{3k}Z_3 + \cdots + a_{kk}Z_k
\end{aligned}$$

Step 2: The inverse relationship between the original variables (Y) and unrotated factors (Z) should be found.

$$\begin{aligned}
Y_1 &= a_{11}Z_1 + a_{21}Z_2 + a_{31}Z_3 + \cdots + a_{k1}Z_k \\
Y_2 &= a_{12}Z_1 + a_{22}Z_2 + a_{32}Z_3 + \cdots + a_{k2}Z_k \\
&\vdots \\
Y_k &= a_{1k}Z_1 + a_{2k}Z_2 + a_{3k}Z_3 + \cdots + a_{kk}Z_k
\end{aligned}$$

Step 3: We can employ the same rules for principal components analysis to select the number of factors to be kept and used for interpretation of the total variance. We can keep the factors that significantly contribute in explaining the total variation. Generally, one can say that only m factors should be retained and presented in the mathematical model.

$$\begin{aligned}
Y_1 &= a_{11}Z_1 + a_{21}Z_2 + a_{31}Z_3 + \cdots + a_{m1}Z_m + e_1 \\
Y_1 &= a_{12}Z_1 + a_{22}Z_2 + a_{32}Z_3 + \cdots + a_{m2}Z_m + e_2 \\
&\vdots \\
Y_k &= a_{1k}Z_1 + a_{2k}Z_2 + a_{3k}Z_3 + \cdots + a_{mk}Z_m + e_k
\end{aligned}$$

Step 4: The principal components $Z_1, Z_2, \ldots, Z_m$ should be scaled by normalizing the eigenvectors to get the unit variance. We can achieve this point easily by dividing each principal component Z_i by its standard deviation (the square root of the eigenvalue λ_i); thus, we divide each component by $\sqrt{\lambda_i}$. Dividing each principal component by the $\sqrt{\lambda_i}$ will keep the factors unrotated.

$$\begin{aligned}
Y_1 &= g_{11}F_1 + g_{12}F_2 + g_{13}F_3 + \cdots + g_{1m}F_m + e_1 \\
Y_2 &= g_{21}F_1 + g_{22}F_2 + g_{23}F_3 + \cdots + g_{2m}F_m + e_1 \\
&\vdots \\
Y_k &= g_{k1}F_1 + g_{k2}F_2 + g_{k3}F_3 + \cdots + g_{km}F_m + e_k
\end{aligned}$$

where $F_i = \frac{Z_i}{\sqrt{\lambda_i}}$ and $g_{ij} = a_{ij}\sqrt{\lambda_i}$.

Step 5: One can employ any technique to rotate the factors and produces desirable results. The model after rotation will be of the form as given below.

$$\begin{aligned}
Y_1 &= d_{11}f_1 + d_{12}f_2 + d_{13}f_3 + \cdots + d_{1m}f_m + e_1 \\
Y_2 &= d_{21}f_1 + d_{22}f_2 + d_{23}f_3 + \cdots + d_{2m}f_m + e_1 \\
&\vdots \\
Y_k &= d_{k1}f_1 + d_{k2}f_2 + d_{k3}f_3 + \cdots + d_{km}f + e_k
\end{aligned}$$

Step 6: The formula in (9.4) can be employed to compute the values for each factor. The rotated factors can be represented by a linear combination of the Y values to calculate the scores.

$$f = \left(\acute{D}D\right)^{-1}\acute{D}Y \tag{9.4}$$

where f refers to the rotated factors, Y refers to the standardized Y values, and D refers to the matrix of the factor loadings after rotation.

EXAMPLE 9.1 ASSESSMENT OF WATER QUALITY FOR TWO ESTUARIES

A researcher wants to investigate the water quality of two estuaries in the Penang state of Malaysia by monitoring 18 parameters. The two locations of interest, with 10 sites in each location, were Kuala Juru (Juru estuary) and Bukit Tambun (Jejawi estuary). The parameters were water temperature (WT) and electrical conductivity (EC), pH, dissolved oxygen (DO), biochemical oxygen demand (BOD), chemical oxygen demand (COD), total nitrates (NO_3^{2-}), and total phosphate (PO_4^{3-}) concentrations, turbidity, and total suspended solids (TSS). The total metal concentration of the metals lead (Pb), zinc (Zn), chromium (Cr), mercury (Hg), copper (Cu), manganese (Mn), iron (Fe) and cadmium (Cd) were measured. The sample data measured from 20 different sites are presented in Table 9.2.

Two packages should be installed and loaded before performing factor analysis. The two packages are `psych` package and `GPArotation` package as shown below:

```
>install.packages ("psych")
>library (psych)
>install.packages ("GPArotation")
>library (GPArotation)
```

One can compute the entries of the correlation matrix of water quality parameters employing R built-in functions. The bivariate correlations between various variables and the *p*-value associated with each correlation to test the significance for the stored data. CSV (`Example9_1`) of the water quality parameters were computed and produced in a matrix form employing the built-in function `corr.test ()`.

```
> corr.test(Example9_1[,2:19])
Call:corr.test(x = Example9_1[, 2:19])
Correlation matrix
         WT    pH    DO   BOD   COD   TSS  Con.  Tur.
WT     1.00 -0.08 -0.15 -0.19 -0.46 -0.35 -0.65 -0.40
pH    -0.08  1.00  0.86  0.24  0.41 -0.07  0.71 -0.37
DO    -0.15  0.86  1.00  0.30  0.35 -0.03  0.63 -0.23
BOD   -0.19  0.24  0.30  1.00  0.32  0.18  0.37  0.24
COD   -0.46  0.41  0.35  0.32  1.00  0.14  0.54 -0.08
TSS   -0.35 -0.07 -0.03  0.18  0.14  1.00  0.22  0.74
Con.  -0.65  0.71  0.63  0.37  0.54  0.22  1.00 -0.03
Tur.  -0.40 -0.37 -0.23  0.24 -0.08  0.74 -0.03  1.00
PO43.  0.06  0.13  0.21  0.41  0.07  0.16  0.12  0.13
NO32. -0.09 -0.17 -0.05  0.05 -0.09  0.62 -0.06  0.53
Cu     0.58  0.07 -0.07 -0.13 -0.10 -0.23 -0.43 -0.34
Zn     0.62 -0.03 -0.14 -0.11 -0.06 -0.21 -0.53 -0.27
Cd     0.77  0.43  0.36 -0.15 -0.20 -0.43 -0.24 -0.61
Cr     0.36  0.12  0.07  0.35 -0.16 -0.01 -0.22 -0.09
Fe     0.80  0.41  0.36 -0.03 -0.13 -0.40 -0.28 -0.57
pb     0.41  0.58  0.42 -0.02  0.05 -0.26  0.07 -0.43
Hg    -0.34 -0.60 -0.68 -0.32 -0.08 -0.12 -0.14  0.11
Mn    -0.76 -0.46 -0.45 -0.03  0.16  0.42  0.23  0.52
       PO43. NO32.   Cu    Zn    Cd    Cr    Fe    pb
WT     0.06 -0.09  0.58  0.62  0.77  0.36  0.80  0.41
pH     0.13 -0.17  0.07 -0.03  0.43  0.12  0.41  0.58
DO     0.21 -0.05 -0.07 -0.14  0.36  0.07  0.36  0.42
BOD    0.41  0.05 -0.13 -0.11 -0.15  0.35 -0.03 -0.02
COD    0.07 -0.09 -0.10 -0.06 -0.20 -0.16 -0.13  0.05
TSS    0.16  0.62 -0.23 -0.21 -0.43 -0.01 -0.40 -0.26
Con.   0.12 -0.06 -0.43 -0.53 -0.24 -0.22 -0.28  0.07
Tur.   0.13  0.53 -0.34 -0.27 -0.61 -0.09 -0.57 -0.43
PO43.  1.00  0.51 -0.11 -0.03  0.10 -0.18  0.15  0.42
NO32.  0.51  1.00 -0.24 -0.23 -0.26 -0.19 -0.19 -0.12
Cu    -0.11 -0.24  1.00  0.92  0.58  0.65  0.67  0.44
```

TABLE 9.2 The Data for Water Quality Parameters for the Two Estuaries (mg/L)

River	WT	pH	DO	BOD	COD	TSS	Con.	Tur.	$PO4^{3-}$	$NO3^{2-}$	Cu	Zn	Cd	Cr	Fe	pb	Hg	Mn
Juru	28.15	7.88	6.73	10.56	1248.00	473.33	42.45	13.05	0.88	12.45	0.05	0.70	1.90	0.06	22.58	0.64	7.54	0.05
Juru	28.20	7.92	6.64	10.06	992.50	461.67	42.75	14.11	0.55	15.05	0.03	0.36	1.92	0.11	22.40	0.13	7.95	0.05
Juru	28.30	7.89	6.36	5.27	775.00	458.33	42.50	17.90	1.01	19.25	0.09	0.44	1.95	0.05	23.12	0.81	9.32	0.05
Juru	28.40	7.84	6.23	14.57	1124.00	473.34	42.35	22.00	1.27	12.45	0.05	0.58	2.02	0.08	23.10	0.72	9.24	0.05
Juru	28.30	7.86	6.29	7.36	1029.50	528.34	40.70	12.36	0.61	12.90	0.39	1.75	1.94	0.15	24.27	0.53	9.38	0.06
Juru	28.35	7.41	5.93	6.01	1265.00	393.34	29.47	25.95	0.88	15.80	0.23	2.13	1.95	0.08	25.87	0.41	9.37	0.06
Juru	28.55	7.40	5.67	4.81	551.00	356.67	29.51	27.70	1.02	25.55	0.07	0.39	1.91	0.08	18.51	0.50	9.42	0.03
Juru	28.75	7.41	5.26	4.96	606.00	603.33	28.45	29.35	0.84	17.90	0.28	2.10	2.00	0.11	29.16	0.48	10.88	0.07
Juru	29.30	7.25	4.19	6.61	730.00	430.00	26.25	35.35	0.73	12.85	0.32	2.72	2.00	0.13	25.72	0.50	13.10	0.08
Juru	29.55	7.18	5.14	5.11	417.00	445.00	25.79	27.75	0.72	17.15	0.09	0.56	1.93	0.06	23.73	0.14	12.40	0.05
Jejawi	27.65	7.27	5.71	16.37	653.00	736.67	37.95	99.95	1.01	21.10	0.03	0.45	1.37	0.12	1.03	0.06	11.23	0.61
Jejawi	27.70	7.53	5.86	6.31	947.50	858.33	37.20	122.9	0.77	21.80	0.01	0.18	1.39	0.04	1.67	0.21	11.43	0.60
Jejawi	27.85	7.45	5.52	7.36	1194.50	823.33	41.50	46.35	1.12	27.60	0.01	0.16	1.35	0.02	0.91	0.07	9.78	0.73
Jejawi	27.85	7.41	5.27	5.87	988.50	821.67	36.75	52.65	0.46	16.50	0.02	0.23	1.56	0.12	1.16	0.14	13.12	0.78
Jejawi	27.80	7.40	4.96	10.51	1074.00	815.00	36.65	82.35	1.02	29.60	0.02	0.20	1.29	0.05	1.50	0.23	12.97	0.74
Jejawi	27.85	7.36	5.16	11.26	911.00	373.33	34.85	36.30	0.73	11.50	0.10	0.21	1.25	0.12	1.30	0.05	14.62	0.71
Jejawi	27.85	7.37	5.08	3.75	632.00	385.00	37.05	33.50	0.71	13.40	0.01	0.24	1.53	0.03	0.98	0.08	15.46	0.72
Jejawi	27.75	7.40	5.23	3.31	787.00	403.33	37.10	36.35	0.58	10.90	0.01	0.17	1.50	0.02	0.92	0.18	16.02	0.71
Jejawi	27.70	7.38	5.30	3.15	879.50	550.00	38.30	30.00	0.75	13.90	0.01	0.19	1.51	0.03	1.54	0.29	19.08	0.77
Jejawi	27.65	7.36	5.17	8.56	1339.00	363.33	38.40	26.15	0.62	12.70	0.02	0.20	1.33	0.02	1.16	0.10	19.05	0.73

```
Zn     -0.03 -0.23  0.92  1.00  0.63  0.55  0.71  0.45
Cd      0.10 -0.26  0.58  0.63  1.00  0.39  0.96  0.75
Cr     -0.18 -0.19  0.65  0.55  0.39  1.00  0.44  0.17
Fe      0.15 -0.19  0.67  0.71  0.96  0.44  1.00  0.73
pb      0.42 -0.12  0.44  0.45  0.75  0.17  0.73  1.00
Hg     -0.42 -0.28 -0.27 -0.28 -0.59 -0.39 -0.64 -0.50
Mn     -0.21  0.14 -0.58 -0.60 -0.94 -0.42 -0.98 -0.73
          Hg    Mn
WT     -0.34 -0.76
pH     -0.60 -0.46
DO     -0.68 -0.45
BOD    -0.32 -0.03
COD    -0.08  0.16
TSS    -0.12  0.42
Con.   -0.14  0.23
Tur.    0.11  0.52
PO43.  -0.42 -0.21
NO32.  -0.28  0.14
Cu     -0.27 -0.58
Zn     -0.28 -0.60
Cd     -0.59 -0.94
Cr     -0.39 -0.42
Fe     -0.64 -0.98
pb     -0.50 -0.73
Hg      1.00  0.71
Mn      0.71  1.00
Sample Size
[1] 20
Probability values (Entries above the diagonal are adjusted for multiple tests.)
         WT   pH   DO  BOD  COD  TSS Con. Tur. PO43. NO32.
WT     0.00 1.00 1.00 1.00 1.00 1.00 0.26 1.00  1.00  1.00
pH     0.72 0.00 0.00 1.00 1.00 1.00 0.06 1.00  1.00  1.00
DO     0.53 0.00 0.00 1.00 1.00 1.00 0.38 1.00  1.00  1.00
BOD    0.42 0.31 0.19 0.00 1.00 1.00 1.00 1.00  1.00  1.00
COD    0.04 0.08 0.13 0.17 0.00 1.00 1.00 1.00  1.00  1.00
TSS    0.13 0.77 0.90 0.45 0.56 0.00 1.00 0.03  1.00  0.43
Con.   0.00 0.00 0.00 0.11 0.01 0.36 0.00 1.00  1.00  1.00
Tur.   0.08 0.10 0.32 0.31 0.73 0.00 0.91 0.00  1.00  1.00
PO43.  0.81 0.57 0.37 0.07 0.78 0.51 0.61 0.58  0.00  1.00
NO32.  0.69 0.47 0.84 0.83 0.71 0.00 0.81 0.02  0.02  0.00
Cu     0.01 0.79 0.76 0.59 0.68 0.33 0.06 0.15  0.63  0.31
Zn     0.00 0.89 0.56 0.64 0.81 0.37 0.02 0.24  0.89  0.33
Cd     0.00 0.06 0.12 0.54 0.40 0.06 0.31 0.00  0.68  0.28
Cr     0.12 0.61 0.76 0.13 0.49 0.98 0.35 0.69  0.46  0.43
Fe     0.00 0.07 0.12 0.89 0.60 0.08 0.22 0.01  0.52  0.41
pb     0.07 0.01 0.06 0.94 0.82 0.27 0.77 0.06  0.07  0.63
Hg     0.15 0.01 0.00 0.18 0.74 0.60 0.55 0.64  0.06  0.23
Mn     0.00 0.04 0.04 0.90 0.51 0.07 0.33 0.02  0.37  0.57
         Cu   Zn   Cd   Cr   Fe   pb   Hg   Mn
WT     0.89 0.44 0.01 1.00 0.00 1.00 1.00 0.01
pH     1.00 1.00 1.00 1.00 1.00 0.88 0.67 1.00
DO     1.00 1.00 1.00 1.00 1.00 1.00 0.13 1.00
BOD    1.00 1.00 1.00 1.00 1.00 1.00 1.00 1.00
```

```
COD   1.00 1.00 1.00 1.00 1.00 1.00 1.00 1.00
TSS   1.00 1.00 1.00 1.00 1.00 1.00 1.00 1.00
Con.  1.00 1.00 1.00 1.00 1.00 1.00 1.00 1.00
Tur.  1.00 1.00 0.55 1.00 1.00 1.00 1.00 1.00
PO43. 1.00 1.00 1.00 1.00 1.00 1.00 1.00 1.00
NO32. 1.00 1.00 1.00 1.00 1.00 1.00 1.00 1.00
Cu    0.00 0.00 0.88 0.28 0.15 1.00 1.00 0.89
Zn    0.00 0.00 0.38 1.00 0.07 1.00 1.00 0.68
Cd    0.01 0.00 0.00 1.00 0.00 0.02 0.84 0.00
Cr    0.00 0.01 0.09 0.00 1.00 1.00 1.00 1.00
Fe    0.00 0.00 0.00 0.05 0.00 0.04 0.35 0.00
pb    0.05 0.05 0.00 0.46 0.00 0.00 1.00 0.04
Hg    0.24 0.23 0.01 0.09 0.00 0.02 0.00 0.07
Mn    0.01 0.01 0.00 0.07 0.00 0.00 0.00 0.00
To see confidence intervals of the correlations, print with the short=FALSE option
```

One can see that most of the water quality variables (parameters) included in the correlation matrix had a significant bivariate relationship with the other variables (small p-value), as presented clearly in the results of probability values (`Probability values (Entries above the diagonal are adjusted for multiple tests)`). Factor analysis was carried out on the data set of the water quality variables measured from the two estuaries (18 variables) to recognize the origin of the differences in the data values and to recognize the variables responsible for the variation in the data values of the water quality of the two estuaries.

The factors were extracted from the correlation matrix employing principal component method. Furthermore, the `Standard deviation` (square roots of the eigenvalues), the `Proportion of variance` (percentage variance accounted for), and the `Cumulative proportion` (cumulative percentage variance) for the unrotated factors were produced employing the two built-in functions `princomp ()` and `summary ()`. The results generated by the two functions were stored in a new place called `variance`.

```
> variance <- princomp (Example9_1[,2:19],cor = TRUE)
> summary (variance)
Importance of components:
                          Comp.1     Comp.2     Comp.3
Standard deviation     2.5993147 1.9527172 1.5441306
Proportion of Variance 0.3753576 0.2118391 0.1324633
Cumulative Proportion  0.3753576 0.5871967 0.7196600
                           Comp.4     Comp.5     Comp.6
Standard deviation     1.24970561 0.96743964 0.96425666
Proportion of Variance 0.08676467 0.05199664 0.05165505
Cumulative Proportion  0.80642471 0.85842135 0.91007640
                          Comp.7    Comp.8     Comp.9
Standard deviation     0.7062347 0.5756577 0.53204500
Proportion of Variance 0.0277093 0.0184101 0.01572622
Cumulative Proportion  0.9377857 0.9561958 0.97192201
                           Comp.10     Comp.11     Comp.12
Standard deviation     0.400480618 0.362574331 0.292227820
Proportion of Variance 0.008910263 0.007303341 0.004744283
Cumulative Proportion  0.980832275 0.988135616 0.992879900
```

```
                         Comp.13     Comp.14     Comp.15
Standard deviation     0.262894673 0.163797642 0.148563082
Proportion of Variance 0.003839645 0.001490537 0.001226166
Cumulative Proportion  0.996719545 0.998210082 0.999436248
                          Comp.16       Comp.17
Standard deviation     0.0884536770 4.021268e-02
Proportion of Variance 0.0004346696 8.983666e-05
Cumulative Proportion  0.9998709175 9.999608e-01
                          Comp.18
Standard deviation     2.657867e-02
Proportion of Variance 3.924587e-05
Cumulative Proportion  1.000000e+00
```

One can see that only four factors can contribute significantly in explaining the total variance in the water quality of the two estuaries' data values, because only four eigenvalues are greater than one: `2.5993147`, `1.9527172`, `1.5441306`, and `1.24970561` as presented in the results of the built-in function `princomp ()`, this criterion is employed because the correlation matrix was used to extract the factors. The first four extracted factors explained 80.64% of the total variance in the data set of water quality, and therefore, the eighteen variables are represented very well by the first four factors. The value produced by dividing the squared value of the standard deviation of a factor by the number of variables involved in the analysis represents the amount of variation of the total variance captured by the factor. For instance, the number of variables (here, 18), the proportion of the variance captured divided by the first factor is $[((2.5993147)^2 / 18) = 0.3753]$. The computed eigenvalues for water quality parameters data can be represented pictorially by producing the scree plot employing the built-in function `screeplot ()`, as given in Fig. 9.1.

The built-in function `$loadings` was employed to generate the loadings for unrotated extracted factors. The loadings for all extracted factors were rounded to two decimals.

```
> round (unclass (variance $ loadings),digits=2)
      Comp.1 Comp.2 Comp.3 Comp.4 Comp.5 Comp.6 Comp.7
WT      0.30   0.22   0.13   0.15   0.03   0.14   0.20
pH      0.16  -0.42  -0.16  -0.03   0.19  -0.11  -0.16
DO      0.13  -0.43  -0.08   0.02   0.27   0.04   0.11
BOD    -0.02  -0.27   0.19  -0.36  -0.35   0.57   0.05
COD    -0.05  -0.29  -0.16  -0.25  -0.44  -0.41   0.51
```

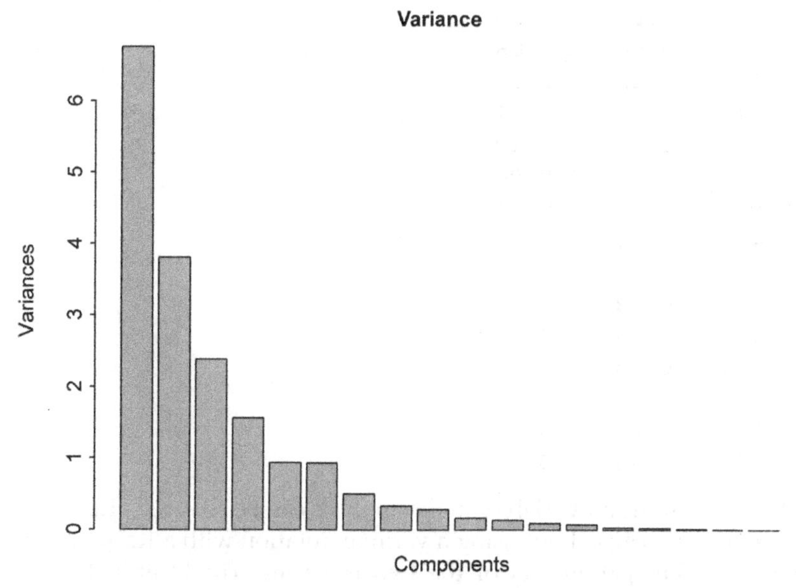

FIG. 9.1 Scree plot showing the eigenvalues extracted from water quality data for the two estuaries.

```
TSS    -0.18  -0.12   0.41  -0.18   0.24  -0.36  -0.08
Con.   -0.10  -0.44  -0.20  -0.07   0.07  -0.04  -0.24
Tur.   -0.24   0.00   0.41  -0.14   0.11  -0.04  -0.22
PO43.   0.04  -0.22   0.36   0.32  -0.53   0.14  -0.18
NO32.  -0.11  -0.08   0.51   0.25   0.08  -0.19   0.24
Cu      0.28   0.17   0.06  -0.32  -0.11  -0.29  -0.14
Zn      0.28   0.19   0.10  -0.27  -0.23  -0.30  -0.02
Cd      0.37  -0.01  -0.02   0.13   0.11  -0.01  -0.01
Cr      0.20   0.05   0.15  -0.56   0.19   0.24  -0.14
Fe      0.38  -0.01   0.04   0.06   0.01  -0.01   0.13
pb      0.29  -0.16   0.02   0.18  -0.21  -0.21  -0.55
Hg     -0.23   0.28  -0.28   0.01  -0.22  -0.03  -0.28
Mn     -0.37   0.06  -0.06  -0.11  -0.07  -0.10  -0.13
       Comp.8 Comp.9 Comp.10 Comp.11 Comp.12 Comp.13 Comp.14
WT       0.17   0.48    0.12    0.25    0.11    0.37    0.04
pH      -0.04   0.12    0.02    0.35    0.24   -0.07    0.39
DO       0.11  -0.42    0.05   -0.16   -0.50    0.27   -0.07
BOD      0.13   0.15    0.03    0.21   -0.15   -0.14   -0.30
COD      0.14   0.09   -0.31   -0.02    0.06    0.11    0.18
TSS      0.16   0.49    0.14   -0.29   -0.26    0.14   -0.10
Con.    -0.16   0.24    0.38    0.15    0.04   -0.21   -0.07
Tur.     0.58  -0.32   -0.06    0.28    0.16   -0.01    0.27
PO43.   -0.16  -0.07    0.23   -0.29   -0.01    0.17    0.41
NO32.   -0.46  -0.03   -0.24    0.35   -0.17   -0.30   -0.06
Cu      -0.29  -0.20    0.30    0.32   -0.13    0.39    0.01
Zn       0.10  -0.17    0.31   -0.12    0.05   -0.48   -0.14
Cd       0.17   0.16   -0.05   -0.28   -0.11   -0.38    0.29
Cr      -0.35   0.06   -0.39   -0.22    0.02    0.03    0.29
Fe       0.13   0.07    0.04    0.10   -0.25   -0.12    0.02
pb       0.09   0.06   -0.50    0.01    0.07    0.12   -0.42
Hg       0.06   0.17   -0.12    0.24   -0.65   -0.10    0.30
Mn      -0.15   0.09    0.06   -0.22    0.13    0.06   -0.03
      Comp.15 Comp.16 Comp.17 Comp.18
WT       0.27    0.29    0.13    0.31
pH      -0.44   -0.06   -0.15    0.36
DO       0.12    0.18    0.10    0.33
BOD     -0.08   -0.26   -0.01    0.15
COD      0.15    0.02    0.04   -0.08
TSS     -0.19   -0.07   -0.22   -0.05
Con.     0.46    0.24    0.16   -0.31
Tur.     0.17    0.03    0.18   -0.10
PO43.   -0.05    0.10   -0.04   -0.04
NO32.    0.12    0.01    0.09    0.14
Cu       0.11   -0.41    0.04   -0.09
Zn      -0.02    0.37   -0.09    0.34
Cd       0.36   -0.55    0.12    0.09
Cr       0.09    0.28    0.04   -0.06
Fe      -0.48    0.14    0.53   -0.45
pb       0.06    0.03    0.05    0.04
Hg       0.01    0.14   -0.07    0.12
Mn      -0.11   -0.12    0.72    0.40
```

The contributions of the less significant variables in the analysis can be reduced after rotation process and result in simpler factors. Factor analysis was carried out using a varimax rotation with a Kaiser normalization to generate the factor model for the water quality parameters of the two estuaries. The built-in function `principal ()` was employed to produce information regarding the loadings for the rotated factors with the communalities and specific variance. Moreover, the variance and cumulative variance captured by each factor after rotation were produced as a result of employing the built-in function `principal ()`. The results of employing the function `principal ()` were stored in a new place called `Hidden`.

```
> Hidden <- principal(Example9_1[,2:19], nfactors = 4, rotate="varimax")
> Hidden
Principal Components Analysis
Call: principal(r = Example9_1[, 2:19], nfactors = 4, rotate = "varimax")
Standardized loadings (pattern matrix) based upon correlation matrix
       RC1    RC2    RC3    RC4   h2    u2  com
WT     0.68  -0.54  -0.09   0.32  0.86 0.137 2.4
pH     0.51   0.80  -0.10  -0.04  0.91 0.092 1.8
DO     0.49   0.77   0.05  -0.10  0.84 0.159 1.7
BOD   -0.08   0.56   0.39   0.31  0.56 0.437 2.5
COD   -0.15   0.69  -0.07   0.01  0.51 0.494 1.1
TSS   -0.38   0.15   0.74   0.09  0.72 0.277 1.6
Con.  -0.11   0.89   0.01  -0.32  0.91 0.093 1.3
Tur.  -0.55  -0.07   0.70   0.04  0.81 0.195 1.9
PO43.  0.42   0.10   0.65  -0.25  0.66 0.336 2.1
NO32.  0.00  -0.16   0.87  -0.23  0.83 0.171 1.2
Cu     0.42  -0.21  -0.20   0.75  0.82 0.178 1.9
Zn     0.44  -0.29  -0.15   0.72  0.82 0.183 2.1
Cd     0.90  -0.08  -0.22   0.26  0.94 0.058 1.3
Cr     0.16   0.07   0.01   0.89  0.83 0.173 1.1
Fe     0.90  -0.06  -0.15   0.37  0.97 0.028 1.4
pb     0.83   0.17  -0.04   0.06  0.71 0.286 1.1
Hg    -0.70  -0.36  -0.42  -0.25  0.85 0.145 2.5
Mn    -0.93   0.01   0.08  -0.30  0.96 0.041 1.2

                       RC1  RC2  RC3  RC4
SS loadings           5.68 3.46 2.71 2.67
Proportion Var        0.32 0.19 0.15 0.15
Cumulative Var        0.32 0.51 0.66 0.81
Proportion Explained  0.39 0.24 0.19 0.18
Cumulative Proportion 0.39 0.63 0.82 1.00

Mean item complexity = 1.7
Test of the hypothesis that 4 components are sufficient.

The root mean square of the residuals (RMSR) is 0.06
with the empirical chi square 25.57 with prob < 1

Fit based upon off diagonal values = 0.98
```

It can be seen clearly that the rotated factors (shown in the results as `RC1 RC2 RC3 RC4`) are considerably more sensible than unrotated factors, as each variable contributes highly to one factor even though it appears with others that have a low contribution (high or low contribution depends on the coefficient (loading) associated with each factor), which assists the interpretation of the results and the identification of the sources of differences in the water quality of the two estuaries. The value produced by the sums of the squares of the factor loadings represents the communalities, for instance, the first communality that, corresponding to the water temperature (`WT`) (`h2`), is computed as $[(0.68)^2+(-0.54)^2+(-0.09)^2+0.32^2=0.8645=0.86]$. One can see that most of the variance of the variables is captured by four extracted factors, as indicated by the high values of the communalities for most of the variables. The specific variance is represented by `u2` in the R outputs and computed employing the formula `u2` = 1 – communality. The correlation between the specific variance and the factors can be measured by the magnitude of the factor loadings (large value indicates high relationship).

The chi-square test was employed to help in deciding the number of factors to be retained in the model. The results of hypothesis testing exhibited that the four factors are enough and sufficient to describe the relation and detect the hidden information.

The value of 0.6 is considered as a cut point for the parameter loading to decide whether the variables have a strong correlation with the factor or not. The relationship between various water quality parameters of the two estuaries and the four extracted factors can be represented in a mathematical equation to show a clear picture of the relationship. The four rotated extracted factors are placed in a mathematical form as in Eqs. (9.5–9.8).

$$F_1 = 0.68\,WT + 0.51\,pH + 0.49\,DO - 0.08\,BOD - 0.15\,COD - 0.38\,TSS - 0.11\,Con. - 0.55\,Tur. + 0.42\,PO4^3 + 0.00\,NO3^2 + 0.42\,Cu + 0.44\,Zn + 0.90\,Cd + 0.16\,Cr + 0.90\,Fe + 0.83\,Pb - 0.70\,Hg - 0.93\,Mn \tag{9.5}$$

$$F_2 = -0.54\,WT + 0.80\,pH + 0.77\,DO + 0.56\,BOD + 0.69\,COD + 0.15\,TSS - 0.89\,Con. - 0.07\,Tur. + 0.10\,PO4^3 - 0.16\,NO3^2 - 0.21\,Cu - 0.29\,Zn - 0.08\,Cd + 0.07\,Cr - 0.06\,Fe + 0.17\,Pb - 0.36\,Hg + 0.01\,Mn \tag{9.6}$$

$$F_3 = -0.09\,WT - 0.10\,pH + 0.05\,DO + 0.39\,BOD - 0.07\,COD + 0.74\,TSS + 0.01\,Con. + 0.70\,Tur. + 0.65\,PO4^3 + 0.87\,NO3^2 - 0.20\,Cu - 0.15\,Zn + 0.22\,Cd + 0.01\,Cr - 0.15\,Fe - 0.04\,Pb - 0.42\,Hg + 0.08\,Mn \tag{9.7}$$

$$F_4 = 0.32\,WT - 0.04\,pH - 0.10\,DO + 0.31\,BOD + 0.01\,COD + 0.09\,TSS - 0.32\,Con. + 0.04\,Tur. - 0.25\,PO4^3 - 0.23\,NO3^2 + 0.75\,Cu + 0.72\,Zn + 0.26\,Cd + 0.89\,Cr + 0.37\,Fe + 0.06\,Pb - 0.25\,Hg - 0.30\,Mn \tag{9.8}$$

One can observe that the results of factor analysis are ready for explanation and draw smart conclusions.

The extracted factors can be labeled to represent a dimension which is different from other factors, the label is usually chosen based on the high loadings associated with each factor. Only 81% of the total variance is captured by only four factors to represent the differences in the data for 18 parameters. 32% of the total variance was captured by F_1 and was negatively correlated with Mn and Hg, but it was positively correlated with Cd, Fe, Pb, and temperature, as presented in the R outputs for factor analysis for water quality parameters. Major industrial pollution is the label of F_1 and includes large-scale manufacturers and industries that discharge pollutants in large quantities and high concentrations directly into the river water. The pollutants produced here are usually rich in toxic chemicals and heavy metals, as well as high concentrations of organic waste.

The second factor captured a lower amount of variance compared to F_1; only 19% of the total variance was captured by F_2 and was positively correlated with conductivity (con.), pH, DO, and COD. This factor represents domestic sewage pollution. The third factor captured only 15% of the total variance and was positively correlated with nitrates, TSS, turbidity, and phosphate. This factor can represent the surface runoff and other agriculture activities, which includes crop farming and animal husbandry. The last factor F_4 captured 15% of the total variance and, because of the positive relationship between this factor and the concentration of Cr, Cu, and An, this factor can represent an industrial pollution source that is related with electroplating, electronic, and other metal processing industries, it is similar to the first one, related to industrial pollution

The values of the extracted factors for each of the n observations for water quality data can be generated employing R built-in functions and commands. The factor scores for the water quality data values can be produced easily by employing the built-in function `$scores`.

```
> round (Hidden $ scores,digits=2)
        RC1   RC2   RC3   RC4
 [1,]  1.09  1.60 -0.37 -0.32
 [2,]  0.45  1.47 -0.62  0.32
 [3,]  1.65  0.51  0.07 -1.25
 [4,]  1.31  1.33  0.31 -0.24
 [5,]  0.52  1.12 -0.79  1.88
 [6,]  0.76 -0.18 -0.28  0.78
 [7,]  1.31 -1.18  0.69 -0.94
 [8,]  0.85 -1.17  0.23  1.08
 [9,]  0.40 -1.68 -0.46  1.92
[10,]  0.76 -2.05 -0.12 -0.37
[11,] -1.09  0.37  1.82  1.01
[12,] -0.78  0.21  1.41 -0.24
```

```
[13,] -0.36  0.30  1.52 -0.98
[14,] -1.30  0.23 -0.05  0.74
[15,] -0.78 -0.11  1.87 -0.28
[16,] -1.31  0.22 -0.68  0.78
[17,] -0.65 -0.65 -0.93 -1.08
[18,] -0.78 -0.36 -1.29 -1.05
[19,] -0.75 -0.32 -0.92 -1.13
[20,] -1.28  0.34 -1.41 -0.61
```

The values of the extracted factors for each of the n observations were plotted with the samples from various estuaries to investigate the behavior of the chosen variables measured from the two estuaries. Shapely and high-resolution graphs (histogram, scatter diagram, bar chart and other types of plots) can be produced by employing R commands and built-in functions to illustrate various relationships. The values of the various factors were used to produce a bar chart employing the built-in function `barplot ()` to understand the behavior of the chosen variables for the two estuaries. The relationship between the values of the factor and the sampling points of different estuaries was investigated. Fig. 9.2 represents a bar chart for the values of the first factor for the two estuaries.

One can see clearly that Jejawi River showed a negative contribution to the industrial pollution factor, while Juru River showed a positive contribution to it. High values of Cd, Fe, water temperature, and Pb are considered as the main source for the positive contribution, whereas the high values of Mn and Hg are considered as the main source for the negative contribution. This conclusion indicated that the two estuaries were different in terms of the chosen variables. Thus, the two estuaries are different in terms of industrial pollution. The values of the second factor (domestic sewage pollution) for all observations are presented in a plot as shown in Fig. 9.3.

Whereas the positive contribution was mainly attributed to the conductivity, pH, DO, COD, and BOD, the negative contribution was mainly due to the water temperature. The parameters correlated with this factor showed distinct behavior from site to site regardless of the region. It may be accounted for the different types of dynamics of the river water at different equilibrium conditions, which results in irregular mixing and equilibrium for the heavy metal distribution in aqueous and solid phase matrices of the estuaries.

Factor 3 (F_3) and factor 4 (F_4) showed different behavior compared to the first two extracted factors; the graphs for the values for all n observations for the two factors are presented in Figs. 9.4 and 9.5.

A scatter plot was produced (Fig. 9.6) employing the values for the first two extracted factors for all observations. One can see that two distinct sets can be recognized in the plot. The graph revealed that separating the two estuaries was due to the first factor, while the fluctuations in the concentrations of the chosen variables was due to the second factor regardless of the estuary (region). This finding indicates that the two estuaries are distinct in terms of the chosen variables.

In summary, it can be said that factor analysis provided sufficient information of the water quality parameters of the surface water of the two estuaries.

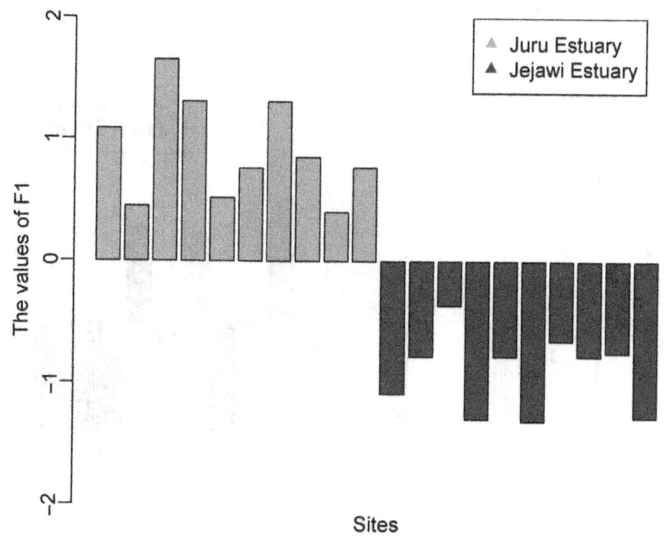

FIG. 9.2 Showing the values of F_1 by the two estuaries.

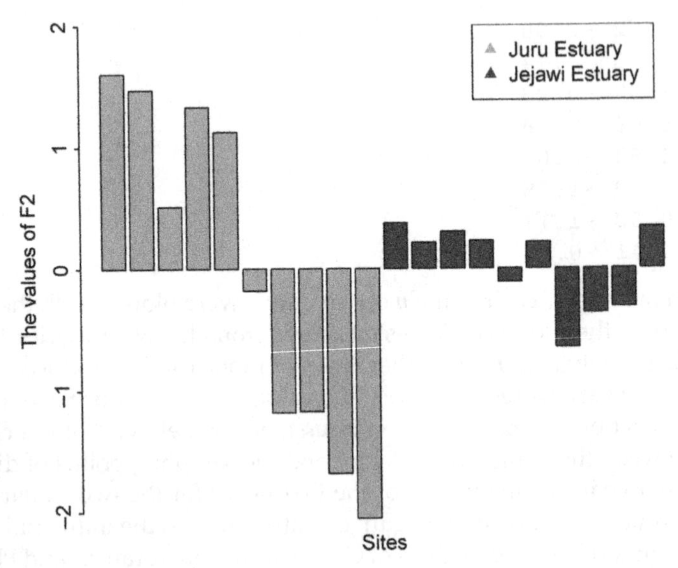

FIG. 9.3 Showing the values of F_2 by the two estuaries.

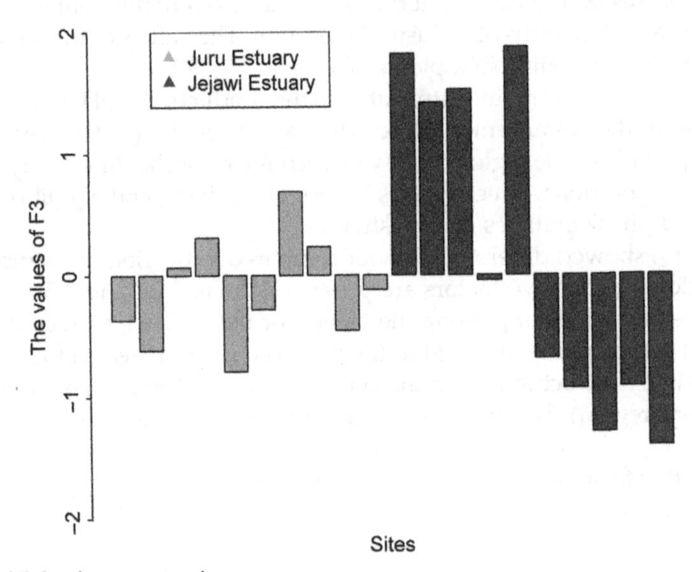

FIG. 9.4 Showing the values of F_3 by the two estuaries.

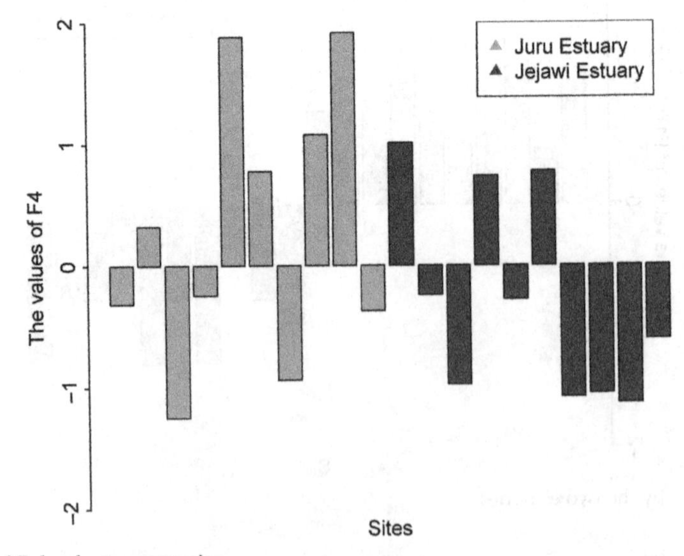

FIG. 9.5 Showing the values of F_4 by the two estuaries.

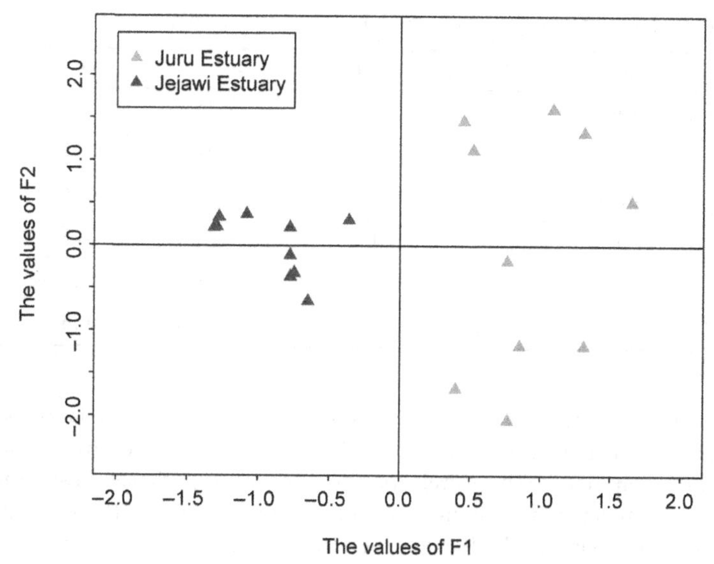

FIG. 9.6 Showing the values of $F_1$and F_2 by the two estuaries.

EXAMPLE 9.2 HEAVY METAL CONCENTRATIONS IN THE SEDIMENT

A researcher wants to assess the concentration of eight heavy metal concentrations in the sediment. The heavy metals are chromium (Cr), cadmium (Cd), zinc (Zn), copper (Cu), lead (Pb), manganese (Mn), iron (Fe), and mercury (Hg). The two locations, with 10 sampling points at each location, were Kuala Juru (the Juru River) and Bukit Tambun (the Jejawi River) in the Penang State of Malaysia. The samples were analyzed for the concentration of heavy metals. A Flame atomic absorption spectrometer (FAAS; Perkin Elmer HGA-600) was employed for the analysis of Cu, Zn, Cd, Fe, Pb, Cr, and Mn; and a cold vapor atomic absorption spectrometer (CV-AAS) method was employed for Hg analysis after sample digestion in acid solution. Blank acid mixtures were digested in the same way. Heavy metals concentrations were measured (mg/L) at 10 different sites. The sample data obtained from 10 different sampling points at each location are given in Table 9.3.

TABLE 9.3 The Data for Heavy Metal Content in Sediment (mg/L)

River	Cu	Zn	Cd	Cr	Fe	Pb	Hg	Mn
Juru	0.63	1.89	1.95	0.15	26.17	0.48	0.1	0.01
Juru	0.73	1.79	1.99	0.17	22.95	0.34	0.12	0.02
Juru	0.35	0.9	1.94	0.09	25.41	0.38	0.12	0.03
Juru	0.76	2.19	1.98	0.19	24.13	0.17	0.13	0.01
Juru	0.6	1.74	1.94	0.14	23.16	0.36	0.1	0.02
Juru	0.36	3.38	1.95	0.14	27.2	0.17	0.14	0.01
Juru	0.63	3.12	1.98	0.12	26.84	0.61	0.12	0.03
Juru	0.52	3.41	1.93	0.14	25.98	0.32	0.12	0.02
Juru	0.55	2.86	1.97	0.19	25.6	0.3	0.12	0.02
Juru	0.47	3.71	1.92	0.13	26.26	0.8	0.11	0.03
Jejawi	0.38	3.27	1.6	0.13	38.72	0.16	0.17	1.63
Jejawi	0.43	2.47	1.69	0.12	38.66	0.18	0.23	1.68
Jejawi	0.39	1.59	1.62	0.17	38.94	0.15	0.15	1.75
Jejawi	0.3	2.12	1.64	0.2	37.74	0.18	0.15	1.75

Continued

TABLE 9.3 The Data for Heavy Metal Content in Sediment (mg/L)—cont'd

River	Cu	Zn	Cd	Cr	Fe	Pb	Hg	Mn
Jejawi	0.49	2.28	1.66	0.28	39.52	0.28	0.18	1.9
Jejawi	0.45	2.24	1.68	0.11	38.49	0.28	0.24	1.67
Jejawi	0.48	1.76	1.76	0.16	38.22	0.15	0.17	1.7
Jejawi	0.46	2.5	1.75	0.23	39.01	0.42	0.22	1.77
Jejawi	0.23	1.8	1.71	0.21	39.61	0.41	0.24	1.85
Jejawi	0.2	1.44	1.73	0.17	38.8	0.15	0.18	1.7

The built-in function `corr.test()` in R was employed to compute the bivariate correlation between various heavy metal concentrations in the sediment. The bivariate correlations for various heavy metal concentrations in the sediments were placed in a matrix form called a correlation matrix. The correlations between various chosen heavy metal concentrations and the *p*-value associated with each correlation to test the significance for the stored data. CSV (`Example9_1`) of various heavy metals in the sediment were computed employing the built-in function `corr.test()`.

```
> corr.test(Example9_2[,2:9])
Call:corr.test(x = Example9_2[, 2:9])
Correlation matrix
      Cu    Zn    Cd    Cr    Fe    Pb    Hg    Mn
Cu  1.00  0.15  0.63 -0.02 -0.66  0.25 -0.51 -0.61
Zn  0.15  1.00  0.16 -0.10 -0.16  0.31 -0.12 -0.25
Cd  0.63  0.16  1.00 -0.29 -0.95  0.46 -0.73 -0.96
Cr -0.02 -0.10 -0.29  1.00  0.35 -0.16  0.24  0.41
Fe -0.66 -0.16 -0.95  0.35  1.00 -0.42  0.84  0.99
Pb  0.25  0.31  0.46 -0.16 -0.42  1.00 -0.31 -0.45
Hg -0.51 -0.12 -0.73  0.24  0.84 -0.31  1.00  0.83
Mn -0.61 -0.25 -0.96  0.41  0.99 -0.45  0.83  1.00
Sample Size
[1] 20
Probability values (Entries above the diagonal are adjusted for multiple tests.)
     Cu   Zn   Cd   Cr   Fe   Pb   Hg   Mn
Cu 0.00 1.00 0.06 1.00 0.04 1.00 0.38 0.09
Zn 0.54 0.00 1.00 1.00 1.00 1.00 1.00 1.00
Cd 0.00 0.51 0.00 1.00 0.00 0.76 0.01 0.00
Cr 0.93 0.67 0.22 0.00 1.00 1.00 1.00 1.00
Fe 0.00 0.50 0.00 0.13 0.00 1.00 0.00 0.00
Pb 0.28 0.19 0.04 0.49 0.07 0.00 1.00 0.81
Hg 0.02 0.63 0.00 0.30 0.00 0.18 0.00 0.00
Mn 0.00 0.29 0.00 0.08 0.00 0.05 0.00 0.00
To see confidence intervals of the correlations, print with the short=FALSE option
```

One can observe that the bivariate correlations between all of the possible combinations of two heavy metals in the sediment were computed and presented in a matrix form (placed in columns 2 to 9). The correlation matrix for the heavy metal concentrations in sediment was tested employing the *p*-values, which were provided under the `Probability values (Entries above the diagonal are adjusted for multiple tests.)`. A strong positive correlation was exhibited between Cu and Cd and a strong negative correlation with Fe, Hg, and Mn. Furthermore, Cd showed a strong negative correlation with Fe, Hg, and Mn. Zn did not show a significant correlation with the other parameters. The chromium (Cr) was positively correlated with Mn (p-value <0.08) while Fe (p-value <0.07) was negatively correlated with Pb, and positively correlated with Hg and Mn. Mn and Hg were positively correlated. It can be said that the concentration of correlated parameters depends on each other and share the origin source.

A factor analysis was performed on the gathered data values for eight heavy metals (8 variables). The factors were extracted by the principal components method employing a correlation matrix of the chosen heavy metals.

The `standard deviation`, the `proportion of variance`, and the `cumulative proportion` for unrotated factors were produced by employing the two built-in functions `princomp ()` and `summary ()`.

```
> variance <- princomp(Example9_2[,2:9],cor=TRUE) # principal components method
> summary (variance)
Importance of components:
                          Comp.1    Comp.2    Comp.3
Standard deviation     2.1249455 1.0488704 0.9901061
Proportion of Variance 0.5644242 0.1375161 0.1225388
Cumulative Proportion  0.5644242 0.7019403 0.8244791
                           Comp.4     Comp.5     Comp.6
Standard deviation     0.81301797 0.67097945 0.50986185
Proportion of Variance 0.08262478 0.05627668 0.03249489
Cumulative Proportion  0.90710383 0.96338051 0.99587540
                            Comp.7      Comp.8
Standard deviation     0.170069744 0.063820873
Proportion of Variance 0.003615465 0.000509138
Cumulative Proportion  0.999490862 1.000000000
```

Only two factors with eigenvalues of more than one were noticed in the factor analysis output ($(2.1249455)^2$ and $(1.0488704)^2$), furthermore, >70% of the differences in the heavy metals data values is captured by only two extracted factors. The significance of the factor can be measured by its eigenvalue; the most significant factor is the one with the highest eigenvalue and captures the highest amount of variations in the data values. The built-in function `screeplot ()` was employed to produce a Scree plot for the extracted factors and presented in Fig. 9.7.

The built-in function `$ loadings` was employed to extract the loadings for the unrotated factors from `variance` (`variance` is the location where the extracted data were stored).

```
> round (unclass (variance $ loadings),digits=2)
  Comp.1 Comp.2 Comp.3 Comp.4 Comp.5 Comp.6 Comp.7 Comp.8
Cu -0.33   0.21  -0.41  -0.21   0.74  -0.29   0.03   0.04
Zn -0.13  -0.77  -0.25  -0.55  -0.12   0.02   0.04  -0.07
Cd -0.44   0.10  -0.02   0.07  -0.02   0.55   0.68  -0.17
Cr  0.18   0.23  -0.86   0.11  -0.36   0.16  -0.09   0.05
Fe  0.46  -0.13  -0.04   0.03   0.09  -0.19   0.62   0.58
Pb -0.25  -0.49  -0.15   0.79   0.11  -0.17  -0.03   0.00
Hg  0.40  -0.19  -0.02   0.07   0.52   0.68  -0.24   0.05
Mn  0.46  -0.03  -0.08   0.05   0.13  -0.23   0.29  -0.79
```

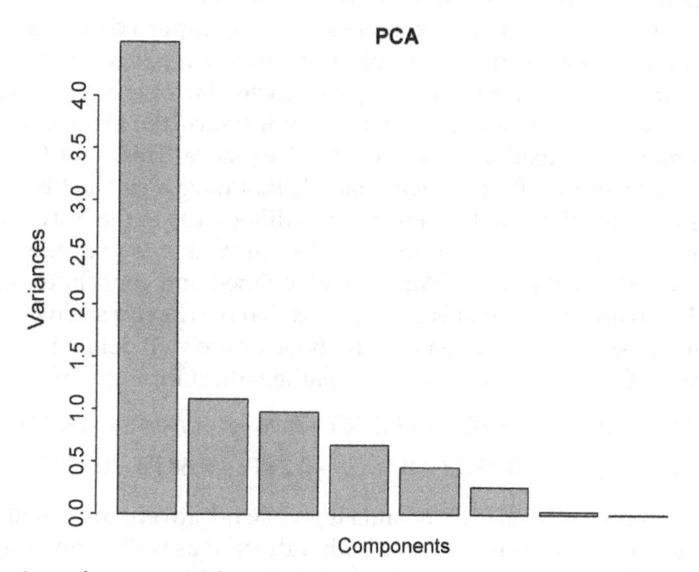

FIG. 9.7 Scree plot showing the eigenvalues extracted from the heavy metal concentrations in the sediments.

One can employ the built-in function `principal ()` to carry out factor analysis, a varimax rotation with a Kaiser normalization was employed to generate the factor model. The loadings, the communalities, and the specific variance for the extracted factors were computed employing R built-in functions and commands. Furthermore, the variance and cumulative variance captured by each factor after rotation were computed as well.

```
> Hidden <- principal(Example9_2[,2:9], nfactors=2, rotate ="varimax", covar = FALSE)
> print (Hidden, digits=2)
Principal Components Analysis
Call: principal(r = Example9_2[, 2:9], nfactors = 2, rotate = "varimax",
    covar = FALSE)
Standardized loadings (pattern matrix) based upon correlation matrix
     RC1   RC2   h2    u2 com
Cu -0.73  0.06 0.54 0.463 1.0
Zn  0.05  0.85 0.73 0.267 1.0
Cd -0.92  0.26 0.90 0.095 1.2
Cr  0.27 -0.36 0.20 0.796 1.8
Fe  0.96 -0.24 0.97 0.028 1.1
Pb -0.30  0.68 0.55 0.453 1.4
Hg  0.86 -0.13 0.75 0.250 1.0
Mn  0.92 -0.34 0.97 0.034 1.3

                       RC1  RC2
SS loadings           4.03 1.58
Proportion Var        0.50 0.20
Cumulative Var        0.50 0.70
Proportion Explained  0.72 0.28
Cumulative Proportion 0.72 1.00

Mean item complexity =  1.2
Test of the hypothesis that 2 components are sufficient.

The root mean square of the residuals (RMSR) is 0.09
with the empirical chi square 9.44 with prob < 0.74

Fit based upon off diagonal values = 0.97
```

Factor analysis for the heavy metal data values was performed employing the built-in function `principal ()` and then the generated results stored in a new location called `Hidden`. The standardized loadings, communalities, and specific variance were generated and appeared in the results as `RC1` and `RC2` to represent the two extracted factors, then the next two column `h2` to represent the communalities and `u2` = 1 – communality to represent the specific variance. The `SS loadings`, `proportion Var` (proportion of variance) captured by each factor and the `Cumulative Var` (cumulative variance) are presented as the last result of the function `principal ()`.

One can observe the amount of variance captured by each extracted factor is improved after rotation because of the reduction in the contributions of the less significant variables; for instance, the first factor captures 56.44% of variance and after the rotation, this amount changed to 50% of the total variance. This reduction in the amount of variation captured because of the reduction of the effect of heavy metals that have a small effect with the first factor.

The two extracted factors are enough to capture most of the differences in the heavy metal data values (>70%) as confirmed by the chi-square test (`Test of the hypothesis that 2 components are sufficient`). Notice that most of the variables associated with each factor are well defined and contribute slightly to the other factors.

The coefficient associated with each factor that is >0.6 is considered to have a strong correlation within each factor. Most of the variables with high loadings associated with each factor are well defined and contribute very little to the other factors. The two extracted factors are presented in a mathematical form and presented in Eqs. (9.9) and (9.10).

$$F_1 = -0.73\,Cu + 0.05\,Zn - 0.92\,Cd + 0.27\,Cr + 0.96\,Fe - 0.30\,Pb + 0.86\,Hg + 0.92\,Mn \tag{9.9}$$

$$F_2 = 0.06\,Cu + 0.85\,Zn + 0.26\,Cd - 0.36\,Cr - 0.24\,Fe + 0.68\,Pb - 0.13\,Hg - 0.34\,Mn \tag{9.10}$$

The first factor was positively correlated with Fe, Mn, and Hg, while negatively correlated with Cd and Cu. This factor may be termed as pseudo anthropogenic factor because both industrial as well as domestic factors contribute to their presence. This factor captured 50% of the differences in the heavy metal data values.

The second factor on the other hand, accounted for 20% of the total variance and is positively correlated with Zn and Pb. Because lead (Pb) and zinc (Zn) are considered as toxic metals, this factor can be termed as an anthropogenic factor because of industrial origin only.

The value produced by the sums of the squares of the loadings represents the communalities; for instance, the first communality that corresponding to the Cu is computed as $[(-0.73)^2+0.06^2=0.5365=0.54]$. One can see that most of the variance for the heavy metals in the sediment is captured by two factors as indicated by the high values of the communalities for most of the heavy metals. The specific variance is represented by `u2` in the R outputs and computed employing the formula `u2` $=1-$ communality. The correlation between the specific variance and the factors can be measured by the magnitude of the factor loadings (large value indicates high relationship). The built-in function `$ scores` was employed to generate the values of the extracted factors for each of the n observations.

```
> round (Hidden $ scores,digits = 2)
        RC1   RC2
 [1,]  -1.21 -0.12
 [2,]  -1.51 -0.76
 [3,]  -1.03 -0.74
 [4,]  -1.39 -0.96
 [5,]  -1.34 -0.53
 [6,]  -0.26  0.78
 [7,]  -0.71  1.59
 [8,]  -0.53  1.04
 [9,]  -0.82  0.19
[10,]  -0.21  2.73
[11,]   1.19  0.73
[12,]   1.07  0.16
[13,]   0.58 -1.13
[14,]   0.80 -0.64
[15,]   0.83 -0.72
[16,]   1.07  0.27
[17,]   0.34 -0.97
[18,]   1.01  0.24
[19,]   1.34 -0.11
[20,]   0.77 -1.05
```

The values of the two extracted factors for each of the n observations were plotted with the samples from various estuaries to understand the behavior of the chosen variables measured from the two rivers. A bar chart (Fig. 9.8) for the values of F_1 for the two estuaries was produced by employing the built-in function `barplot ()` to investigate the relationship between the values of the factor and the sampling points of different estuaries.

One can observe that a positive contribution was shown by the samples selected from all of the Juru Estuary to the first factor (pseudo anthropogenic factor), whereas a negative contribution was exhibited by the samples selected from all of the Jejawi Estuary. This may be due to changes in discharges along the study area. A positive contribution was imputed to the concentrations of Fe, Mn, and Hg, whereas a negative contribution was imputed to the concentrations of Cd and Cu. Thus, the two estuaries were different with regard to these heavy metals. This also indicates the pattern of distribution of polluting industries, agricultural, and orchid plantation activities surrounding these two estuaries.

The values of F_2 for all observations are presented in a plot as shown in Fig. 9.9. The positive contribution was mainly due to the concentrations of Zn and Pb, and the negative contribution was primarily due to the concentration of Cr, Fe, Hg, and Mn in the sediment; a smaller effect was observed from Cu. The concentrations of heavy metals exhibited different behavior from site to site, regardless of the region.

A scatter plot was produced (Fig. 9.10) employing the values for the first two factors for all observations. One can see that two distinct sets can be recognized in the scatter plot. The graph revealed that separating the two estuaries was due to the first factor, while the fluctuations in the concentrations of the chosen variables was due to the second factor regardless of the estuary (region). This finding indicates that the two estuaries are distinct in terms of the chosen variables.

Factor analysis has provided a satisfactory picture of the concentration of heavy metals in the sediment. Factor analysis provides valuable information about the source of the differences in the heavy metals, as >70% of the total variance was explained by only two factors.

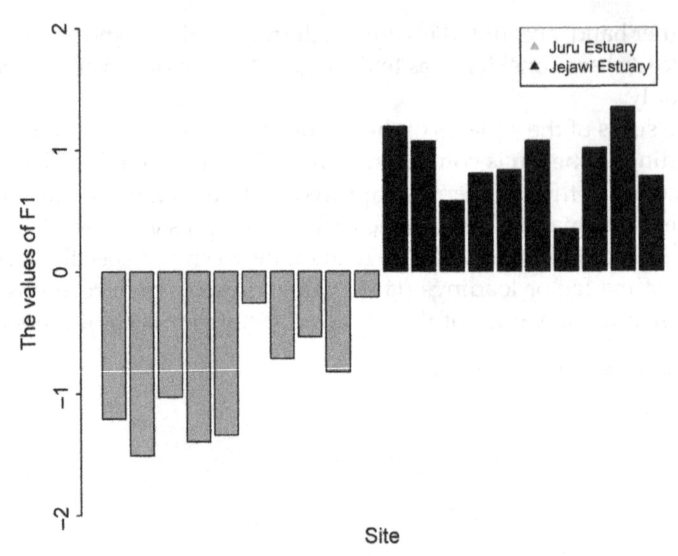

FIG. 9.8 Showing the values of F_1 by the two estuaries.

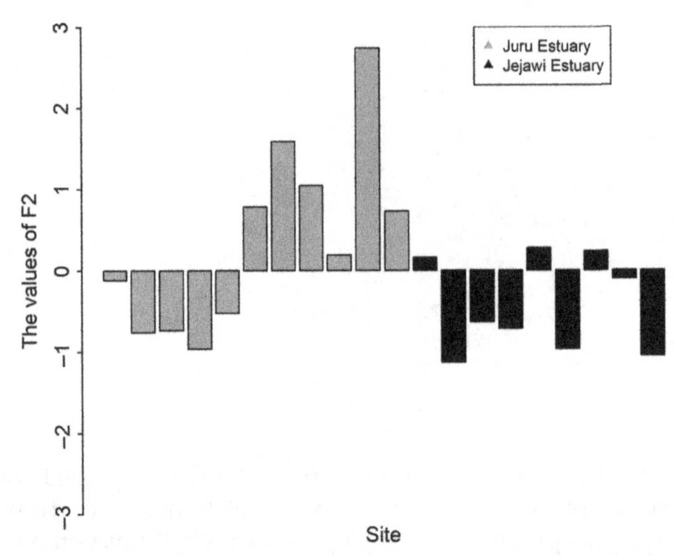

FIG. 9.9 Showing the values of F_2 by the two estuaries.

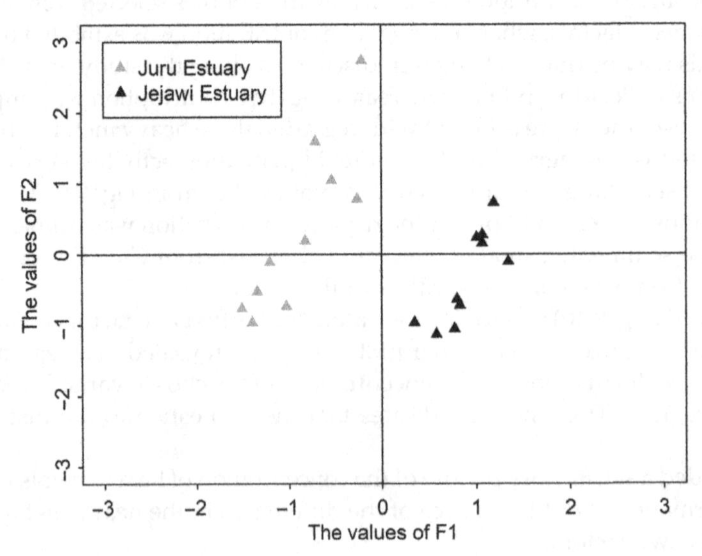

FIG. 9.10 Showing the values of F_1 and F_2, by the two estuaries.

Further Reading

Alkarkhi, A.F.M., Alqaraghuli, W.A.A., 2019. Easy Statistics For Food Science With R, first ed. Academic Press.
Alkarkhi, A.F.M., Anees, A., Azhar, M.E., 2009. Assessment of surface water quality of selected estuaries of malaysia-multivariate statistical techniques. Environmentalist 29, 255–262.
Alkarkhi, A.F.M., Ismail, N., Ahmed, A., Easa, A.M., 2009. Analysis of heavy metal concentrations in sediments of selected estuaries of malaysia—a statistical assessment. Environ. Monit. Assess. 153, 179–185.
Bryan, F.J.M., 1991. Multivariate Statistical Methods: A Primer. Chapman & Hall, Great Britain.
Daniel & Hocking, 2013. Blog Archives, High Resolution Figures In R [Online]. R-Bloggers. Available, https://www.r-bloggers.com/Author/Daniel-Hocking.
Documentation, R., Factor Analysis. [Online]. Available: https://stat.ethz.ch/R-manual/R-devel/library/stats/html/factanal.html. [(Accessed 7 July 2018)].
James, H. S. Exploratory Factor Analysis With R [Online]. Available: Exploratory Factor Analysis With R [Accessed 7 July 2018].
Johnson, R.A., Wichern, D.W., 2002. Applied Multivariate Statistical Analysis. Prentice Hall, New Jersey.
Rencher, A.C., 2002. Methods of Multivariate Analysis. J. Wiley, New York.
Robert, I.K., 2017. Quick-R-principal components and factor analysis [Online]. In: Datacamp. Available: https://www.statmethods.net/advstats/factor.html. [(Accessed 7 July 2018)].
Sanchez, G., 2012. 5 Functions To Do Principal Components Analysis In R [Online]. Available: http://www.gastonsanchez.com/visually-enforced/how-to/2012/06/17/PCA-in-R/. [(Accessed 16 June 2018)].
William, R., 2018. How To: Use The Psych Package For Factor Analysis And Data Reduction [Online]. Department Of Psychology—Northwestern University. Available, http://personality-project.org/r/psych/HowTo/factor.pdf. [(Accessed 7 July 2018)].

CHAPTER

10

Discriminant Analysis

LEARNING OBJECTIVES

After careful consideration of this chapter, you should be able:

- *To explain discriminant analysis (DA).*
- *To explain the procedure of using a discriminant function for classification process.*
- *To understand how to employ discriminant analysis to environmental science data.*
- *To describe how to interpret the extracted discriminant function (DA).*
- *To describe the built-in functions and commands in R for discriminant function.*
- *To comprehend the result of R output for discriminant function and how to employ it in the analysis.*
- *To draw smart conclusions and write useful reports.*

10.1 INTRODUCTION

Multivariate analysis consists of many methods used to analyze data values obtained from several variables (k) measured on each experimental unit. Researchers are usually interested in identifying the contribution of each chosen variable in separating various groups. Discriminant analysis (DA) is usually employed to identify the contribution of each chosen variable in separating the groups through finding a linear combination from the k measured variables; the minimum number of linear combinations used is one. Furthermore, a linear or quadratic function can be employed to predict or allocate a new individual to one of predefined groups. For example, this method could be used to separate data for two or more rivers based on the water quality parameters or heavy metals concentrations. Researchers can use discriminant analysis if the groups are already defined prior to the project.

A mathematical model can be generated as the end result of discriminant analysis; this model can be employed for the prediction of group elements. Moreover, one can employ the model to identify the contribution of various variables and to study the relationship between various chosen variables and the observations.

10.2 DISCRIMINANT ANALYSIS IN R

R statistical software provides built-in functions and commands to carry out discriminant analysis easily and efficiently and generate the required results with detailed information. The `MASS` package should be installed from R library and loaded to your computer before starting the analysis. There are two options to install and load the package, the first option employs two calls; the first call is `install packages ("mass")` and the second call is `library (MASS)`. The second option is to employ the **Packages** button on the upper-left row of the R menu, and then choose **Load package**. Moreover, the package can be installed and loaded in the RStudio from **Tools** on the upper-left row of RStudio. Linear and quadratic classifications can be carried out employing R built-in functions and commands. Furthermore, R provides more built-in functions to extract information related to discriminant analysis. We provided the built-in functions and commands with an easy explanation along with the structure of each call.

Applied Statistics for Environmental Science with R
https://doi.org/10.1016/B978-0-12-818622-0.00010-1

1. Linear discriminant analysis can be carried out by employing the built-in function `lda ()`. Consider there are two or more sets and k responses $Y_1, Y_2, \ldots, Y_k$, measured from each set. The built-in function `lda ()` can generate information regarding the `Prior probabilities of groups`, `group means`, and the `Coefficients of linear discriminants`. The structure of this call is:

```
lda (Set ~ Y1 + Y2+ ... + Yk, data = data frame)
```

The grouping variable is represented by `Set` in the discriminant function call, which is a categorical variable. The frequencies in each group will be employed as the prior probabilities for the classification purpose. We can change the default setting of prior probabilities if required. Equal prior probabilities for r sets can take the vector form as $(1/r, 1/r, \ldots, 1/r)$.

```
lda (Set ~ Y1 + Y2 + ... + Yk, data = Data frame, prior = c (1/r, 1/r,..., 1/r))
```

Desired values can be placed instead of the equal prior probabilities.

2. Quadratic discriminant analysis can be carried out by employing the built-in function `qda ()`. Consider there are r sets with equal prior probabilities, then the structure of the call is:

```
qda (Sate ~ Y1 + Y2 + ...+ Yk, data = Data frame, prior = c (1/r, 1/r,..., 1/r))
```

One can see that the structure of the calls for performing both the linear and quadratic discriminant analysis are the same and the only difference is to employ the built-in function `qda ()` instead of the built-in function `lda ()`.

3. The coefficients of the discriminant function (DF) associated with each variable can be extracted by employing the character `$`.

```
Data frame $ LDi
```

where `i` represents the number of extracted discriminant function.

4. The produced results of the function `lda()` can be employed by the built-in function `predict ()` to generate the posterior probabilities (the probability of assigning observations to predefined groups) of each sample for each group (`$ posterior`), the classification of the groups (`$ class`), and the values for each discriminant function (`$ x`). The structure of this call is:

```
predict (data frame)
```

5. The scores of each extracted discriminant function can be generated by employing the built-in function `$x`. The structure of this call is:

```
Data frame $ x
```

6. The classification of the groups can be generated by employing the built-in function `$class`. The structure of this call is:

```
Data frame $ class
```

7. Posterior probabilities can be produced by employing the built-in function `$posterior`. The structure of this call is:

```
Data frame $ posterior
```

8. The classification table can be generated employing the built-in function `table ()`. The structure of this call is:

```
table (data frame $ grouping variable, Scores $ class)
```

9. The percentages of the correct classifications and misclassifications can be generated by employing the built-in function `prop.table ()`. The structure of this call is:

```
prop.table (data frame,1 or 2)
```

Calculating the percentages of correct classifications and misclassifications could be based on the row or the column: the numbers 1 or 2 appearing in the call `prop.table ()` refer to the method of calculation: 1 refers to the percentage that is computed using the row, and 2 refers to the percentage that is computed using the column.

10. The built-in function `diag (prop.table())` can be employed to generate the percentage of only correct classifications. The structure of this call is:

```
diag (prop.table (data frame,1 or 2))
```

11. The built-in function `sum (diag (prop.table()))` can be employed to generate the total percentage for the correct classification, which represents the proportion of cases that lie along the diagonal (the overall predictive accuracy). The structure of this call is:

```
sum (diag (prop.table (data frame,1 or 2)))
```

10.3 CONFIGURATION OF DISCRIMINANT ANALYSIS DATA

Consider there are r normally distributed populations. r samples were chosen, one sample from each of the r populations of sizes $n_1, n_2, \ldots, n_r$, with k response variables measured from each sample. The configuration of data for a discriminant analysis is presented in Table 10.1.

The description of the symbol Y_{ijl} in the Table 10.1 can be clarified as: i takes the values $i=1, 2, 3, \ldots, n_i$, which represents the ith observation in the table, while the population (sample) is represented by $j=1, 2, 3, \ldots, r$, and the variable measured from each population is represented by l ($l=1, 2, 3, \ldots, k$).

A linear combination will be formed of the k measured variables from the data in Table 10.1 to be employed to separate the r populations; the linear combination is called the discriminant function.

10.4 THE CONCEPT OF DISCRIMINANT FUNCTION

Researchers are usually interested in identifying the contribution of each measured variable in separating various populations (groups). The relative contribution shows the importance of a variable in separating various populations.

Consider there are r populations and k measured variables ($Y's$) of interest. Let us consider the case where there are two populations and then we generalize the case for r populations.

The k measured variables ($Y's$) can be employed to produce a function that is a linear combination of the k measured variables, so that the distance between the two mean vectors is maximized. The linear combination of the k variables presented in (10.1) is called the discriminant function (DF), also called canonical discriminant function, and can be employed in separating the two populations.

$$Z = a_1 Y_1 + a_2 Y_2 + \ldots + a_k Y_k \tag{10.1}$$

where the coefficient associated with each variable a_i needs to be estimated employing the sample data that show the contribution of each response variable in separating the two populations.

We can generalize the situation of the two populations to include r populations. Consider there are r populations and k response variables measured from each population with a sample size of n_i (the sample size selected from the ith population). The k response variables can be employed to form discriminant functions as presented below.

$$Z_i = a_{i1} Y_1 + a_{i2} Y_2 + \ldots + a_{ik} Y_k$$

TABLE 10.1 The General Configuration of Data for a Discriminant Analysis

Individuals	Variable				Population
	Y_1	Y_2	...	Y_k	
1	Y_{111}	Y_{112}	...	Y_{11k}	
2	Y_{211}	Y_{212}	...	Y_{21k}	
⋮	⋮	⋮	...	⋮	Population 1
n_1	Y_{n_111}	Y_{n_112}	...	Y_{n_11k}	
1	Y_{121}	Y_{122}	...	Y_{12k}	
2	Y_{221}	Y_{222}	...	Y_{22k}	
⋮	⋮	⋮	...	⋮	Population 2
n_2	Y_{n_221}	Y_{n_222}	...	Y_{n_22k}	
⋮	⋮	⋮	...	⋮	
1	Y_{1r1}	Y_{1r2}	...	Y_{1rk}	
2	Y_{2r1}	Y_{2r2}	...	Y_{2rk}	
⋮	⋮	⋮	...	⋮	Population r
n_r	Y_{n_rr1}	Y_{n_rr2}	...	Y_{n_rrk}	

The maximum differences among the r populations are captured by the first discriminant function and is represented by Z_1.

$$Z_1 = a_{11}Y_1 + a_{12}Y_2 + \ldots + a_{1k}Y_k \tag{10.2}$$

The second maximum differences are captured by the second discriminant function, which is represented by Z_2. The amount of differences that are represented by Z_2 is not captured by Z_1, and the discriminant functions are uncorrelated.

$$Z_2 = a_{21}Y_1 + a_{22}Y_2 + \ldots + a_{2k}Y_k \tag{10.3}$$

The smaller of k and $r-1$ represents the maximum number of discriminant functions that can be employed for separating different populations. Moreover, the discriminant functions are uncorrelated; Z_i is uncorrelated with $Z_1, Z_2, \ldots, Z_{i-1}$.

10.4.1 Common Steps for Computing the Discriminant Function

The common steps for computing the discriminant function need to compute the coefficients (a_i) of the linear combination. The value of a_i can be computed by employing the entries of a multivariate analysis of variance, the common steps for estimating the coefficients are:

Step 1: The sum of squares for between populations (B), and the sum of squares for within (W) should be computed first.

Step 2: Employ the matrix $W^{-1}B$ to compute the eigenvalues (λ) and then compute the eigenvector associated with each eigenvalue to represent a_i. The discriminant function that corresponds to the first eigenvalue represents the highest contribution in separating the population means; thus, the contribution of each discriminant function can be placed in order such that $\lambda_1 > \lambda_2 > \ldots > \lambda_r$.

Step 3: Keep the significant eigenvalues and discard insignificant ones because only significant eigenvalues contribute significantly in separating the population means. Several tests are available for this purpose, such as the Wilks and chi-square tests. Various statistical packages provide the significance test along with the results of the analysis for discriminant analysis.

Step 4: Researchers usually employ standardize discriminant function to help in interpreting the contribution of each chosen response variable, especially when the variables are not commensurate.

Step 5: Compute the values of each significant discriminant function for each of n observations to be used in interpreting the contribution of the chosen response variables in separating the r populations. The values can be used to build a plot such as a bar-chart or a scatter-plot for the values of the first two discriminant functions. We can deliver three important points regarding the discriminant analysis as follows:

- The differences among the r populations are most probably captured by the first two or three discriminant functions.
- The contribution of each discriminant function can be computed by $\frac{\lambda_i}{\sum_{i=1}^{s}\lambda_i}$ which represents the relative importance of each discriminant function.
- Researchers should note that not always helpful results can be produced by discriminant analysis.

10.5 ALLOCATION

The Mahalanobis distance between each observation and the population center can be used for assigning or allocating observations to one of the selected populations. Assigning the individuals can be carried out either by employing linear functions or quadratic functions and based on the variance–covariance matrix of the selected populations. Furthermore, classification to the predefined populations can be performed with regard to an independent variable; this variable is a categorical variable, such as kind of chemicals or location.

Consider there are r populations. r samples were selected, one sample from each of the chosen r populations with n_i sample size and k response variables measured from each sample. The steps for assigning observations to a population are illustrated below.

Step 1: The mean vector for the selected response variables of each sample should be computed:

$$(\overline{Y}_1, \overline{Y}_2, \ldots, \overline{Y}_k)$$

Step 2: The Mahalanobis distance should be computed.
Step 3: Employ linear or quadratic classification functions as follows:

(i) If the populations have equal covariance matrices employ linear classification function.

$$D_i^2 = (Y - \overline{Y}_i)' \, S_p^{-1} \, (Y - \overline{Y}_i)'$$

where S_p is the pooled variance.

(ii) If the populations have unequal covariance matrices employ quadratic classification function.

$$D_i^2 = (Y - \overline{Y}_i)' \, S_i^{-1} \, (Y - \overline{Y}_i)'$$

where S_i is the sample covariance matrix for ith population.

Step 4: Allocate the observation Y to the population for which D_i^2 has the smallest value.

Correct classification and misclassification are usually computed as proportions to evaluate the classification procedure through the ability of the procedure to predict sample membership. Consider correct classification is represented by p_1, then misclassification is represented by $p_2 = 1 - p_1$. Various statistical packages produce these proportions with the classification results.

If no restriction, then equal probabilities should be employed to all populations; otherwise, the researcher can use other prior probabilities.

EXAMPLE 10.1 ASSESSMENT OF HEAVY METALS IN THE SURFACE WATER

A researcher wants to investigate the concentrations of seven heavy metals (copper (Cu), lead (Pb), zinc (Zn), cadmium (Cd), chromium (Cr), arsenic (As), and mercury (Hg)) in the surface water obtained from two locations. The two locations, with 20 sampling points at each location, were Kuala Juru (the Juru River) and Bukit Tambun (the Jejawi River) in the Penang State of Malaysia. Flame atomic absorption spectrometer (FAAS; Perkin Elmer HGA-600) was employed for the analysis of Cu, Zn, Cd, Pb, Cr, Cd, and Cu and cold vapor atomic absorption spectrometer (CV-AAS) method was employed for Hg. The sample data for the concentration of heavy metals in the surface water are given in Table 10.2.

TABLE 10.2 The Concentration of Heavy Metals in the Surface Water for Juru and Jejawi Rivers (mg/L)

Location	Cu	Pb	Zn	Cd	Cr	As	Hg
Juru	0.075	0.290	0.112	0.135	0.080	5.980	0.011
Juru	0.048	0.412	0.108	0.123	0.034	1.767	0.037
Juru	0.067	0.467	0.058	0.132	0.102	2.881	0.017
Juru	0.055	0.349		0.134	0.091	3.801	0.038
Juru	0.049	0.367	0.082	0.124	0.130	3.266	0.020
Juru	0.042	0.345	0.103	0.144	0.034	1.666	0.031
Juru	0.085	0.301	0.089	0.129	0.111	3.835	0.015
Juru	0.176	0.244	0.100	0.119	0.068	5.276	0.032
Juru	0.030	0.243	0.079	0.152	0.122	2.202	0.042
Juru	0.017	0.256	0.103	0.159	0.148	3.865	0.008
Juru	0.013	0.420	0.085	0.174	0.131	2.920	0.017
Juru	0.002	0.380	0.071	0.192	0.136	3.515	0.014
Juru	0.004	0.294	0.087	0.207	0.141	3.015	0.004
Juru	0.008	0.263	0.088	0.176	0.150	3.635	0.016
Juru	0.104	0.285	0.079	0.174	0.143	3.210	0.009
Juru	0.016	0.341	0.083	0.209	0.147	3.800	0.025
Juru	0.022	0.311	0.101	0.283	0.145	3.585	0.015
Juru	0.029	0.204	0.088	0.271	0.146	3.280	0.004
Juru	0.062	0.221	0.110	0.255	0.124	3.465	0.008
Juru	0.064	0.133	0.115	0.125	0.125	2.725	0.010
Jejawi	0.012	0.210	0.058	0.278	0.115	3.840	0.009
Jejawi	0.047	0.179	0.052	0.219	0.177	3.245	0.014
Jejawi	0.036	0.276	0.067	0.193	0.160	3.295	0.009
Jejawi	0.061	0.339	0.062	0.157	0.160	3.665	0.004
Jejawi	0.036	0.186	0.038	0.126	0.279	2.974	0.010
Jejawi	0.050	0.385	0.050	0.135	0.162	3.424	0.013
Jejawi	0.050	0.229	0.043	0.163	0.233	2.827	0.006
Jejawi	0.044	0.247	0.048	0.088	0.229	3.004	0.009
Jejawi	0.055	0.324	0.056	0.083	0.208	3.760	0.008
Jejawi	0.054	0.249	0.056	0.045	0.191	3.350	0.005
Jejawi	0.035	0.276	0.055	0.102	0.218	3.275	0.006

TABLE 10.2 The Concentration of Heavy Metals in the Surface Water for Juru and Jejawi Rivers (mg/L)—cont'd

Location	Cu	Pb	Zn	Cd	Cr	As	Hg
Jejawi	0.045	0.195	0.058	0.096	0.245	3.195	0.004
Jejawi	0.046	0.263	0.054	0.093	0.232	3.575	0.003
Jejawi	0.049	0.151	0.042	0.144	0.188	3.545	0.009
Jejawi	0.049	0.230	0.053	0.141	0.211	3.710	0.015
Jejawi	0.050	0.246	0.074	0.130	0.216	3.655	0.007
Jejawi	0.046	0.213	0.042	0.119	0.198	3.605	0.008
Jejawi	0.046	0.214	0.050	0.161	0.200	3.500	0.004
Jejawi	0.044	0.269	0.044	0.136	0.208	3.820	0.002
Jejawi	0.045	0.266	0.081	0.111	0.245	3.585	0.015

The location of the two rivers represents the categorical variable of interest; the Juru River and the Jejawi River, where the surface water samples were obtained and the chosen heavy metals were measured. The function `lda ()` in R was employed to carry out the discriminant analysis for the stored data. `CSV (EXAMPLE10_1)` of heavy metals in the surface water of the two rivers. The results of the function `lda ()` should be stored in a new location called `LDis` and prepared to be used for other objectives, such as to study the behavior of selected parameters regarding each river. Because the categorical variable has only two options (Juru River or Jejawi River), only one discriminant function was generated to describe the differences between the two rivers in terms of chosen heavy metals; this function captured 100% of the differences between the two rivers. The results of discriminant analysis for the chosen heavy metals in the surface water are presented below.

```
install.packages("MASS")
library(MASS)
> LDis <- lda( Location ~ Cu + pb + Zn + Cd + Cr + As + Hg, data = Example10_1)
> LDis
Call:
lda(Location ~ Cu + pb + Zn + Cd + Cr + As + Hg, data = Example10_1)
Prior probabilities of groups:
Jejawi Juru
 0.5 0.5
Group means:
             Cu        pb        Zn       Cd       Cr       As       Hg
Jejawi 0.0450 0.24735 0.053975 0.13600 0.20375 3.442425 0.008019
Juru   0.0484 0.30630 0.091200 0.17085 0.11540 3.384225 0.018761
Coefficients of linear discriminants:
              LD1
Cu -0.3314727
pb  5.1616279
Zn 62.9586460
Cd  5.7205524
Cr -6.8285906
As -0.1027578
Hg 24.7198052
```

One can see that the package `MASS` should be installed and loaded prior to performing the analysis as appeared in the results. The function `lda ()` was shown in the third row to tell us the starting of the discriminant analysis for the stored data of heavy metals and then store the output generated by this function in a place called `LDis`. The proportion of each river within the data set (equal probabilities (0.5) for both Juru and Jejawi Rivers) was displayed as `Prior probabilities of groups`; the proportions represent the prior probabilities of each river. More information was generated, such as the mean value of each chosen heavy metal below the message `Group means`. The coefficient associated with each heavy metal for the linear discriminant function appear as `Coefficients of linear discriminants:LD1`. One can represent the discriminant function for the chosen heavy metal in the surface water in terms of a mathematical equation. as in Eq. (10.4):

$$Z_1 = -0.331\,Cu + 5.161\,pb + 62.958\,Zn + 5.720\,Cd - 6.828\,Cr - 0.102\,As + 24.719\,Hg \tag{10.4}$$

The chosen heavy metals contribute in explaining the differences between the two rivers. The importance of each heavy metal can be identified based on the coefficient associated with each heavy metal in the discriminant function (10.4).

One can observe that each heavy metal influences the discriminant function (Z_1) in (10.4) either positively or negatively. The highest effect on Z_1 was exhibited by *Zn* and captures most of the differences between the two rivers. On the other hand, the second highest effect was seen in *Hg*. The effect of each heavy metal concentration in the surface water can be placed in order: $Zn > Hg > Cr > Cd > Pb > Cu > As$. More information can be generated by the built-in function `predict ()`, such as assigning the chosen samples to the respective river `$ class`, the posterior probabilities (`$posterior`) of each sample to each river, and the values of the discriminant function for each observation (`$x`).

```
> O.Vs <- predict (DA)
> O.Vs          # scores for each sample
$`class`
 [1] Juru Juru Juru Juru Juru Juru Juru Juru Juru Juru Juru Juru
[13] Juru Juru Juru Juru Juru Juru Juru Juru Jejawi Jejawi Jejawi Jejawi
[25] Jejawi Jejawi Jejawi Jejawi Jejawi Jejawi Jejawi Jejawi Jejawi Jejawi Jejawi Jejawi
[37] Jejawi Jejawi Jejawi Jejawi
Levels: Jejawi Juru
$posterior
         Jejawi         Juru
1  6.009458e-05 9.999399e-01
2  9.640183e-08 9.999999e-01
3  1.500258e-01 8.499742e-01
4  2.222642e-04 9.997777e-01
5  9.563194e-03 9.904368e-01
6  1.245061e-06 9.999988e-01
7  7.701155e-03 9.922988e-01
8  2.607969e-04 9.997392e-01
9  8.021835e-03 9.919782e-01
10 1.458230e-03 9.985418e-01
11 6.885880e-04 9.993114e-01
12 4.340799e-02 9.565920e-01
13 1.020176e-02 9.897982e-01
14 1.360677e-02 9.863932e-01
15 1.039739e-01 8.960261e-01
16 2.115802e-03 9.978842e-01
17 2.814406e-05 9.999719e-01
18 1.370707e-02 9.862929e-01
19 3.675837e-05 9.999632e-01
20 7.193591e-04 9.992806e-01
21 8.118114e-01 1.881886e-01
22 9.963449e-01 3.655127e-03
23 6.988438e-01 3.011562e-01
24 8.998190e-01 1.001810e-01
25 9.999987e-01 1.285597e-06
26 9.783900e-01 2.161002e-02
27 9.999573e-01 4.272215e-05
28 9.999416e-01 5.842217e-05
29 9.982551e-01 1.744905e-03
30 9.997583e-01 2.417385e-04
31 9.993523e-01 6.477211e-04
32 9.999066e-01 9.336007e-05
```

```
33 9.998364e-01 1.635866e-04
34 9.999844e-01 1.563566e-05
35 9.991107e-01 8.893324e-04
36 9.533936e-01 4.660644e-02
37 9.999769e-01 2.313045e-05
38 9.997677e-01 2.323410e-04
39 9.999414e-01 5.855486e-05
40 7.779580e-01 2.220420e-01

$x
          LD1
1   2.6120929
2   4.3415349
3   0.4661142
4   2.2605427
5   1.2470455
6   3.6539689
7   1.3057477
8   2.2175663
9   1.2946968
10  1.7546674
11  1.9565248
12  0.8311623
13  1.2295006
14  1.1511737
15  0.5788347
16  1.6544600
17  2.8159684
18  1.1491725
19  2.7442026
20  1.9447675
21 -0.3928604
22 -1.5071220
23 -0.2262306
24 -0.5899570
25 -3.6453586
26 -1.0246647
27 -2.7037945
28 -2.6196785
29 -1.7063565
30 -2.2379655
31 -1.9729785
32 -2.4936891
33 -2.3429358
34 -2.9739362
35 -1.8877178
36 -0.8111555
37 -2.8686941
38 -2.2486239
39 -2.6190687
40 -0.3369563
```

It can easily be seen that if the generated information by the built-in function `predict` is stored in `O.Vs`, then the results start with the function `$ class`, which was employed to identify `Juru` or `Jejawi` rivers. The second generated result is the posterior probabilities of each observation with regard to each river `$posterior`. The value of each sample for the discriminant function was produced as the last result (`$x`) of the built-in function `predict ()`.

We can try employing the value of the discriminant function Z_1 regarding each sample to investigate the relationship between various samples of the surface water and the generated discriminant function. A bar chart of the values of the discriminant function Z_1 for various samples obtained from the Juru and Jejawi Rivers was produced by employing the built-in function `barplot ()` as presented in Fig. 10.1. The surface water samples are represented by the horizontal axes (20 sampling points of each river), and the values of the discriminant function is represented by the height of the Z_1 for all samples. A negative contribution to the discriminant function was exhibited in Fig. 10.1 by all of the water samples obtained from the Jejawi River. In contrast, a positive contribution was exhibited by all of the water samples obtained from the Juru River. The positive contribution was mainly attributed to the high concentration of Zn, Hg, and Pb in the surface water, and the negative contribution was mainly attributed to the high concentration of Cr, Cu, and As in the surface water. The difference in the contribution of various samples to the discriminant function can be attributed to the differences in the industrial and agricultural activities operating close to the study areas, such as rubber manufacturing, pulp, paper, electroplating, and metal finishing industrial sectors.

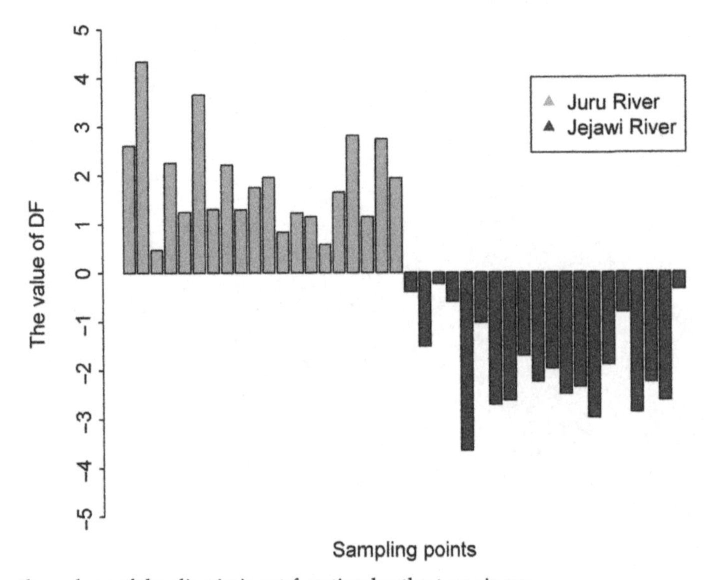

FIG. 10.1 A bar chart showing the values of the discriminant function by the two rivers.

EXAMPLE 10.2 CLASSIFICATION OF THE HEAVY METALS IN THE SURFACE WATER

Produce a table for the percentages of correct classifications and misclassifications and a table for the number of samples for the data of heavy metals concentrations in the surface water samples obtained from the Juru and Jejawi Rivers, along with the results of the discriminant function obtained in Example 10.1.

The discriminant function Z_1 in (10.1) was employed to clarify the classification steps for the samples of the heavy metals concentrations in the surface water obtained from the Juru River and Jejawi River. R's built-in functions can easily be employed to perform classification analysis. A table for the number of correct classifications and misclassifications for all of the observations was generated by the built-in function `table ()`. Moreover, the percentages of correct classifications and misclassifications for the classification operation with the data on the heavy metals concentration in the surface water was generated by the built-in function `prop.table (data frame,1)`; the number 1 in the call indicates that the calculation was carried out for the row percentage. The results of employing the two built-in functions `table ()` and `prop.table (data frame, 1)` for the heavy metals concentration in the surface water obtained from various locations of the Juru and Jejawi Rivers are shown below.

```
> Table <- table (Example10_1 $ Location, Scores $ class) #classification Table
> Table

         Jejawi Juru
Jejawi       20    0
Juru          0   20
```

```
> prop.table (Table, 1)  # 1 means row percentage, 2 column percentage#correct and
misclassification percentage
        Jejawi Juru
 Jejawi      1    0
 Juru        0    1
> diag (prop.table (Table, 1)) #correct Classification percentage
Jejawi Juru
     1    1
> sum (diag (prop.table (Table))) # total percent correct
[1] 1
```

All the samples of the surface water (20 samples) obtained from the Jejawi River were correctly assigned into their population as exhibited by the output of the function `table ()`; this result represents 100% of the Jejawi River samples, and 0% misclassifications samples. On the other hand, 100% of the Juru River samples were correctly assigned into the Juru River population.

In summary, one can say that all the samples of the surface water collected from Juru and Jejawi Rivers were correctly assigned into their respective populations as generated by the function `sum (diag (prop.table (Table, 1)))`. Furthermore, the results of the classification also showed that significant differences existed between these two rivers, which are expressed in terms of one discriminant function.

EXAMPLE 10.3 AMOUNT OF THE SOLID WASTE IN THE PALM OIL MILLS

The amount of solid waste generated (tons/month) in five different palm oil mills in Malaysia were analyzed for empty fruit bunches (EFB), potash ash, fiber, and shell. The five mills were assessed based on the production cycle, which is divided into several sectors: Sector A (preliminary process) covers the sterilization process and stripping and digestion process while Sector B (oil clarification and purification process) includes screening, decentering, and centrifuging processes. Sector C (nut and fiber separation process) includes depericarping, hydrocyclone, and nut cracking processes. The sample data for the amount of the solid waste in the palm oil mills generated in 12 months are presented in Table 10.3.

Five different palm oil mills represent the categorical variable of interest, which is the location of the palm oil mill. The researchers want to identify the parameters responsible for the differences between various palm oil mills. Discriminant

TABLE 10.3 The Amount of Solid Waste Generated in Palm Oil Mills (tons)

Location	EFB	Potash	Fiber	Shell	Location	EFB	Potash	Fiber	Shell
1	5520	120	3240	1320	3	2870	62	1685	686
1	5500	119	3227	1314	3	2760	60	1620	660
1	5432	118	3188	1300	3	2815	61	1652	673
1	5765	125	3382	1378	3	2843	62	1669	680
1	5720	124	3357	1368	3	2732	59	1409	653
1	5500	120	3227	1315	3	2650	58	1555	634
1	5530	120	3243	1321	4	3312	72	1944	792
1	5500	120	3227	1315	4	3257	71	1912	779
1	5420	118	3182	1296	4	3395	74	1993	812
1	5480	119	3217	1311	4	3477	76	2041	832
1	5395	117	3165	1290	4	3456	75	2028	826
1	5495	119	3224	1313	4	3422	74	2009	818
2	1656	36	972	396	4	3395	74	1993	812
2	1601	35	940	383	4	3477	76	2041	832
2	1573	34	923	376	4	3560	77	2090	851

Continued

TABLE 10.3 The Amount of Solid Waste Generated in Palm Oil Mills (tons)—cont'd

Location	EFB	Potash	Fiber	Shell	Location	EFB	Potash	Fiber	Shell
2	1711	37	1004	409	4	3533	77	2074	845
2	1656	36	972	396	4	3367	73	1976	805
2	1546	34	907	370	4	3422	74	2009	818
2	1573	34	923	376	5	3588	78	2106	858
2	1562	34	917	374	5	3698	80	2171	884
2	1794	39	1053	429	5	3665	80	2151	876
2	1711	37	1004	409	5	3643	79	2138	971
2	1656	36	972	396	5	3709	81	3177	887
2	1546	34	907	370	5	3776	82	2216	903
3	2760	60	1620	660	5	3754	82	2203	898
3	3036	66	1782	726	5	3665	80	2151	876
3	2926	64	1717	670	5	3643	79	2138	971
3	2953	64	1733	706	5	3709	81	3177	887
3	2926	64	1717	670	5	3671	80	2155	879
3	3036	66	1782	726	5	3560	77	2089	851

analysis (DA) was carried out for the stored data. CSV (`Example10_3`) of the solid waste generated in palm oil mills employing the built-in function `lda ()` in R.

```
> LDis <- lda(Location ~ EFB+ Potash+ Fibre+ Shell, data = Example10_3)
> LDis
Call:
lda(Location ~ EFB + Potash + Fibre + Shell, data = Example10_3)
Prior probabilities of groups:
  1   2   3   4   5
0.2 0.2 0.2 0.2 0.2
Group means:
        EFB    Potash      Fibre     Shell
1 5521.417 119.91667 3239.9167 1320.0833
2 1632.083  35.50000  957.8333  390.3333
3 2858.917  62.16667 1661.7500  678.6667
4 3422.750  74.41667 2009.1667  818.5000
5 3673.417  79.91667 2322.6667  895.0833

Coefficients of linear discriminants:
                 LD1          LD2           LD3          LD4
EFB     0.0007618302 -0.028938637 -0.0711391936 -0.031584129
Potash -0.4714822349  0.718413069  3.3946245787  1.196704635
Fibre   0.0006413023  0.004076789 -0.0007475991 -0.004250796
Shell  -0.0069800148  0.045539654 -0.0087490849  0.033887137

Proportion of trace:
   LD1    LD2    LD3    LD4
0.9978 0.0022 0.0000 0.0000
```

One can observe that the outputs of employing the built-in function `lda ()` show the call to carry out discriminant analysis and the name of the location where the generated data will be stored; the new location is called `LDis`. The proportion of each mill within the data set (equal probabilities (0.2) for all mills) appeared as `prior probabilities of groups`, the proportions represent the prior probabilities of each palm oil mill. Furthermore, the mean value of each chosen variable (`EFB, Potash, Fiber, Shell`) was shown below the message (`Group means`). The coefficients associated with each selected variable for the linear discriminant function appeared as `Coefficients of linear discriminants`. Two discriminant functions captured 100% of the total variance as presented under `Proportion of trace (LD1 LD2)`, and are responsible for all of the differences among different palm oil mills. In contrast, the other discriminant functions captured 0% of the differences. The two discriminant functions for the chosen variables for the palm oil mills can be represented mathematically as in Eqs. (10.5) and (10.6).

$$Z_1 = 0.00076\,EFB - 0.47148\,Potash + 0.00064\,Fibre - 0.00698\,Shell \quad (10.5)$$

$$Z_2 = -0.02893\,EFB + 0.71841\,Potash + 0.00407\,Fibre + 0.04553\,Shell \quad (10.6)$$

The importance of the chosen variables can be assessed through the coefficients associated with each variable in the discriminant functions Eqs. (10.5) and (10.6).

One can observe that the results of the first discriminant function (Z_1) exhibited that potash had the highest contribution in separating the five mills and captured most of the differences between the five mills. The importance of each chosen variable in separating various mills can be placed in order: Fiber < EFB < Shell < Potash; 99.78% of the differences between the mills was captured by Z_1, and very little of the total variance was captured by the second discriminant function (Z_2), the amount of differences captured by Z_2 is 0.22%. Furthermore, the classification of each sample (`$class`), posterior probabilities (`$posterior`), and the values of the samples (`$x`) of each discriminant function for different palm oil mills can be generated by employing the built-in function `predict ()`.

```
> O.Vs <- predict (LDis)
> O.Vs  # scores for each sample
$`class`
 [1] 1 1 1 1 1 1 1 1 1 1 1 1 2 2 2 2 2 2 2 2 2 2 2 2 3 3 3 3 3 3 3 3 3 3 3 3 4 4 4 4 4 4 4 4 4 4
[47] 4 4 5 5 5 5 5 5 5 5 5 5 5 4
Levels: 1 2 3 4 5

$posterior
            1            2             3             4             5
1  1.000000e+00 0.000000e+00 1.246094e-179 1.157978e-111  5.806396e-87
2  1.000000e+00 0.000000e+00 1.231841e-173 4.785737e-107  3.798713e-83
3  1.000000e+00 0.000000e+00 2.006805e-167 4.626806e-102  1.590713e-78
4  1.000000e+00 0.000000e+00 1.284786e-210 4.084355e-136 1.348659e-108
5  1.000000e+00 0.000000e+00 1.636775e-204 2.623073e-131 2.365239e-104
6  1.000000e+00 0.000000e+00 2.082379e-179 2.045032e-111  1.342974e-86
7  1.000000e+00 0.000000e+00 1.233692e-179 1.033039e-111  3.928118e-87
8  1.000000e+00 0.000000e+00 2.082379e-179 2.045032e-111  1.342974e-86
9  1.000000e+00 0.000000e+00 3.469484e-167 7.751107e-102  2.904910e-78
10 1.000000e+00 0.000000e+00 1.437384e-173 6.571517e-107  8.190025e-83
11 1.000000e+00 0.000000e+00 2.970829e-161  3.035036e-97  2.131712e-74
12 1.000000e+00 0.000000e+00 1.336056e-173 5.336221e-107  4.632444e-83
13 0.000000e+00 1.000000e+00  3.564561e-37  1.545370e-79 3.240208e-104
14 0.000000e+00 1.000000e+00  5.187672e-40  1.133675e-83 7.107206e-109
15 0.000000e+00 1.000000e+00  7.223140e-43  7.126330e-88 8.614148e-114
16 0.000000e+00 1.000000e+00  2.449286e-34  2.106574e-75  1.477226e-99
17 0.000000e+00 1.000000e+00  3.564561e-37  1.545370e-79 3.240208e-104
18 0.000000e+00 1.000000e+00  7.624014e-43  8.425949e-88 1.572227e-113
19 0.000000e+00 1.000000e+00  7.223140e-43  7.126330e-88 8.614148e-114
20 0.000000e+00 1.000000e+00  7.587884e-43  8.034062e-88 1.205893e-113
21 0.000000e+00 1.000000e+00  1.208703e-28  4.568181e-67  5.556577e-90
22 0.000000e+00 1.000000e+00  2.449286e-34  2.106574e-75  1.477226e-99
```

```
23  0.000000e+00  1.000000e+00  3.564561e-37  1.545370e-79 3.240208e-104
24  0.000000e+00  1.000000e+00  7.624014e-43  8.425949e-88 1.572227e-113
25 3.329062e-192  1.605042e-32  1.000000e+00  1.446121e-11  3.889870e-22
26 3.631262e-155  1.394384e-49  9.989022e-01  1.097797e-03  9.938026e-11
27 2.287176e-170  5.784929e-43  9.999995e-01  4.813450e-07  3.135515e-16
28 1.990922e-167  6.888717e-44  9.999975e-01  2.500986e-06  1.305255e-14
29 2.287176e-170  5.784929e-43  9.999995e-01  4.813450e-07  3.135515e-16
30 3.631262e-155  1.394384e-49  9.989022e-01  1.097797e-03  9.938026e-11
31 1.070804e-179  3.432923e-38  1.000000e+00  5.672781e-09  1.714630e-18
32 3.329062e-192  1.605042e-32  1.000000e+00  1.446121e-11  3.889870e-22
33 6.024889e-186  2.335893e-35  1.000000e+00  2.868898e-10  2.580926e-20
34 6.078566e-180  3.252419e-38  1.000000e+00  6.354645e-09  2.964938e-18
35 1.107987e-196  1.730889e-30  1.000000e+00  1.236420e-12  2.643347e-24
36 1.034985e-204  7.504275e-27  1.000000e+00  3.686490e-14  8.824698e-26
37 4.751342e-123  1.453120e-71  1.196930e-05  9.996835e-01  3.045707e-04
38 5.208282e-128  1.980799e-67  2.374511e-04  9.996715e-01  9.106532e-05
39 1.964523e-113  6.667811e-80  2.710383e-08  9.947431e-01  5.256909e-03
40 7.599927e-104  2.710862e-88  5.559711e-11  9.134687e-01  8.653128e-02
41 1.980154e-108  5.212705e-84  1.411758e-09  9.846714e-01  1.532859e-02
42 3.883882e-113  7.898433e-80  3.041800e-08  9.965882e-01  3.411813e-03
43 1.964523e-113  6.667811e-80  2.710383e-08  9.947431e-01  5.256909e-03
44 7.599927e-104  2.710862e-88  5.559711e-11  9.134687e-01  8.653128e-02
45  1.232488e-98  2.236856e-92  2.921064e-12  8.343022e-01  1.656978e-01
46  5.729537e-99  1.735505e-92  2.392140e-12  7.653571e-01  2.346429e-01
47 4.331396e-118  1.065252e-75  6.029112e-07  9.989815e-01  1.017919e-03
48 3.883882e-113  7.898433e-80  3.041800e-08  9.965882e-01  3.411813e-03
49  3.371874e-94  8.213287e-97  7.778776e-14  4.937412e-01  5.062588e-01
50  4.472031e-85 7.243361e-106  3.207427e-17  7.986128e-02  9.201387e-01
51  1.142000e-85 4.116471e-106  1.863119e-17  5.132232e-02  9.486777e-01
52  1.116785e-86 1.591270e-110  1.492427e-21  6.665427e-05  9.999333e-01
53  8.064399e-90 3.331967e-107  6.197073e-20  1.351642e-04  9.998648e-01
54  9.689596e-77 1.803558e-115  4.053986e-21  4.575019e-03  9.954250e-01
55  3.884229e-77 1.133238e-115  2.639490e-21  3.245720e-03  9.967543e-01
56  1.142000e-85 4.116471e-106  1.863119e-17  5.132232e-02  9.486777e-01
57  1.116785e-86 1.591270e-110  1.492427e-21  6.665427e-05  9.999333e-01
58  8.064399e-90 3.331967e-107  6.197073e-20  1.351642e-04  9.998648e-01
59  1.617674e-85 3.306704e-106  1.662099e-17  4.916205e-02  9.508379e-01
60  1.251837e-98  2.209489e-92  2.913665e-12  8.349328e-01  1.650672e-01

$x
           LD1          LD2          LD3          LD4
1  -22.6245227 -0.305097135  0.312982681  0.134395719
2  -22.1347339 -0.770973651 -1.596644729 -0.578688797
3  -21.6423467 -0.318109352 -0.002160594  0.063688509
4  -25.1090614 -0.582793809 -0.756602840 -0.258351956
5  -24.6180939 -0.556284492 -0.843782882 -0.266372227
6  -22.6131961 -0.007020928  1.789230764  0.651902975
7  -22.6219605 -0.536713480 -0.409401137 -0.160310828
8  -22.6131961 -0.007020928  1.789230764  0.651902975
9  -21.6274164 -0.177465062  0.890991663  0.332654293
10 -22.1354435 -0.369587770 -0.140137613 -0.006159656
11 -21.1440020 -0.514955550 -0.659949383 -0.205506390
12 -22.1334869 -0.684050489 -1.229956879 -0.441902898
13  19.0313332 -0.157700501 -0.173928882 -0.018625920
```

```
14  19.5311333 -0.006961302  0.481763459  0.217289267
15 20.0192422 -0.303175535 -0.847010921 -0.260006165
16 18.5315331 -0.308439700 -0.829621223 -0.254541106
17 19.0313332 -0.157700501 -0.173928882 -0.018625920
18 20.0302920  0.139701108  1.138203399  0.457455250
19 20.0192422 -0.303175535 -0.847010921 -0.260006165
20 20.0209743 -0.100390573 -0.042496028  0.045149763
21 17.5436240 -0.162964667 -0.156539183 -0.013160861
22 18.5315331 -0.308439700 -0.829621223 -0.254541106
23 19.0313332 -0.157700501 -0.173928882 -0.018625920
24 20.0302920  0.139701108  1.138203399  0.457455250
25  7.1296601 -0.199813825 -0.034811292  0.025094548
26  4.1542418 -0.210342156 -0.000031895  0.036024665
27  5.3626011 -1.279130171  1.574572932 -0.504508264
28  5.1421508 -0.355817190 -0.673113935 -0.205355580
29  5.3626011 -1.279130171  1.574572932 -0.504508264
30  4.1542418 -0.210342156 -0.000031895  0.036024665
31  6.1307012 -0.497215435 -1.346943573 -0.450986621
32  7.1296601 -0.199813825 -0.034811292  0.025094548
33  6.6298600 -0.350553024 -0.690503633 -0.210820638
34  6.1417510 -0.054338792  0.638270747  0.266474793
35  7.4933564 -1.286925066 -1.218551452  0.372453604
36  8.1286189  0.097587784  1.277320989  0.501175718
37  1.1788235 -0.220870487  0.034747502  0.046954782
38  1.6786236 -0.070131288  0.690439843  0.282869969
39  0.1909145 -0.075395454  0.707829542  0.288335027
40 -0.7983977  0.094941427  1.452798374  0.565550198
41 -0.3093708 -0.341996452 -0.385689842 -0.115950188
42  0.1798646 -0.518272097 -1.277384779 -0.429126387
43  0.1909145 -0.075395454  0.707829542  0.288335027
44 -0.7983977  0.094941427  1.452798374  0.565550198
45 -1.3078445 -0.523536262 -1.259995080 -0.423661329
46 -1.2967947 -0.080659619  0.725219241  0.293800086
47  0.6790234 -0.371609686 -0.620944839 -0.188960404
48  0.1798646 -0.518272097 -1.277384779 -0.429126387
49 -1.7965948 -0.231398818  0.069526900  0.057884899
50 -2.7955536 -0.528800428 -1.242605381 -0.418196270
51 -2.7776800 -0.019678429  1.189932667  0.437998833
52 -2.9943963  4.171827397 -1.461073927  3.210683376
53 -2.6344456  4.109156035  0.591156141 -3.743556752
54 -3.7828570 -0.300479020 -0.202087893 -0.035779329
55 -3.7730542  0.055474460  1.416438578  0.544896186
56 -2.7776800 -0.019678429  1.189932667  0.437998833
57 -2.9943963  4.171827397 -1.461073927  3.210683376
58 -2.6344456  4.109156035  0.591156141 -3.743556752
59 -2.7914838 -0.040384131  0.733859854  0.333152281
60 -1.3084858 -0.527613051 -1.259247481 -0.419410532
```

It can easily be seen that the results generated by the built-in function `predict ()` were stored in the location `O.Vs` to be employed for further analysis, such as to build a plot or to serve other objectives.

One can employ the value of the discriminant function concerning each sample to investigate the relationship between various samples of the solid waste and the generated discriminant function. A bar chart of the values of Z_1 for various samples obtained from various palm oil mills was produced by employing the built-in function `barplot ()`, as presented in Fig. 10.2. The solid waste samples are represented by the horizontal axes (12 bars of each mill), and the values of the discriminant function is represented by the height of the Z_1 for all samples. The second palm oil mill (Mill 2) showed the highest positive effect to the first discriminant function, and the third palm oil mill (Mill 3) exhibited the second highest effect to the first discriminant function. The highest negative effect was exhibited by the first palm oil mill (Mill 1), and the fifth palm oil mill (Mill 5) showed negative effect that was smaller than the first palm oil mill. High EFB and fiber is the main source for the positive effect, whereas the negative effect is mainly attributed to potash; much less of the contribution was due to the shell. The second discriminant function showed a different pattern of the effect of chosen parameters, as presented in Fig. 10.3; the positive contribution is mainly due to potash. Furthermore, most palm oil mills exhibited a negative contribution, which is mainly due to the high value of EFB and less to the other parameters. This difference in the behavior of different palm oil mills indicates that different mills have different properties and are affected by the activities surrounding the location or other environmental reasons.

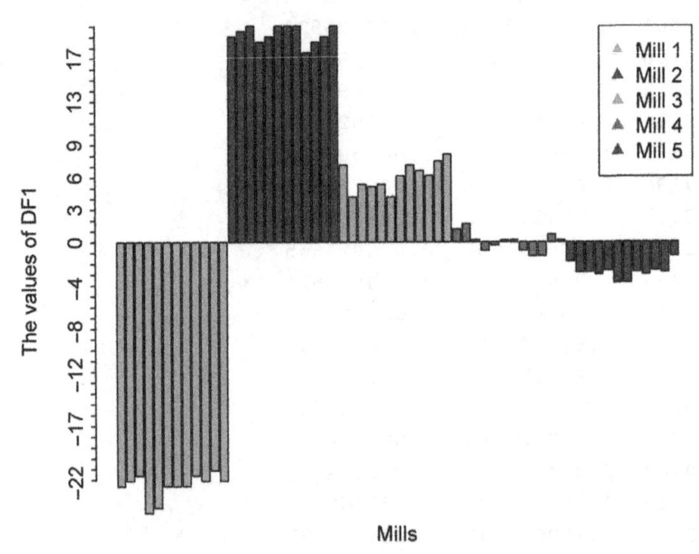

FIG. 10.2 Showing the values for the first discriminant function for palm oil mills.

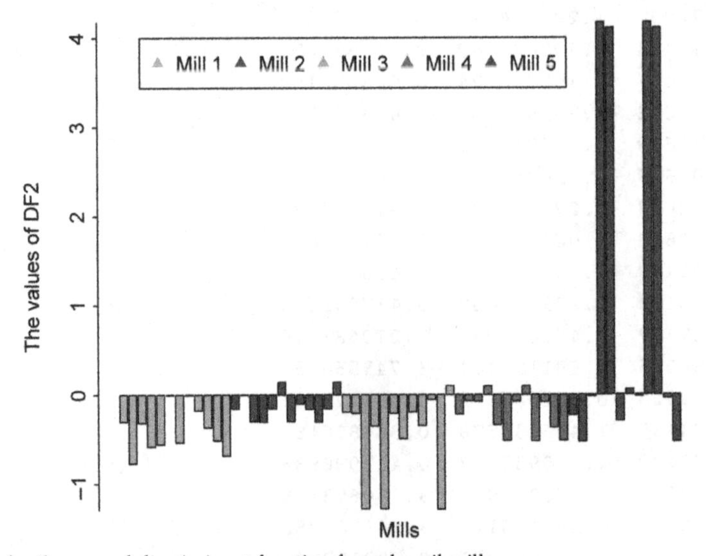

FIG. 10.3 Showing the values for the second discriminant function for palm oil mills.

The values of the first two discriminant functions called for further analysis to obtain a clear picture of the behavior of different palm oil mills with respect to the first two discriminant functions. A plot of the values of

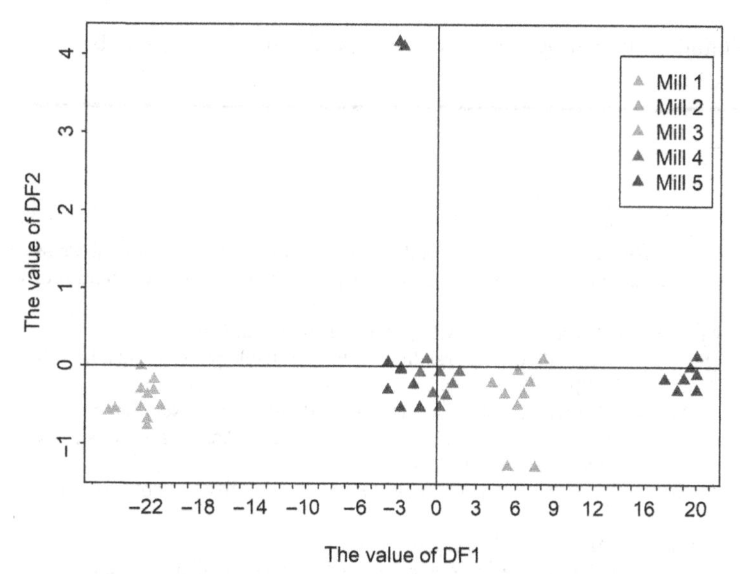

FIG. 10.4 Showing the values for the first and second discriminant functions for palm oil mills.

the first two discriminant functions was generated by employing the built-in function `plot ()`, as presented in Fig. 10.4 The difference is very clear in the behavior of different palm oil mills with regard to the first two discriminant functions.

EXAMPLE 10.4 CLASSIFICATION OF THE AMOUNT OF THE SOLID WASTE IN THE PALM OIL MILLS

Produce a table for the percentages of correct classifications and misclassifications and construct a table for the number of samples regarding the data of solid waste samples obtained from various palm oil mills along with the results of the discriminant function obtained in Example 10.3.

The rule of classification was carried out on the data set of the solid waste generated from various palm oil mills, which included five palm oil mills and four parameters. Employing the built-in functions `table ()`, `prob.table (data frame, 1)`, and `sum (diag (prop.table (data frame, 1)))` will produce the information presented below.

```
> Table <- table (Example10_3 $ Location, Scores $ class)
> Table
   1  2  3  4  5
1 12  0  0  0  0
2  0 12  0  0  0
3  0  0 12  0  0
4  0  0  0 12  0
5  0  0  0  1 11
> prop.table (Table, 1)
           1          2          3          4          5
1 1.00000000 0.00000000 0.00000000 0.00000000 0.00000000
2 0.00000000 1.00000000 0.00000000 0.00000000 0.00000000
3 0.00000000 0.00000000 1.00000000 0.00000000 0.00000000
4 0.00000000 0.00000000 0.00000000 1.00000000 0.00000000
5 0.00000000 0.00000000 0.00000000 0.08333333 0.91666667
> diag (prop.table (Table, 1))
        1         2         3         4         5
1.0000000 1.0000000 1.0000000 1.0000000 0.9166667
> sum (diag (prop.table (Table))) # total percent correct
[1] 0.9833333
```

We see that 98.33% of the samples gathered from various palm oil mills were correctly assigned (allocated) into their respective palm oil mill, and a 1.6667% misclassification rate is shown by the results of the built-in functions. This strong result indicates that each palm oil mill is affected by various conditions. The difference could be due to various locations or other environmental reasons.

Further Reading

Alkarkhi, A.F.M., Alqaraghuli, W.A.A., 2019. Easy Statistics for Food Science with R, first ed. Academic Press.

Alkarkhi, A.F.M., Anees, A., Norli, I., Azhar, M.E., 2008. Multivariate analysis of heavy metals concentrations in river estuary. Environ. Monit. Assess. 143, 179–186.

Bryan, F.J.M., 1991. Multivariate Statistical Methods: A Primer. Chapman & Hall, Great Britain.

Daniel & Hocking, 2013. Blog Archives, High Resolution Figures In R. [Online]. R-Bloggers. Available: https://www.r-bloggers.com/author/daniel-hocking/. [(Accessed 26 February 2019)].

Johnson, R.A., Wichern, D.W., 2002. Applied Multivariate Statistical Analysis. Prentice Hall, New Jersey.

Kabacoff, R.I., 2017. Quick-R, Accessing The Power Of R, Discriminant Function Analysis [Online]. Datacamp. Available: https://www.statmethods.net/advstats/factor.html. [(Accessed 26 February 2019)].

Ngu, C.C., Nik Norulaini, N.A.R., Alkarkhi, A.F.M., Anees, A., Omar, M., K, A., 2010. Assessment of major solid wastes generated in palm oil Mills. Int. J. Environ. Technol. Manage. 13.

Rencher, A.C., 2002. Methods of Multivariate Analysis. J. Wiley, New York.

Thiagogm, G.M., 2014. Computing And Visualizing Lda In R [Online]. R-Bloggers. Available: https://www.r-bloggers.com/computing-and-visualizing-lda-in-r/. [(Accessed 26 February 2019)].

CHAPTER

11

Clustering Approaches

LEARNING OBJECTIVES

After careful consideration of this chapter, you should be able:

- *To explain the concept of cluster analysis (CA).*
- *To explain the procedure of using cluster analysis for assigning observations.*
- *To know how to present the steps of cluster analysis pictorially in a graph called a dendrogram.*
- *To understand how to apply cluster analysis to environmental science data.*
- *To understand how to interpret the generated dendrogram.*
- *To describe the built-in functions and commands in R for cluster analysis.*
- *To comprehend the result of R output for cluster analysis and how to employ it in the analysis.*
- *To draw smart conclusions and recommendations.*

11.1 WHAT IS CLUSTER ANALYSIS?

Cluster analysis (CA) or clustering is a statistical technique employed to sort a set of observations (individuals) into different groups called clusters; each cluster represents a collection of observations (individuals) that are close to each other, and the observations are similar within each cluster and dissimilar with other clusters. Similarity and dissimilarity is represented by the distance between various observations. A good clustering should generate a high quality cluster with similar observations in each cluster and dissimilar with other clusters. In cluster analysis the number of clusters and the number of individuals in each cluster are not predefined. Clustering can be performed by various algorithms; the algorithms are significantly different in the way of clustering and forming the clusters. Cluster analysis is an iterative process to form various clusters. Researchers from various fields have employed cluster analysis to solve a wide variety of research issues. Cluster analysis is useful for classifying and recognizing the correct groups. For instance, an environmentalist wants to investigate the water quality of a river, he collects a large number of of water samples from various sampling points distributed along the river, and he wants to know how many different sampling points can be recognized in this river. If the sampling points can be sorted in clusters, then he can select a sample from each cluster to investigate the water quality, so long as the sampling points are similar within the same cluster and differ with other clusters. Researchers have employed cluster analysis to deal with a variety of issues in various fields such as food, marketing, medicine, science, engineering, and others.

11.2 CLUSTER ANALYSIS IN R

Variety of built-in functions and commands are offered by R to carry out cluster analysis and generate beautiful and lovely plots. The built-in functions are presented in an easy and simple way along with sufficient explanation of the structure and the use of each function.

Applied Statistics for Environmental Science with R
https://doi.org/10.1016/B978-0-12-818622-0.00011-3

1. The Euclidean distance matrix for all possible combination of two observations in a data set can be computed by employing the built-in function `dist ()` in R. The results of this function should be kept in a new location to be used for other objectives.

```
dist (data frame)
```

2. After computing the distance matrix, the built-in function `hclust ()` should be employed to carry out cluster analysis (hierarchical analysis). The structure of this built-in function is:

```
hclust (Data frame)
```

Cluster analysis can be carried out in R employing various methods such as "complete", "ward", "single", and "average"; the complete method is the default setting in R. Below is the structure of performing cluster analysis with the desired method employed to carry out the analysis.

```
hclust (Data frame, method = " average")
```

3. It is better to present the results of clustering graphically employing a dendrogram. A dendrogram is a graph used to represent the results of clustering in a pictorial form and show how the clusters are shaped. The dendrogram can be generated by employing the built-in function `plot ()` in R.

```
plot (Data frame)
```

We can place a label to the dendrogram call as shown below.

```
plot (Data frame, labels = as.character (file name $ variable name))
```

4. A given number of clusters can be recognized by employing the built-in function `rect.hclust ()`.

```
rect.hclust (Data frame, g)
```

where g represents the number of clusters to be recognized.
Or we can add the argument `h = height` to cut off the dendrogram at a given height (h).

```
rect.hclust (Data frame, h = height)
```

5. Saving the cluster numbers in a new place can be done by employing the built-in function `cutree ()`.

```
Cluster_number <- cutree (Data frame, n)
```

6. Non-hierarchical clustering can be carried out by employing the built-in function `kmeans ()`. We need to specify the number of clusters in the structure of the call.

```
Kmeans (Data frame, No. of clusters)
```

7. Calculating the cluster means can be done by employing the built-in function `aggregate ()`

```
Aggregate (Data frame, by = list ( $cluster), FUN = mean)
```

11.3 MEASURES OF DISTANCE

The procedure of clustering depends on measuring similarity in order to group a set of data into various clusters. Similarity is represented in terms of Euclidean distance in clustering approaches. The Euclidean distance function is usually used to measure the distance between the individuals.

Consider A and B are two points in the plane or 3-dimensional space, the distance between $A=(X_1, X_2, X_k)$ and $B=(Y_1, Y_2, \ldots, Y_k)$ is presented in Eq. (11.1).

$$d=\sqrt{(X_1-Y_1)^2+(X_2-Y_2)^2+\ldots+(X_k-Y_k)^2} \tag{11.1}$$

where d is called the Euclidean distance.

We are usually interested in standardizing the variables in order to take into consideration the correlation (or covariances) between various selected variables. The distance formula in (11.1) will take the form in (11.2) after standardizing the variables and is called the statistical distance.

$$d = \sqrt{\frac{(X_1 - Y_1)^2}{S_{11}} + \frac{(X_2 - Y_2)^2}{S_{22}} + \ldots + \frac{(X_k - Y_k)^2}{S_{kk}}} \tag{11.2}$$

11.4 CLUSTERING PROCEDURES

There are two types of procedures for performing cluster analysis; the first type is hierarchical approach and the second type is nonhierarchical approach.

11.4.1 Hierarchical Clustering Procedure

Hierarchical clustering consists of two methods or techniques; the two methods are agglomerative and divisive methods. Distance matrix should be computed first and then clusters formed sequentially. The result of hierarchical clustering ends with a figure called a dendrogram, as presented in Fig. 11.1 for linkage method, including single linkage, complete linkage and average linkage.

11.4.1.1 Agglomerative Method

The agglomerative method considers each individual as a cluster starts, and then, based on the distance, the close individuals are grouped in a new cluster. We continue grouping similar individuals in one cluster that are different from other clusters. Every time a new cluster is formed and the distance should be computed between all possible pairs, the number of clusters loses one and the number of individuals in each cluster becomes higher. The result of this procedure ends with only one group to represent all individuals.

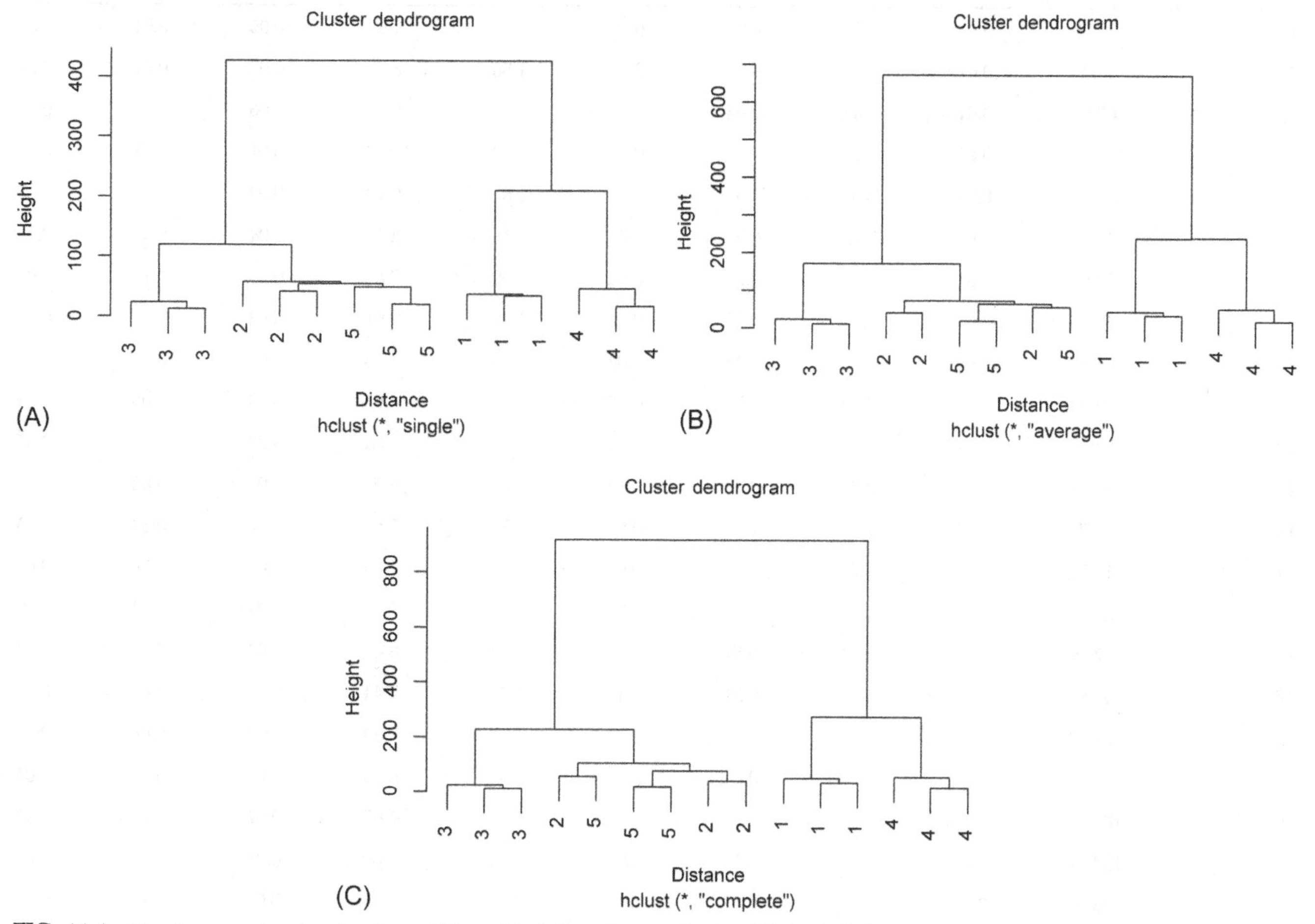

FIG. 11.1 Dendrogram showing the steps of hierarchical clustering employing: (A) single linkage; (B) complete linkage; (C) average linkage.

11.4.1.2 *Divisive Method*

The divisive method starts with a single cluster representing all observation, then this cluster is divided into two different clusters; at each step, one cluster should be divided into two clusters. We continue until consistency (stability) is accomplished.

In summary, it can be said that the agglomerative method is more desirable than the divisive method. The hierarchical methods are single linkage (nearest neighbor), complete linkage (farthest neighbor), average linkage, median, ward's method, and the flexible data method.

11.4.2 Non-Hierarchical Clustering Procedure

The second approach of clustering analysis is known as a non-hierarchical procedure, and uses different criteria for grouping observations into the clusters. This approach consists of a number of methods used for clustering, such as partitioning, k-means, and mixture methods.

EXAMPLE 11.1 INORGANIC ELEMENTS IN PARTICULATE MATTER IN THE AIR

The concentrations of nine inorganic elements (Al, Zn, Fe, Cu, Ca, Na, Mn, Ni, and Cd) in particulate matter (PM_{10}) in the air of an equatorial urban coastal location during 2009 were studied during summer and winter monsoon seasons using high-volume sampling techniques. Atomic absorption spectrophotometry was used to analyze the samples. The sample data for the inorganic elements are taken from Table 1.4. The sample data for the concentration of inorganic elements are reproduced in Table 11.1 (representing the summer season only).

TABLE 11.1 The Results of Inorganic Elements for Summer Season in Particulate Matter in the Air (μg/m^3)

Parameter	PM_{10}	Al	Zn	Fe	Cu	Ca	Na	Mn	Ni	Cd
1	37.71	0.01	2.32	0.97	0	1.37	7.94	0.05	0.01	0.01
2	48.03	0.01	2.05	0.75	0	1.56	8.46	0.05	0.01	0.04
3	67.87	0.01	3.43	0.44	0	1.41	9.98	0.06	0.1	0.05
4	39.01	0.01	4.24	0.45	0	1.12	12.93	0.01	0.07	0
5	38.33	0.01	3.51	0.5	0.02	1.6	9.92	0.02	0	0.05
6	29.7	0.01	2.78	0.58	0.02	1.38	12.13	0.04	0.06	0.02
7	53.66	0.01	2.4	0.73	0.02	1.59	7.81	0.23	0.11	0.02
8	132.28	0.01	2.34	0.55	0.02	1.59	9.01	0.03	0	0.01
9	66.31	0.01	2.13	0.45	0.02	1.23	8.76	0.02	0	0.05
10	69.2	0.01	2.13	0.5	0.03	1.95	8.66	0.03	0.08	0.05
11	78.17	0.01	1.96	0.56	0.04	1.35	8.88	0.03	0.02	0.07
12	31.63	0.01	2.21	0.42	0.04	1.24	9.02	0.02	0.08	0.05
13	66.73	0.01	1.46	0.51	0.03	1.33	8.53	0.02	0.13	0.03
14	113.56	0.01	2.07	0.6	0.03	1.51	7.65	0.02	0.06	0.05
15	123.4	0.01	1.75	0.57	0.03	1.75	7.45	0.02	0.12	0.04
16	72.39	0.01	1.61	0.57	0.03	1.54	8.4	0.02	0.07	0.02
17	51.85	0.01	1.16	0.51	0.18	1.72	7.44	0.01	0.09	0
18	77.59	0.01	1.66	0.81	0.04	2.04	7.89	0.03	0.08	0.03
19	30.3	0.01	1.09	0.61	0.03	1.8	6.28	0.02	0.05	0.03
20	100.4	0.01	1.82	0.47	0.04	1.7	7.05	0.02	0.05	0.02
21	132.98	0.01	1.86	0.52	0.03	1.74	7.69	0.01	0.1	0.03
22	126.38	0.01	0.34	0.02	0	0.01	1.36	0.01	0.18	0
23	31.82	0.01	1.41	0.52	0.03	1.93	7.04	0.02	0.04	0
24	110.53	0.01	1.25	0.67	0.04	2.29	7.39	0.02	0.03	0

TABLE 11.1 The Results of Inorganic Elements for Summer Season in Particulate Matter in the Air (μg/m³)—cont'd

Parameter	PM_{10}	Al	Zn	Fe	Cu	Ca	Na	Mn	Ni	Cd
25	38.41	0.01	1.55	0.79	0.12	2.21	7.74	0.02	0.06	0.03
26	124.26	0.01	2.24	0	0	1.56	9.65	0.07	0.06	0.01
27	53.62	0.01	1.97	0.01	0	1.39	7.07	0.03	0.04	0.03
28	23.3	0.01	1.28	0.02	0	1.8	6.39	0.03	0	0
29	67.34	0.01	0.91	0.02	0	0.8	7.14	0	0.08	0
30	22.58	0.03	1.29	0.02	0.01	0.92	7.21	0.02	0.05	0.02
31	54.52	0.03	1.37	0.04	0.01	1.48	7.98	0.01	0.03	0.03
32	112.83	0.01	1.88	0.03	0	1.46	8.44	0.01	0.02	0
33	175.28	0.01	1.68	0.03	0	1.38	8.13	0	0.03	0
34	47.3	0.01	2.2	0.03	0	1.13	11.39	0.01	0.03	0.01
35	57.08	0.01	1.69	0.04	0	1.04	8.34	0.02	0.07	0
36	15.16	0.01	1.58	0.04	0.01	1.77	6.69	0.05	0.08	0.02
37	272.98	0.01	1.88	0.04	0.01	1.6	7.05	0.04	0.08	0
38	101.82	0.01	1.86	0.02	0.01	1.79	6.02	0.05	0.13	0
39	59.9	0.01	1.73	0.02	0.01	1.7	7.25	0.05	0	0.02
40	31.07	0.01	1.92	0.02	0.02	1.4	4.92	0.05	0.03	0.04
41	107.43	0.01	2.03	0.02	0.02	1.48	5.19	0.04	0.06	0
42	30.51	0.01	2.42	0.03	0.03	2.08	5.7	0.07	0	0.04
43	84.02	0.01	2.1	0.22	0.01	1.61	5.85	0.03	0	0
44	7.37	0.01	2.63	0.35	0.02	1.24	3.9	0.03	0	0
45	7.65	0.01	1.38	0.34	0.01	1.11	4.2	0.04	0	0
46	15.05	0.01	1.56	0.21	0.02	0.35	2.47	0.03	0	0
47	84.13	0.01	1.89	0.04	0.02	0.41	4.44	0.02	0	0.02

The Euclidean distance for all possible combinations between different inorganic elements in particulate matter in the air should be computed to produce the distance matrix as the first step in the clustering procedure.

The built-in function `dist ()` was employed to compute the distances for all possible combinations between different observations for the stored data `.CSV` (`Example11_1`) of different inorganic elements in particulate matter in the air sample data. The results of employing the function `dist ()` are stored in a place called (`Euclidean`).

```
> Euclidean <- round (dist (Example11_1[-1]), digits = 2)
> Euclidean
      1       2      3      4      5      6      7      8      9
2   589.18
3   578.77  13.82
4   587.52  2.95   11.76
5   589.13  0.23   13.76  2.84
6   580.95  10.37  6.08   8.24   10.27
7   547.93  53.19  40.11  50.93  53.12  43.55
8   589.07  0.28   13.66  2.82   0.33   10.22  53.05
9   588.91  0.40   13.58  2.75   0.39   10.11  52.93  0.38
10  589.14  0.15   13.75  2.87   0.24   10.30  53.12  0.27   0.39
```

The function `dist ()` generated the distance matrix for the inorganic elements in particulate matter in the air and the results stored in a new location called `Euclidean`. The built-in function `round ()` was employed to round the values in the distance matrix to two decimals. The first column and the first row of the results showed the numbers 1–10, these represent the parameters $PM_{10} = 1$,

Al=2, Zn=3, Fe=4, Cu=5, Ca=6, Na=7, Mn=8, Ni=9, and Cd=10. The next step is to call the stored data for the Euclidean distance (`Euclidean`) and start conducting clustering analysis for the inorganic elements in particulate matter in the air environment. The built-in function `hclust ()` was employed to perform cluster analysis using the distance matrix computed earlier.

```
> Cluster <- hclust (Distance, method = "single")
> Cluster

Call:
hclust(d = Distance, method = "single")

Cluster method : single
Distance       : euclidean
Number of objects: 10
```

The single linkage (nearest neighbor) method was employed to carry out cluster analysis, as appeared in the results generated by the function `hclust ()` and the results stored in a place called `Cluster`. We employed the single linkage method to illustrate the use of distances in all stages of clustering analysis and show how the agglomeration occurs. Researchers can choose other methods to conduct cluster analysis. The single linkage technique uses the smallest distance to form the clusters.

The Euclidean distance matrix for the inorganic elements in particulate matter in the air environment has been presented in Table 11.2. The smallest distance in Table 11.2 is **0.15** (bolded) which corresponds to the distance between the parameter Al=2 and Cd=10 (the highlighted column and row). Thus, the first cluster will represent two elements Al and Cd, C1=Al, Cd].

TABLE 11.2 Distance Matrix for the Inorganic Elements in Particulate Matter in the Air Environment

	1	2	3	4	5	6	7	8	9	10
1	0.00	589.18	578.77	587.52	589.13	580.95	547.93	589.07	588.91	589.14
2	589.18	0.00	13.00	2.95	0.23	10.37	53.19	0.28	0.40	**0.15**
3	578.77	13.00	0.00	11.76	13.76	6.08	40.11	13.66	13.58	13.75
4	587.52	2.95	11.76	0.00	2.84	8.24	50.93	2.82	2.75	2.87
5	589.13	0.23	13.76	2.84	0.00	10.27	52.12	0.33	0.39	0.24
6	580.95	10.37	6.08	8.24	10.27	0.00	43.55	10.22	10.11	10.30
7	547.93	53.19	40.11	50.93	52.12	43.55	0.00	52.05	52.93	53.12
8	589.07	0.28	13.66	2.82	0.33	10.22	52.05	0.00	0.38	0.27
9	588.91	0.40	13.58	2.75	0.39	10.11	52.93	0.38	0.00	0.39
10	589.14	**0.15**	13.75	2.87	0.24	10.30	53.12	0.27	0.39	0.00

A new table of distances for different inorganic elements in particulate matter in the air environment is given in Table 11.3. This table has one fewer column and one fewer row than Table 11.2.

TABLE 11.3 The Distances After Forming the First Cluster

	1	C1	3	4	5	6	7	8	9
1	0.00	589.14	578.77	587.52	589.13	580.95	547.93	589.07	588.91
C1	589.14	0.00	13.00	2.87	**0.23**	10.30	53.12	0.27	0.39
3	578.77	13.00	0.00	11.76	13.76	6.08	40.11	13.66	13.58
4	587.52	2.87	11.76	0.00	2.84	8.24	50.93	2.82	2.75
5	589.13	**0.23**	13.76	2.84	0.00	10.27	52.12	0.33	0.39
6	580.95	10.30	6.08	8.24	10.27	0.00	43.55	10.22	10.11
7	547.93	53.12	40.11	50.93	52.12	43.55	0.00	52.05	52.93
8	589.07	0.27	13.66	2.82	0.33	10.22	52.05	0.00	0.38
9	588.91	0.39	13.58	2.75	0.39	10.11	52.93	0.38	0.00

The second step is to find the smallest distance in the new table (Table 11.3) to form the second cluster; it can be observed that the smallest distance is **0.23** (bolded) between C1 and Cu = 5 (the highlighted column and row). These points will form the second cluster C2 = [C1, Cu]. The new table of distances is given in Table 11.4, and as before, it has one fewer column and one fewer row from the table that preceded it, namely, Table 11.3.

TABLE 11.4 The Distances After Forming the First and Second Clusters

	1	C2	3	4	6	7	8	9
1	0.00	589.13	578.77	587.52	580.95	547.93	589.07	588.91
C2	589.13	0.00	13.00	2.84	10.27	52.12	**0.27**	0.39
3	578.77	13.00	0.00	11.76	6.08	40.11	13.66	13.58
4	587.52	2.84	11.76	0.00	8.24	50.93	2.82	2.75
6	580.95	10.27	6.08	8.24	0.00	43.55	10.22	10.11
7	547.93	52.12	40.11	50.93	43.55	0.00	52.05	52.93
8	589.07	**0.27**	13.66	2.82	10.22	52.05	0.00	0.38
9	588.91	0.39	13.58	2.75	10.11	52.93	0.38	0.00

The next step is to search the smallest distance in Table 11.4. The smallest distance is **0.27** (bolded) between cluster 2 (C2) and Mn = 8 (the highlighted column and row), which are combined to form the third cluster (C3), C3 = [C2, Mn]. The new table of distances is given in Table 11.5.

TABLE 11.5 The Distances After Forming the First Three Clusters

	1	C3	3	4	6	7	9
1	0.00	589.07	578.77	587.52	580.95	547.93	588.91
C3	589.07	0.00	13.00	2.82	10.22	52.05	**0.38**
3	578.77	13.00	0.00	11.76	6.08	40.11	13.58
4	587.52	2.82	11.76	0.00	8.24	50.93	2.75
6	580.95	10.22	6.08	8.24	0.00	43.55	10.11
7	547.93	52.05	40.11	50.93	43.55	0.00	52.93
9	588.91	**0.38**	13.58	2.75	10.11	52.93	0.00

The fourth cluster C4 is formed from the third cluster (C3) and Ni = 9: C4 = [C3, Ni] (the highlighted column and row) because the smallest distance in Table 11.5 is **0.38**. The new table of distances is given in Table 11.6, and as before, it has one fewer column and one fewer row from the table that preceded it, namely, Table 11.5.

TABLE 11.6 The Distances After Forming the First Four Clusters

	1	C4	3	4	6	7
1	0.00	588.91	578.77	587.52	580.95	547.93
C4	588.91	0.00	13.00	**2.75**	10.11	52.05
3	578.77	13.00	0.00	11.76	6.08	40.11
4	587.52	**2.75**	11.76	0.00	8.24	50.93
6	580.95	10.11	6.08	8.24	0.00	43.55
7	547.93	52.05	40.11	50.93	43.55	0.00

The fifth cluster C5 is formed from the fourth cluster (C4) and Fe = 4: C5 = [C3, Fe] (the highlighted column and row) because the smallest distance in Table 11.6 is **2.75**.

We continue employing the same steps to form other clusters until all parameters are in one cluster. The tables of distances for the other clusters are given in Tables 11.7–11.10.

TABLE 11.7 The Distances After Forming the First Five Clusters

	1	C5	3	6	7
1	0.00	587.52	578.77	580.95	547.93
C5	587.52	0.00	11.76	8.24	50.93
3	578.77	11.76	0.00	**6.08**	40.11
6	580.95	8.24	**6.08**	0.00	43.55
7	547.93	50.93	40.11	43.55	0.00

TABLE 11.8 The Distances After Forming the First Six Clusters

	1	C5	C6	7
1	0.00	589.07	578.77	547.93
C5	589.07	0.00	**8.24**	50.93
C6	578.77	**8.24**	0.00	40.11
7	547.93	52.05	40.11	0.00

TABLE 11.9 The Distances After Forming the First Seven Clusters

	1	C7	7
1	0.00	578.77	547.93
C7	578.77	0.00	**40.11**
7	547.93	**40.11**	0.00

TABLE 11.10 The Distances After Forming the First Eight Clusters

	1	C8
1	0.00	**547.93**
C8	**547.93**	**0**

The last cluster is cluster 9, which consists of cluster 8 (C8) and PM_{10} as shown in Table 11.10; C9 = [C8, PM_{10}] ends with one cluster for all inorganic elements in the air environment.

The steps of clustering analysis for inorganic elements in particulate matter in the air environment can be summarized in a graph called a dendrogram. The function `plot ()` was used to produce a dendrogram (Fig. 11.2) for the inorganic elements in particulate matter in the air environment.

```
> plot (Cluster,labels = Example11_1 $ Parameter, font.main = 1, cex.lab = 0.9, cex.main=.9,
font.sub = 1, cex.sub = .9, cex.axis = 0.9, cex=0.9)
```

The steps for the single linkage (nearest neighbor) method are presented in Fig. 11.2. The first cluster was formed by observation 2 (Al) and observation 4 (Cd), and the second cluster was formed by C1 and observation 5 (Cu). Then, the third cluster is formed from C2 and Mn and so on for other clusters.

In summary, we determined that inorganic elements in particulate matter in the air environment are close to each other. However, PM_{10} is different from other elements, as can easily be observed in Fig. 11.2. Therefore, only two groups or clusters were identified; the first cluster is of PM_{10}, and the second cluster is of inorganic elements in particulate matter in the air environment.

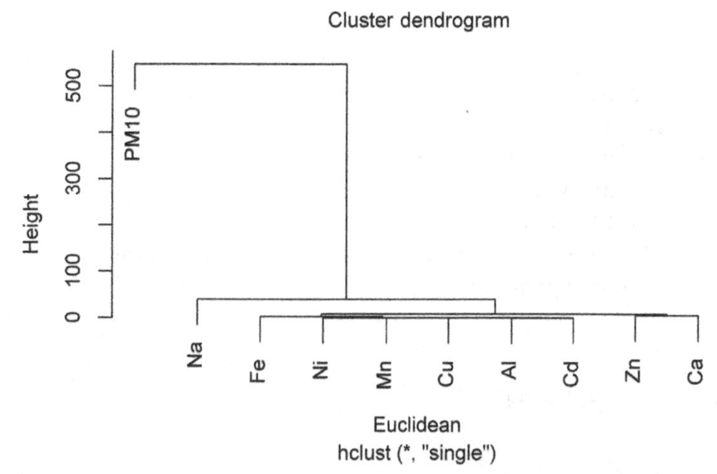

FIG. 11.2 Dendrogram showing the steps of clustering for inorganic elements in particulate matter in the air environment.

EXAMPLE 11.2 SIMILARITY OF SAMPLING POINTS

A researcher wants to investigate the similarity of selected sampling points in the Juru River by monitoring eight physiochemical parameters of the surface water. Ten sampling points were selected and analyzed for the physiochemical parameters: the temperature, pH, dissolved oxygen (DO), biochemical oxygen demand (BOD), chemical oxygen demand (COD), total suspended solids (TSS), electrical conductivity (EC), and turbidity (Tur.). The sample data for the physiochemical parameters of surface water are given in Table 11.11.

TABLE 11.11 The Results of Selected Physiochemical Parameters for the Juru River

Site	TC	pH	DO	BOD	COD	TSS	EC	Tur.
1	28.15	7.88	6.73	10.56	1248.00	473.33	42.45	13.05
2	28.20	7.92	6.64	10.06	992.50	461.67	42.75	14.11
3	28.35	7.41	5.93	6.01	1265.00	393.34	29.47	25.95
4	28.40	7.84	6.23	14.57	1124.00	473.34	42.35	22.00
5	28.30	7.86	6.29	7.36	1029.50	528.34	40.70	12.36
6	28.30	7.89	6.36	5.27	775.00	458.33	42.50	17.90
7	28.55	7.40	5.67	4.81	551.00	356.67	29.51	27.70
8	28.75	7.41	5.26	4.96	606.00	603.33	28.45	29.35
9	29.30	7.25	4.18	6.61	730.00	430.00	26.25	35.35
10	29.55	7.18	5.14	5.11	417.00	445.00	25.79	27.75

Cluster analysis for the physiochemical parameters of the surface water should start with computing the Euclidean distances between various sampling points to produce the distance matrix. The built-in function `dist ()` was employed to produce the distance matrix for the stored data`.CSV` (`Example11_2`) of the physiochemical parameters of the surface water and prepare the results to be employed by the built-in function `hclust ()` to conduct cluster analysis for various sampling points of the Juru River.

```
> Euclidean <- round (dist (Example11_2[2:9]),digits=2)
        1      2      3      4      5      6      7      8      9
2  255.77
3   83.93 281.53
4  124.39 132.33 162.90
5  225.35  76.35 272.03 110.01
6  473.29 217.61 494.53 349.47 264.03
7  706.99 454.24 714.94 585.01 508.72 246.53
8  655.41 412.21 691.66 534.38 430.61 223.42 252.73
9  520.56 265.80 536.35 397.01 316.41  58.34 193.63 213.22
10 831.80 576.18 849.58 707.85 618.52 358.78 160.54 246.58 313.46
> Cluster <- hclust (Euclidean,"single")
> Cluster

Call:
hclust(d = Euclidean, method = "single")

Cluster method   : single
Distance         : euclidean
Number of objects: 10
```

The distance matrix was stored in a new place called "`Euclidean`" and ready to be used for further analysis. One can observe that the numbers in first row and first column of the generated results are used to represent the sites, i.e., 1 = site 1, 2 = site 2, 3 = site 3, 4 = site 4, 5 = site 5, 6 = site 6, 7 = site 7, 8 = site 8, 9 = site 9, and 10 = site 10. All of the possible distances between various sites are presented in Table 11.12.

TABLE 11.12 Distance Matrix for Different Sites

	Site									
	1	**2**	**3**	**4**	**5**	**6**	**7**	**8**	**9**	**10**
1	0.00	255.77	83.93	124.39	225.35	473.29	706.99	655.41	520.56	831.80
2	255.77	0.00	281.53	132.33	76.35	217.61	454.24	412.21	265.80	576.18
3	83.93	281.53	0.00	162.90	272.03	494.53	714.94	691.66	536.35	849.58
4	124.39	132.33	162.90	0.00	110.01	349.47	585.01	534.38	397.01	707.85
5	225.35	76.35	272.03	110.01	0.00	264.03	508.72	430.61	316.41	618.52
6	473.29	217.61	494.53	349.47	264.03	0.00	246.53	223.42	**58.34**	358.78
7	706.99	454.24	714.94	585.01	508.72	246.53	0.00	252.73	193.63	160.54
8	655.41	412.21	691.66	534.38	430.61	223.42	252.73	0.00	213.22	246.58
9	520.56	265.80	536.35	397.01	316.41	**58.34**	193.63	213.22	0.00	313.46
10	831.80	576.18	849.58	707.85	618.52	358.78	160.54	246.58	313.46	0.00

We can use the distance matrix presented in Table 11.2 to carry out cluster analysis employing the single linkage (nearest neighbor) method. The single linkage starts searching for the smallest distance in Table 11.2, which is **58.34** (bolded) to form the first cluster. The smallest distance represents the distance between site 6 and site 9 (the highlighted column and row in Table 11.12). Thus, the two sites (6 and 9) represent the elements of first cluster (C1), C1 = [Site 6, Site 9]. The next step is to compute the distance matrix for the new table (after combining site 6 and site 9 to represent C1). The new distances are presented in Table 11.13.

TABLE 11.13 The Distances After Forming the First Cluster

	1	2	3	4	5	C1	7	8	10
1	0.00	255.77	83.93	124.39	225.35	473.29	706.99	655.41	831.80
2	255.77	0.00	281.53	132.33	**76.35**	217.61	454.24	412.21	576.18
3	83.93	281.53	0.00	162.90	272.03	494.53	714.94	691.66	849.58
4	124.39	132.33	162.90	0.00	110.01	349.47	585.01	534.38	707.85
5	225.35	**76.35**	272.03	110.01	0.00	264.03	508.72	430.61	618.52
C1	473.29	217.61	494.53	349.47	264.03	0.00	193.63	213.22	313.46
7	706.99	454.24	714.94	585.01	508.72	193.63	0.00	252.73	160.54
8	655.41	412.21	691.66	534.38	430.61	213.22	252.73	0.00	246.58
10	831.80	576.18	849.58	707.85	618.52	313.46	160.54	246.58	0.00

A new search for the smallest value should start in the Table 11.13; the smallest value is **76.35** (bolded), which represents the distance between site 2 and site 5 (the highlighted column and row in Table 11.13). These two sites are merged to represent the second cluster (C2), C2 = [Site 2, Site 5]. The new table of distances will have three rows and three columns because site 2 and site 5 were combined to form the second cluster (C2), as shown in Table 11.14.

TABLE 11.14 The Distances After Forming the First and Second Clusters

	1	C2	3	4	C1	7	8	10
1	0.00	225.35	**83.93**	124.39	473.29	706.99	655.41	831.80
C2	225.35	0.00	272.03	110.01	217.61	454.24	412.21	576.18
3	**83.93**	272.03	0.00	162.90	494.53	714.94	691.66	849.58
4	124.39	110.01	162.90	0.00	349.47	585.01	534.38	707.85
C1	473.29	217.61	494.53	349.47	0.00	193.63	213.22	313.46
7	706.99	454.24	714.94	585.01	193.63	0.00	252.73	160.54
8	655.41	412.21	691.66	534.38	213.22	252.73	0.00	246.58
10	831.80	576.18	849.58	707.85	313.46	160.54	246.58	0.00

The same steps should be employed to form other clusters, including computing the distances and selecting the smallest distance value in each round. The tables for distances for other clusters are presented in Tables 11.15–11.20.

TABLE 11.15 The Distances After Forming the First Three Clusters

C3	0.00	225.35	124.39	473.29	706.99	655.41	831.80
C2	225.35	0.00	**110.01**	217.61	454.24	412.21	576.18
4	124.39	**110.01**	0.00	349.47	585.01	534.38	707.85
C1	473.29	217.61	349.47	0.00	193.63	213.22	313.46
7	706.99	454.24	585.01	193.63	0.00	252.73	160.54
8	655.41	412.21	534.38	213.22	252.73	0.00	246.58
10	831.80	576.18	707.85	313.46	160.54	246.58	0.00

TABLE 11.16 The Distances After Forming the First Four Clusters

	C3	*C4*	*C1*	7	8	10
C3	0.00	**124.39**	473.29	706.99	655.41	831.80
C4	**124.39**	0.00	217.61	454.24	412.21	576.18
C1	473.29	217.61	0.00	193.63	213.22	313.46
7	706.99	454.24	193.63	0.00	252.73	160.54
8	655.41	412.21	213.22	252.73	0.00	246.58
10	831.80	576.18	313.46	160.54	246.58	0.00

TABLE 11.17 The Distances After Forming the First Five Clusters

C5	0.00	217.61	454.24	412.21	576.18
C1	217.61	0.00	193.63	213.22	313.46
7	454.24	193.63	0.00	252.73	**160.54**
8	412.21	213.22	252.73	0.00	246.58
10	576.18	313.46	**160.54**	246.58	0.00

TABLE 11.18 The Distances After Forming the First Six Clusters

C5	0.00	217.61	454.24	412.21
C1	217.61	0.00	**193.63**	213.22
C6	454.24	**193.63**	0.00	246.58
8	412.21	213.22	246.58	

TABLE 11.19 The Distances After Forming the First Seven Clusters

C5	0.00	217.61	412.21
C7	217.61	0.00	**213.22**
8	412.21	**213.22**	0.00

TABLE 11.20 The Distances After Forming the First Eight Clusters

C5	0.00	**217.61**
C9	**217.61**	0.00

The last cluster is formed between cluster 5 (C5) and cluster 9, as the smallest distance is **217.61** (Table 11.20). The smallest distances for all clusters are given in Table 11.21.

TABLE 11.21 The Smallest Distances for All Steps of Clustering Physiochemical Parameters of Surface Water

Stage	Smallest distance
1	58.345
2	76.345
3	83.928
4	110.013
5	124.389
6	160.541
7	193.632
8	213.224
9	217.612

The steps of the clustering analysis can be summarized in a pictorial form to show the steps of clustering and to convey the similarities among various sampling points. A dendrogram was produced (Fig. 11.3) employing the built-in function `plot ()`.

```
plot(Cluster, labels = Example11_2 $ Site,font.main = 1, cex.lab = 0.9, cex.main=.9,
font.sub = 1, cex.sub = .9, cex.axis = 0.9, cex=0.9)
```

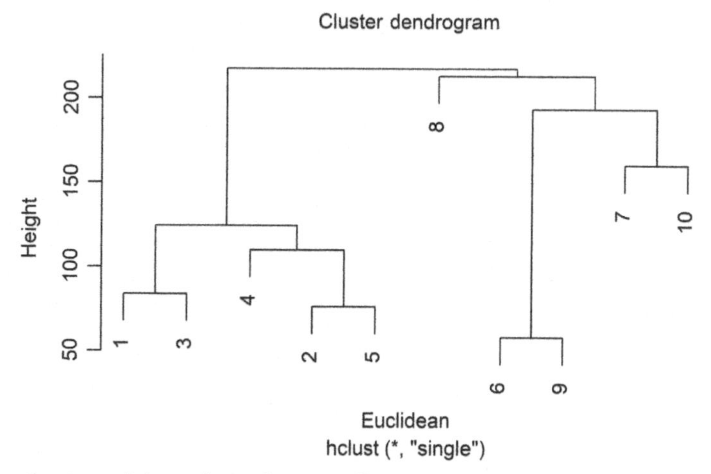

FIG. 11.3 Dendrogram showing the steps of cluster for various sampling points in Juru River.

In summary, it can be concluded that sites 1–5 are close to each other, sites 7 and 10 are close to each other, and sites 6 and 9 are close to each other. Therefore, only four clusters can be identified. The first cluster has five elements which represent five sites (1–5), the second cluster represents two sites (7 and 10), the third represents two sites (6 and 9), and the fourth cluster is of site 8. The differences could be attributed to different environmental factors such as different activities surrounding the sampling points such as industrial activities or residential areas.

Note:

- Researchers can choose whether to classify the variables or the observations according to the research objective.
- The number of clusters is determined by the researchers. Knowledge and experience regarding the research area play an important role to decide the number of clusters.

Further Reading

Alkarkhi, A.F.M., Alqaraghuli, W.A.A., 2019. Easy Statistics for Food Science with R, first ed. Academic Press.

Alkarkhi, A.F.M., Anees, A., Azhar, M.E., 2009. Assessment of surface water quality of selected estuaries of Malaysia-multivariate statistical techniques. The Environmentalist 29, 255–262.

Analytics, P., 2017. Exploring Assumptions of K-Means Clustering Using R [Online]. R-Bloggers. Available: https://www.r-bloggers.com/exploring-assumptions-of-k-means-clustering-using-r/. [(Accessed 8 July 2018)].

Bryan, F.J.M., 1991. Multivariate Statistical Methods: A Primer. Chapman & Hall, Great Britain.

Chi, Y. 2017. R Tutorial—An R Introduction To Statistics, Hierarchical Cluster Analysis [Online]. Available: http://www.r-tutor.com/gpu-computing/clustering/hierarchical-cluster-analysis [Accessed 28 February 2019].

Daniel & Hocking. 2013. Blog Archives, High Resolution Figures In R [Online]. R-Bloggers. Available: [Accessed 26 February 2019].

Johnson, R.A., Wichern, D.W., 2002. Applied Multivariate Statistical Analysis. Prentice Hall, New Jersey.

Rencher, A.C., 2002. Methods of Multivariate Analysis. J. Wiley, New York.

Robert, I.K., 2017. Quick-R, Cluster Analysis [Online]. Datacamp. Available: https://www.statmethods.net/advstats/cluster.html. [(Accessed 28 February 2019)].

Yusup, Y., Alkarkhi, A.F.M., 2011. Cluster analysis of inorganic elements in particulate matter in the air environment of an equatorial urban coastal location. Chem. Ecol. 27, 273–286.

Appendix

CHAPTER 1

Standard Normal Curve

```
jpeg ("Figure 1.1.jpeg", res = 600, height = 12, width = 16, units = 'cm')
x = seq (-4, 4, 0.1)
y = dnorm (x)
par (mar = c (3, 1, 2, 1))
plot (x, y, type = "l", lwd = 2, lty = 1, col = "black", xaxt = "n",yaxt = "n",
  ylab = "", xlab = "") # plot the curve
x = seq (0, 0, length = 200)
y = dnorm (x)
segments (x, rep(0, length (x)), x, dnorm (x, 0, 1), col = "blue", lwd = 2, lty = 4)
mtext (c ("μ"), side = 1, line = 1, col = "black", at = c (0,0))
box (which = "plot", lty = "solid", col = "blue" )
dev.off ()    #developer
```

CHAPTER 2

Example 2.17

```
T <- c (28, 31, 25, 27, 33, 23)
jpeg ("Figure 2.9.jpeg", res = 600, height = 16, width = 16, units = 'cm')
par (mar = c (7, 4, 5, 2))
plot (T, ylab = "", xlab = "", pch = 17, cex = 0.9, xaxt = "n", yaxt = "n")
axis (1, seq (1 : 5), cex.axis = 0.9, tck = -0.01, mgp = c (3, .3, 0))
axis (2, seq (1 : 10), cex.axis = 0.9, tck = -0.01, mgp = c (3, .3, 0))
mtext (side = 2, "Temperature", line = 1.5, cex = 0.9 )
mtext (side = 1, "index", line = 1.5, cex = 0.9 )
dev.off ()
```

Example 2.18

```
T <- c (28, 31, 25, 27, 33, 23)
jpeg ("Figure 2.10.jpeg", res = 600, height = 16, width = 16, units = 'cm')
par (mar = c (7, 4, 5, 2))
plot (T, type = "o", col = "blue", ylab = "", xlab = "", pch = 17, cex = 0.9, xaxt =
  "n", yaxt = "n")
axis (1, seq (1 : 5), cex.axis = 0.9, tck = -0.01, mgp = c (3, .3, 0))
axis (2, seq (1 : 10), cex.axis = 0.9, tck = -0.01, mgp = c (3, .3, 0))
mtext (side = 2, "Temperature", line = 1.5, cex = 0.9 )
mtext (side = 1, "index", line = 1.5, cex = 0.9 )
dev.off ()
```

CHAPTER 4

Example 4.1

```
mean (Example4_1 $ pH)
## key in the data
```

```
pH<- c (7.88, 7.92, 7.41, 7.84, 7.86, 7.89, 7.4, 7.41, 7.25, 7.18)
mean (pH)
##more details
pH <- c (7.88, 7.92, 7.41, 7.84, 7.86, 7.89, 7.4, 7.41, 7.25, 7.18)
s = sum (pH)
s                   #print the sum (s)
n =length (pH)
n                   #print number of values (n)
Average = s / n
Average                #print the Average
```

Example 4.2

```
mean (Example4_2 $ Solid)
## key in the data
Solid <- c (2550, 2540, 2509, 2662, 2642, 2540, 2553, 2540, 2504, 2531, 2491, 2537)
mean (Solid)
#more details
Solid <- c (2550, 2540, 2509, 2662, 2642, 2540, 2553, 2540, 2504, 2531, 2491, 2537)
s = sum (Solid)
s                   #print s
n = length (Solid)
n                   #print n
Average = s / n
Average                #print Average
```

Example 4.3

```
Average <- sapply (Example4_3, mean)
Average
```

Example 4.4

```
Average <- sapply (Example4_4, mean)
Average
```

Example 4.5

```
mean (Example4_5 $ pH)
var (Example4_5 $ pH)
sd (Example4_5 $ pH)
```

Example 4.6

```
mean (Example4_6 $ Solid)
var (Example4_6 $ Solid)
sd (Example4_6 $ Solid)
```

Example 4.7

```
round (cov (Example4_7), digits = 3)
round (cor (Example4_7), digits = 3)
#Test the correlation
Library (psych)
Library (GPArotation)
corr.test (Example4_7)
```

Example 4.8

```
round (cov (Example4_8), digits = 2)
round (cor (Example4_8), digits = 2)
#Test the correlation
Library (psych)
Library (GPArotation)
corr.test (Example4_8)
```

Example 4.9

```
jpeg ("FIGURE 4.1.jpeg", res = 600,height = 16, width = 16, units = 'cm')
par (mar = c (6.5, 4, 5, 2))
plot (Example4_9 $ Cr, Example4_9 $ Mn, xlab = "", ylab = "",
cex=0.9, pch = 15, col = "darkgreen", cex.axis = 0.9, xaxt = "n", yaxt = "n")
axis (1, seq (0, .3, .07), cex.axis = 0.9, tck = -0.01, mgp = c(3, .3, 0))
axis (2, seq (1, 2, .09), cex.axis = 0.9, tck = -0.01, mgp = c(3, .3, 0))
mtext (c ("Cr"), side = 1, line = 1.5, cex = 0.9)
mtext (c ("Mn"), side = 2, line = 1.5, cex = 0.9)
dev.off ()
```

Figure 4.2

#Figure4_2a

```
jpeg ("Figure4_2a.jpeg", res = 600,height = 16, width = 16, units = 'cm')
par (mar = c (6.5, 4, 5, 2))
plot (Figure4.2a $ x, Figure4.2a $ y, xlab="", ylab= "", cex = 0.9, pch = 16, col =
  "Orange", cex.axis = 0.9, tck=-0.01, mgp = c (3, 0.3, 0), xaxt = "n", yaxt = "n")
axis (1, seq (1, 10, 2), cex.axis = 0.9, tck = -0.01, mgp = c(3, .3, 0))
axis (2, seq (0, 100, 20), cex.axis = 0.9, tck = -0.01, mgp = c(3, .3, 0))
mtext (c ("x"), side = 1, line = 1.5, cex = 0.9)
mtext (c ("Y"), side = 2, line = 1.5, cex = 0.9)
dev.off ()
```

#Figure4_2b

```
jpeg ("Figure4_2b.jpeg", res = 600,height = 16, width = 16, units = 'cm')
par (mar = c (6.5, 4, 5, 2))
plot (Figure4.2b $ x, Figure4.2b $ y, xlab ="", ylab = "", cex = 0.9, pch = 16, col =
  "orange", cex.axis = 0.9, tck = -0.02, mgp = c (3, .3, 0), xaxt = "n", yaxt = "n")
axis (1, seq (0, 12, 2), cex.axis = 0.9, tck = -0.01, mgp = c (3, .3, 0))
axis (2, seq (40, 100, 10), cex.axis = 0.9, tck = -0.01, mgp = c (3, .3, 0))
mtext (c ("x"), side = 1, line = 1.5, cex = 0.9)
mtext (c ("Y"), side = 2, line = 1.5, cex = 0.9)
dev.off ()
```

Example 4.10

```
jpeg ("FIGURE 4.3.jpeg", res = 600,height = 16, width = 14, units = 'cm')
par (mar = c(6.5, 4, 5, 2))
pairs (~ pH + DO + BOD + EC +TSS, data = Example4_10, col ="darkgreen", cex = 0.9, pch =
  12, cex.axis = 0.9, tck = -0.03, mgp = c (3, .3, 0))
dev.off ()
```

Example 4.11

```
round (dist (Example4_11), digits = 2)
```

Example 4.12

```
round (dist (Example4_12), digits = 2)
```

CHAPTER 5

#Example 5.2 (a)

```
alpha = 0.01
z1 = round (qnorm (1 - alpha / 2) , digits = 2) # critical value
z1
c.vs = c (-z1, z1)            # critical value for two sided
c.vs #Pring critical values
jpeg ("FIGURE 5.1a.jpeg", res = 600,height = 14, width = 16, units = 'cm')
par (mar = c (5, 6, 3, 2))
x = seq (-3, 3, 0.1)          # graph the function
y = dnorm (x)
plot (x, y, type = "l", col = "blue", xaxt="n", xlab = "", yaxt = "n", cex = 0.9, ylab =
  "Probability")
axis (2, seq (0, 1, .1), col.axis = "black", cex.axis = 0.9, tck = -0.01, mgp = c (3,
  .3, 0))
x1 = seq (-3, -z1, length = 200)
y1 = dnorm (x1)
polygon (c (-3, x1, -z1), c (0, y1, 0), col = "gray")
axis (1, at = -z1, col.axis = "black", cex.axis = 0.9, tck = -0.01, mgp = c (3, .3, 0))
arrows (-2.7, 0.1, -2.7, 0.01, xpd = FALSE, length = 0.05)
text (-2.7, 0.12, alpha / 2, col = "blue", cex = 0.9)
text (0, 0.1, "Nonrejection region", cex = 0.9)
x2 = seq (z1, 3, length = 200)
y2 = dnorm (x2)
polygon (c (z1, x2, 3), c (0, y2, 0), col = "gray")
axis (1, at = z1, col.axis = "black", cex.axis = 0.9, tck = -0.01, mgp = c (3, .3, 0))
arrows (2.7, 0.1, 2.7, 0.01, xpd = FALSE, length = 0.05)
text (2.7, 0.12, alpha / 2, col = "blue", cex = 0.9)
dev.off ()
```

#Example 5.2 (b)

```
alpha = 0.01
z1 = round (qnorm (1 - alpha), digits = 3) # critical value
z1
c.vs = c (-z1, z1)            # critical value for two sided
c.vs #Pring critical values
jpeg ("FIGURE 5.1b.jpeg", res = 600, height = 14, width = 16, units = 'cm')
par (mar = c (5, 6, 3, 2))
x = seq (-3, 3, 0.1)          # graph the function
y = dnorm (x)
plot (x, y, type = "l", col = "blue", xaxt = "n", yaxt = "n", yaxt = "n", xlab = "", cex =
  0.9, ylab = "Probability")
```

```
axis (2, seq (0, 1, .1), col.axis = "black", cex.axis = 0.9, tck = -0.01, mgp = c (3,
  .3, 0))
text (-0.1, 0.1,"Nonrejection region", cex = 0.9)
axis(2,seq (0, 1, .1), col.axis = "black", cex.axis = 0.9, tck = -0.01, mgp = c(3,
  .3, 0))
x2 = seq (z1, 3, length = 200)
y2 = dnorm (x2)
polygon (c (z1, x2, 3), c(0, y2, 0), col = "gray")
axis (1, at = z1, col.axis = "black", cex.axis = 0.9, tck = -0.01, mgp = c (3, .3, 0))
arrows (2.5, 0.1, 2.5, 0.01, xpd = FALSE, length = 0.05)
text (2.5, 0.12, alpha, col = "blue", cex = 0.9)
dev.off ()
```

#Example 5.2 (c)

```
alpha = 0.01
z1 = round (qnorm (1 - alpha) , digits = 3) # critical value
z1
c.vs = c (-z1, z1)              # critical value for two sided
c.vs   #Pring critical values
jpeg ("FIGURE 5.1c.jpeg", res = 600,height = 14, width = 16, units = 'cm')
par (mar = c (5, 6, 3, 2))
x = seq (-3, 3, 0.1)            # graph the function
y = dnorm (x)
plot (x, y, type = "l", col = "blue", xaxt = "n", yaxt = "n", xlab = "", ylab =
  "Probability")
axis (2, seq (0, 1, .1), col.axis = "black", cex.axis = 0.9, tck = -0.01, mgp = c(3,
  .3, 0))
text (0.2, 0.1, "Nonrejection region", cex = 0.9)
x2 = seq (-z1, -3, length = 200)
y2 = dnorm (x2)
polygon (c (-z1, x2, -3), c(0, y2, 0), col = "gray")
axis (1, at = -z1, col.axis = "black", cex.axis = 0.9, tck = -0.01, mgp = c (3, .3, 0))
arrows (-2.5, 0.1, -2.5, 0.01, xpd = FALSE, length = 0.05)
text (-2.5, 0.12, alpha, col = "blue", cex = 0.9)
dev.off ()
```

Example 5.3

```
average = 65
m0 = 77
s = 44
n = 75
z = (average - m0) / ( s / sqrt (n))
z
alpha = 0.05
z1 = round (qnorm (1 - alpha / 2), digits = 2) # The Z critical value
c.v = c (-z1, z1)
c.v
x = seq (-3, 3, 0.1)
y = dnorm (x)
jpeg ("FIGURE 5.2.jpeg", res = 600, height = 14, width = 16, units = 'cm')
par (mar = c (5, 6, 3, 2))
```

```
plot (x, y, type = "l", col = "blue", xaxt="n", xlab = "",yaxt = "n", cex = 0.9, ylab =
  "Probability")
axis (2, seq (0, 1, .1), col.axis = "black", cex.axis = 0.9, tck = -0.01, mgp = c (3,
  .3, 0))
x1 = seq (-3, -z1, length = 200)
y1 = dnorm (x1)
polygon (c (-3, x1, -z1), c (0, y1, 0), col = "gray")
axis (1, at = -z1, col.axis = "black", cex.axis = 0.9, tck = -0.01, mgp = c (3, .3, 0))
arrows (-2.3, 0.1, -2.3, 0.01, xpd = FALSE, length = 0.05)
text (-2.2, 0.12, alpha / 2, col = "black", cex = 0.9)
text (0, 0.12, "Nonrejection region", cex = 0.9)
x2 = seq (z1, 3, length = 200)
y2 = dnorm (x2)
polygon (c (z1, x2, 3), c (0, y2, 0), col = "gray")
axis (1, at = z1, col.axis = "black", cex.axis = 0.9, tck = -0.01, mgp = c(3, .3, 0))
axis (2, seq (0, 1, 0.1), cex.axis = 0.9, tck = -0.01, mgp = c (3, .3, 0))
arrows (2.4, 0.1, 2.4, 0.01, xpd = FALSE, length = 0.05)
text (2.5, 0.12, alpha / 2, col= "black", cex =0.9)
text (-2.6, 0.17, round (z , digits = 2), col = "blue", cex = 0.9)
arrows (-2.6, 0.15, -2.6, 0.001, xpd = FALSE, length = 0.05)
dev.off ()
```

Example 5.4

```
average = 465
m0 = 450
v = 85
s = sqrt (v)
n = 40
z = (average - m0) / (s / sqrt (n))
z
alpha = 0.01
z1 = round (qnorm (1 - alpha / 2), digits = 2) # The Z critical value
c.v = c (-z1, z1)
c.v
x = seq (-3, 3, 0.1)
y = dnorm (x)
jpeg ("FIGURE 5.3.jpeg", res = 600, height = 14, width = 16, units = 'cm')
par (mar = c (5, 6, 3, 2))
plot (x, y, type = "l", col = "blue", xaxt= "n", xlab = "", yaxt = "n", cex = 0.9, ylab =
  "Probability")
axis (2, seq (0, 1, .1), col.axis = "black", cex.axis = 0.9, tck = -0.01, mgp = c(3,
  .3, 0))
x1 = seq (-3, -z1, length = 200)
y1 = dnorm (x1)
polygon (c (-3, x1, -z1), c(0, y1, 0),col = "gray")
axis (1, at = -z1, col.axis = "black", cex.axis = 0.9, tck = -0.01, mgp = c (3, .3, 0))
arrows (-2.7, 0.1, -2.7, 0.005, xpd = FALSE, length = 0.05)
text (-2.6, 0.12, alpha / 2, col = "black", cex = 0.9)
text (0, 0.12, "Nonrejection region", cex = 0.9)
x2 = seq (z1, 3, length = 200)
y2 = dnorm (x2)
polygon (c (z1, x2, 3), c (0, y2, 0), col = "gray")
axis (1, at = z1, col.axis = "black", cex.axis = 0.9, tck = -0.01, mgp = c (3, .3, 0))
```

```
arrows (2.7, 0.1, 2.7, 0.005, xpd = FALSE, length = 0.05)
text (2.6, 0.12, alpha / 2, col = "black", cex = 0.9)
text (2.9, 0.17, round (z, digits = 2), col = "blue", cex = 0.9)
arrows (2.9, 0.15, 2.9, 0.001, xpd = FALSE, length = 0.05)
axis (2, seq (0,1,0.1), cex.axis = 0.9, tck = -0.01, mgp = c (3, .3, 0))
dev.off ()
```

T graph

```
x <- seq (-4, 4, length = 200)
y <- dnorm (x)
d.f <- c (1, 3, 15)
colors <- c ("yellow", "blue", "darkgreen","black")
labels <- c ("d.f = 1", "d.f = 3", "d.f = 15", "normal")
jpeg ("FIGURE 5.4.jpeg", res = 600,height = 14, width = 16, units = 'cm')
par (mar = c (5, 6, 3, 2))
plot (x, y, type = "l", lty = 2, ylab = "", xlab = "", xaxt= "n", yaxt = "n")
for (i in 1 : 5) {
 lines (x, dt(x, d.f [i]), lwd = 1, col = colors [i])
}
legend ("topright", inset =.05, labels, lwd = 1, lty = c (1, 1, 1, 1, 1), col = colors)
axis (1, seq (-4, 4, 1), cex.axis = 0.9, tck = -0.01, mgp = c (3, .3, 0))
axis (2, seq (0, 1, .1), cex.axis = 0.9, tck = -0.01, mgp = c (3, .3, 0))
mtext (c ("Probability"), side = 2, line = 1.5, cex = 0.9)
mtext (c ("X value"), side = 1, line = 1.5, cex = 0.9)
dev.off ()
```

Example 5.5

```
Average = 0.23
m0 = 0.16
s = 0.03
n = 11
t = (Average - m0) / (s / sqrt (n))
t
alpha = 0.05
c.v = qt (1 - alpha / 2, df = n-1)
t1 = round (c.v, digits = 2)
c.v = c(-t1, t1)
c.v
x = seq (-3, 3, 0.1)
y = dnorm (x)
jpeg ("FIGURE 5.5.jpeg", res = 600, height = 14, width = 16, units = 'cm')
par (mar = c (5, 6, 3, 2))
plot (x, y, type = "l", col = "blue", xaxt="n", xlab = "", yaxt ="n", cex = 0.9, ylab =
  "Probability")
axis (2, seq (0, 1, .1), col.axis = "black", cex.axis = 0.9, tck = -0.01, mgp = c (3,
  .3, 0))
x1 = seq (-3, -t1, length = 200)
y1 = dnorm (x1)
polygon (c (-3, x1, -t1), c (0, y1, 0),col = "gray")
axis (1, at = -t1, col.axis = "black", cex.axis = 0.9, tck = -0.01, mgp = c (3, .3, 0))
arrows (-2.7, 0.1, -2.7, 0.004, xpd = FALSE, length = 0.05)
text (-2.7, 0.12, alpha / 2, col = "black", cex = 0.9)
```

```
text (0, 0.12, "Nonrejection region", cex = 0.9)
x2 = seq (t1, 3, length = 200)
y2 = dnorm (x2)
polygon (c (t1, x2, 3), c (0, y2, 0), col = "gray")
axis (1, at = t1, col.axis = "black", cex.axis = 0.9, tck = -0.01, mgp = c (3, .3, 0))
arrows (2.7, 0.1, 2.7, 0.004, xpd = FALSE, length= 0.05)
text (2.7, 0.12, alpha / 2, col = "black", cex = 0.9)
text (2.9, 0.17, round (t, digits = 2), col = "blue", cex = 0.9)
arrows (3, 0.15, 3, 0.001, xpd = FALSE, length = 0.05)
dev.off ()
```

Example 5.6

```
Average = 0.092
m0 = 0.02
s = 0.014
n = 8
t = (Average - m0) / (s / sqrt (n))
t
alpha = 0.10
c.v = qt (1 - alpha / 2, df = n-1)
t1 = round (c.v, digits = 2)
c.v = c (-t1, t1)
c.v
x = seq (-3, 3, 0.1)
y = dnorm (x)
jpeg ("FIGURE 5.6.jpeg", res = 600, height = 14, width = 16, units = 'cm')
par (mar = c (5, 6, 3, 2))
plot (x, y, type = "l", col = "blue", xaxt = "n", yaxt = "n", xlab = "", ylab = "")
x1 = seq (-3, -t1, length = 200)
y1 = dnorm (x1)
polygon (c (-3, x1, -t1), c (0, y1, 0), col = "gray")
axis (1, at = -t1, col.axis = "black", tck = -0.01, mgp = c(3, .3, 0), cex.axis = 0.9)
arrows (-2.7, 0.1, -2.7, 0.004, xpd = FALSE, length = 0.05)
text (-2.7, 0.12, alpha/2, col = "black", cex = 0.9)
text (0, 0.12, "Nonrejection region", cex = 0.9)
x2 = seq (t1, 3, length = 200)
y2 = dnorm (x2)
polygon (c (t1, x2, 3), c (0, y2, 0), col = "gray")
axis (1, at = t1, col.axis = "black", tck = -0.01, mgp = c(3, .3, 0), cex.axis = 0.9)
arrows (2.7, 0.1, 2.7, 0.004, xpd = FALSE, length = 0.05)
text (2.7, 0.12, alpha / 2, col = "black", cex = 0.9)
text (2.9, 0.17, round (t, digits = 2), col = "blue", cex = 0.9)
arrows (3, 0.15, 3, 0.001, xpd = FALSE, length = 0.05)
axis (2, seq (0, 1, 0.1), cex.axis = 0.9, tck = -0.01, mgp = c(3, .3, 0), cex =0.9)
mtext (c ("Probability"), side = 2, line = 1.5, cex = 0.9)
dev.off ()
```

Example 5.7

```
t.test (Example5_7)
t.test (Example5_7, mu = 0.12)
t.test (Example5_7, mu = 0.12, alternative = "less")
t.test (Example5_7, mu = 0.12, alternative = "greater")
```

Example 5.8

```
# Compute the mean vector
Ms <- round (sapply (Example5_8, mean), digits = 3)
Ms
Mp <- c (0.67, 15, 102, 27, 1450)
# Compute the difference
D = Ms - Mp
D
#Compute the transpose
Dt = t (D)
Dt
# Compute the number of observations
n <- length (Example5_8 $ Cd)
n
# Calculate the covariance matrix
cov = round (cov (Example5_8), digits = 3)
cov
# Compute the inverse
INV = solve (cov)
INV
#Compute Hotelling's test value
TSQ = Dt %*% INV %*% D * n
TSQ
```

Example 5.9

```
#calculate the mean vector (observed)
Ms <- round (sapply (Example5_9, mean), digits = 3)
Ms
# proposed (Hypothesized) mean vector
Mp<- c (0.25, 0.10, 4.50)
#Compute the difference
D = Ms - Mp
D
# Compute the transpose for the difference
Dt = t (D)
Dt
# Compute the number of values
n <- length (Example5_9 $ Cd)
n
# Ccompute the covariance matrix
cov = round (cov (Example5_9), digits = 3)
cov
# computee the inverse
IN = solve (cov)
IN
# Ccompute Hotelling's test value
TSQ = Dt %*% IN %*% D * n
TSQ
```

Example 5.10

```
T.v<- t.test (Example5_10 $ Juru, Example5_10 $ Jejawi)
T.v
```

```
T.v $ statistic
#Part 2 Calculate the critical values for the graph
alpha = 0.05
n1 = length (Example5_10 $ Juru)      #calculate the sample size for sample 1
n2 = length (Example5_10 $ Jejawi)    #calculate the sample size for sample 2
c.v = qt (1-alpha / 2, df = n1 + n2 - 2)    # calculate the t critical value
t1 = round (c.v, digits = 3)
t1
c.v1 = c (-t1, t1)
c.v1
jpeg ("FIGURE 5.7.jpeg", res = 600, height = 16, width = 16, units = 'cm')
x = seq (-3, 3, 0.1)            # graph the function
y = dnorm (x)
par (mar = c (6, 7, 5, 2))
plot (x, y, type = "l", col = "blue", xaxt = "n", yaxt = "n", ylab = "", xlab = "")
x1 = seq (-3, -t1, length = 200)
y1 = dnorm (x1)
polygon (c (-3, x1, -t1), c(0, y1, 0), col = "gray")
axis (1, at = -t1, col.axis = "black", cex.axis = 0.9, tck = -0.01, mgp = c (3, .3, 0))
arrows (-2.7, 0.1, -2.7, 0.004, xpd = FALSE, length = 0.05)
text (-2.7, 0.12, alpha / 2, col = "black", cex = 0.9)
text (0, 0.12, "Nonrejection region", cex = 0.9)
x2 = seq (t1, 3, length = 200)
y2 = dnorm (x2)
polygon (c (t1, x2, 3), c (0, y2, 0), col = "gray")
axis (1, at = t1, col.axis = "black", cex.axis = 0.9, tck = -0.01, mgp = c(3, .3, 0))
arrows (2.7, 0.1, 2.7, 0.004, xpd = FALSE, length = 0.05)
text (2.7, 0.12, alpha / 2, col = "black", cex = 0.9)
text (2.9, 0.17, round (T.v $ statistic, digits = 2), col = "blue", cex = 0.9)
arrows (3, 0.15, 3, 0.001, xpd = FALSE, length = 0.05)
axis (2, seq (0, 1, 0.1), cex.axis = 0.9, tck = -0.01, mgp = c(3, .3, 0))
mtext (c ("Probability"), side = 2, line = 1.5, cex = 0.9)
dev.off ()
```

Example 5.11

```
#Compute the t test statistic for PM10 in air for two seasons
T.v <- t.test (Example5_11 $ Dry, Example5_11 $ Wet)
T.v
T.v $ statistic
# calculate the critical value
n1 = length (Example5_11 $ Dry)
n2 = length (Example5_11 $ Wet)
df = n1 + n2 - 2
df
alpha = 0.05
t1 = round (qt(1 - alpha / 2, df = n1 + n2 - 2), digits = 3)
t1
c.v = c (-t1, t1)
jpeg ("FIGURE 5.8.jpeg", res = 600, height = 16, width = 16, units = 'cm')
par(mar = c(6, 6, 5, 2))
x = seq (-3, 3, 0.1)            # graph the function
y = dnorm (x)
plot (x, y, type = "l", col = "black", xaxt = "n", yaxt = "n", ylab = "", xlab = "")
```

```
x1 = seq (-3, -t1, length = 200)
y1 = dnorm (x1)
polygon (c (-3, x1, -t1), c (0, y1, 0), col = "gray")
axis (1, at = -t1, col.axis = "black", cex.axis = 0.9, tck = -0.01, mgp = c (3, .3, 0))
arrows (-2.7, 0.1, -2.7, 0.001, xpd = FALSE, length = 0.05)
text (-2.7, 0.12, alpha / 2, col = "blue", cex = 0.9)
text (0, 0.12, "Nonrejection region", cex = 0.9)
x2 = seq (t1, 3, length = 200)
y2 = dnorm (x2)
polygon (c (t1, x2, 3), c (0, y2, 0), col = "gray")
axis (1, at = t1, col.axis = "black", cex.axis = 0.9, tck = -0.01, mgp = c (3, .3, 0))
arrows (2.7, 0.1, 2.7, 0.001, xpd = FALSE, length = 0.05)
text (2.7, 0.12, alpha/2, col = "blue", cex = 0.9)
text (1.24, 0.13, round (T.v $ statistic, digits = 2), col = "blue", cex =0.9)
arrows (1.24, 0.11, 1.24, 0.001, xpd = FALSE, length = 0.05)
axis (2, seq (0, 1, 0.1), cex.axis = 0.9, tck = -0.01, mgp = c (3, .3, 0))
mtext (c ("Probability"), side = 2, line = 1.5, cex = 0.9)
dev.off ()
```

#Brief data _ Input information

```
MA = 70.336
MB = 58.658
n1 = 28
n2 = 28
df = n1 + n2 - 2
M = MA - MB
sA = 36.0569
sB = 34.313
pv = sqrt (((n1 - 1)*(sA ^ 2) + (n2 - 1)*(sB ^ 2)) / df)
pv
A = sqrt ((1 / n1) + (1 / n2))
T.v = M / (pv * A)
T.v
alpha = 0.05
c.v = qt(1 - alpha / 2, df = n1 + n2 - 2)
t = round (c.v, digits = 3)
c.vs = c (-t, t)
c.vs
```

Example 5.12

```
#compute the mean vector
MV1<- sapply (Example5_12 [1 : 10, 2 : 6], mean)
MV1
MV2 <- sapply (Example5_12 [11 : 20, 2 : 6], mean)
MV2
#Compute the variance-covariance matrix
round (cov (Example5_12 [1 : 10, 2 : 6]), digits = 3) #Juru River
round (cov (Example5_12 [11:20,2:6]), digits = 3) #Jejawi River
# Calculate the number of observations
n1 = length (Example5_12 [1 : 10, 2 : 6] $ Cu)
n2 = length (Example5_12 [11 : 20, 2 : 6] $ Cu)
#Calculate the pooled variance
```

```
spv = (((n1 - 1) * cov (Example5_12 [1 : 10, 2 : 6])) + ((n2 - 1) * cov (Example5_12
  [11 : 20, 2 : 6]))) / (n1 + n2 - 2)
round (spv, digits = 3)
D = MV1 - MV2              # Compute the difference
D1 = t (D)      # Compute the transpose
INV = solve (spv)    # Compute the inverse
#Compute the value of the Hotelling's test statistic
TSQ = ((n1 * n2) / (n1 + n2)) * (D1 %*% INV %*% D)
TSQ                 #print the result of Hotelling's test
```

Example 5.13

```
#compute the mean vector
MV1 <- sapply (Example5_13 [1 : 10,], mean)
MV1                              #Print m1
MV2 <- sapply (Example5_13 [11 : 20,], mean)
MV2
#Compute the variance-covariance matrix
round (cov (Example5_13 [1 : 10,]), digits = 3)
round (cov (Example5_13 [11 : 20,]), digits = 3)
# Compute the number of observations
n1 = length (Example5_13 [1 : 10,] $ DO)
n1
n2 = length (Example5_13 [11 : 20,] $ DO)
n2
#Compute the pooled variance
spv = (((n1 - 1) * cov (Example5_13 [1:10,])) + ((n2 - 1) * cov (Example5_13
  [11 : 20,]))) / (n1 + n2 - 2)
round (spv, digits = 3)              # Compute the difference
D = MV1 - MV2
D1 = t(D)     # Compute the transpose
INV = solve (spv)    # Compute the inverse
#Compute the value of the Hotelling's test statistic
TSQ = ((n1 * n2) / (n1 + n2)) * (D1 % * % INV% * %D)
TSQ             #print the result of Hotelling's test
```

CHAPTER 6

Example 6.1

```
ANOVA <- aov (TSS ~ Location, data = Example6_1)
ANOVA
summary (ANOVA)
```

Example 6.2

```
ANOVA <- aov (PM10 ~ Location, data = Example6_2)
ANOVA
summary (ANOVA)
```

Example 6.3

```
pH = as.factor (Example6_3 $ pH)
Shaking = as.factor (Example6_3 $ Shaking)
COD = Example6_3 $ COD
```

```
aov (COD ~ pH * Shaking) # analysis of variance with interaction
summary (aov (COD ~ pH * Shaking))
jpeg ("Figure 6.1.jpeg", res = 600, height = 16, width = 18, units = 'cm')
par (mar = c (7, 7, 5, 4))
interaction.plot ( pH, Shaking, COD, col = c ("darkgreen", "blue", "black"), legend =
  T, cex.axis = .9, tck = -0.01, mgp = c(3, .3, 0))
dev.off ()
```

Example 6.4

```
pH = factor (Example6_4 $ pH, levels = c (3, 6, 9))
Pectin = factor (Example6_4 $ Pectin, levels = c (1, 4.5, 8))
Turbidity = Example6_4 $ Turbidity
aov (Turbidity ~ pH*Pectin)        # Analysis of variance with interaction
summary (aov (Turbidity ~ pH * Pectin))
jpeg ("Figure 6.2.jpeg", res = 600, height = 16, width = 18, units = 'cm')
par(mar = c (7, 7, 5, 4))
interaction.plot (pH, Pectin, Turbidity, col = c ("darkgreen","blue", "black"),
  legend = T, cex.axis = .9, tck = -0.01, mgp = c (3, .3, 0))
dev.off ()
```

Example 6.5

```
PM10 = Example6_5 $ PM10      #Define and extract PM10 from the file
PM2.5 = Example6_5 $ PM2.5
PM1 = Example6_5 $ PM1
Location = as.factor (Example6_5 $ Point)
Responses <- cbind (PM10, PM2.5, PM1)     #To combine the responses in one place
MANOVA <- manova (Responses ~ Location)  #Perform multivariate analysis of variance
summary (MANOVA, test = "Wilks")
summary (MANOVA, test = "Hotelling-Lawley")
summary (MANOVA, test = "Roy")
summary (MANOVA, test = "Pillai")summary (MANOVA, test = "Pillai")
```

Example 6.6

```
location = as.factor (Example6_6 $ Location)
Responses <- cbind (Example6_6 $ Cd, Example6_6 $ Pb, Example6_6 $ Zn, Example6_6 $
  Cu, Example6_6 $ Fe)
MANOVA <- manova (Responses ~ location)
summary (MANOVA, "Wilks")
summary (MANOVA, "Hotelling-Lawley")
summary (MANOVA, "Pillai")
summary (MANOVA, "Roy")
```

Example 6.7

```
PM10 = Example6_7 $ PM10
PM2.5 = Example6_7 $ PM2.5
PM1 = Example6_7 $ PM1
Responses <- cbind (PM10, PM2.5, PM1)
Time = as.factor (Example6_7 $ Time)
Location = as.factor (Example6_7 $ Location)
MANOVA <- manova (Responses ~ Time * Location)
summary (MANOVA, test = "Wilks")
```

```
summary (MANOVA, test = "Hotelling-Lawley")
summary (MANOVA, test = "Pillai")
summary (MANOVA, test = "Roy")
```

Example 6.8

```
Temperature = Example6_8 $ Temperature
Conductivity = Example6_8 $ Conductivity
pH = Example6_8 $ pH
TDS = Example6_8 $ TDS
DO = Example6_8 $ DO
BOD = Example6_8 $ BOD
COD = Example6_8 $ COD
Turbidity = Example6_8 $ Turbidity
TSS = Example6_8 $ TSS
Responses <- cbind (Temperature, Conductivity, pH, TDS, DO, BOD, COD, Turbidity, TSS)
Season = as.factor (Example6_8 $ Season)
Location = as.factor (Example6_8 $ Location)
MANOVA <- manova (Responses ~ Season * Location)
summary (MANOVA, test = "Wilks")
summary (MANOVA, test = "Pillai")
summary (MANOVA, test="Hotelling-Lawley")
summary (MANOVA, test = "Roy")
```

CHAPTER 7

Example 7.1

```
Fitting = lm (y ~ x, data = Example7_1) #Fit the regression model
Fitting
anova (lm (y ~ x, data = Example7_1))   #Perform analysis of variance
summary (lm(y ~ x, data = Example7_1))
```

Example 7.2

```
Fitting = lm (Cd ~ Cu, data = Example7_2)
Fitting
anova (lm (Cd ~ Cu, data = Example7_2))
summary (lm (Cd ~ Cu, data = Example7_2))
```

Example 7.3

```
Fitting = lm (y ~ x, data = Example7_1)
Fitting
anova (lm (y ~ x, data = Example7_1))
summary (lm (y ~ x, data = Example7_1))
Fitting = (lm (y ~ x, data = Example7_1))  #Step 1: Fitting the regression model
newdata = data.frame (x = 0.03)            #Step 2: Define the required value
predict (Fitting, newdata, interval = "predict")   # Predict
```

Example 7.4

```
Cu_B = Example7_4B $ Cu
Cd_B = Example7_4B $ Cd
Fitting_B = lm (Cd ~ Cu, data = Example7_4B) #Fitting Before changing
```

```
Fitting_B
jpeg ("FIGURE 7.1a.jpeg", res = 600, height = 16, width = 18, units = 'cm')
par (mar = c (7, 5, 5, 2))
plot (Cu_B , Cd_B , pch = 19, col= "blue", tck = -0.01, mgp = c(3, .3, 0), cex = .9)
  ##before
abline (Fitting_B, col = "black")   #draw regression line for before
dev.off ()
Cu_A = Example7_4A $ Cu
Cd_A = Example7_4A $ Cd
Fitting_A = lm (Cd1 ~ Cu1, data = Example7_4A) # Fitting After changing
Fitting_A
jpeg ("FIGURE 7.1b.jpeg", res = 600, height = 16, width = 18, units = 'cm')
par (mar = c (7, 5, 5, 2))
plot (Cu_A, Cd_A, pch = 19, col = "blue", tck = -0.01, mgp = c(3, .3, 0), cex = .9)
  ##after
abline (Fitting_A, col = "black")   #draw regression line for after
dev.off ()
```

Example 7.5

```
Fitting = lm (y ~ x, data = Example7_5)
Fitting
anova (lm (y ~ x, data = Example7_5))
summary (lm (y ~ x, data = Example7_5))
predicted = round (predict (Fitting), digits = 3)
predicted
#or fitted
round (fitted (Fitting), digits = 3)
round (resid (Fitting), digits = 3) #residual
Mean = round (mean (Example7_5 $ y), digits = 3)
Mean
(Example7_5 $ y - Mean) ^ 2
sum ((Example7_5 $ y - Mean) ^ 2)
sum (round ((Example7_5 $ y - Mean) ^ 2, digits = 3)) # total variance
((predicted - Mean) ^ 2)
sum ((predicted - Mean) ^ 2) #Explained variation
Example7_5 $ y - predicted #Error
round ((Example7_5 $ y - predicted) ^ 2, digits = 3)
sum ((Example7_5 $ y - predicted) ^ 2)    #Unexplained variation
```

CHAPTER 8

Example 8.1

```
install.packages ("psych")
library (psych)
install.packages ("GPArotation")
library (GPArotation)
# calculate the variance
round (sapply (Example8_1 [2 : 5], var), digits = 2)
#calculate the correlation
corr.test (Example8_1 [,2 : 5])
#principal component analysis
Components <- princomp (Example8_1 [2 : 5], cor = TRUE)
summary (Components)
```

#Figure 8.1

```
jpeg ("Figure 8.1.jpeg", res = 600, height = 16, width = 16, units = 'cm')
par (mar = c (8, 6, 8, 3))
screeplot (Components, npcs = 4, type = "barplot", xaxt = "n", yaxt = "n")
mtext (c("Components"), side = 1, line = 1.5, cex = 0.9)
axis(2, seq (0, 4, 1), cex.axis = 0.9, tck = -0.01, mgp = c (3, .3, 0))
dev.off ()
#To get the loadings
round (unclass (Components $ loadings), digits = 3)
# To get the scores of principal components
round (Components $ scores, digits = 3)
#Bar plot for the scores
A = as.data.frame (Components $ scores)
```

#Figure 8.2

```
jpeg ("Figure 8.2.jpeg", res = 600,height = 16, width = 16, units = 'cm')
par (mar = c (8, 6, 8, 3))
barplot (A[,1], xaxt = "n", yaxt = "n", ylim = c (-5, 5), col = c (rep ('gray', 12),
  rep ('blue', 12), rep ('darkkhaki', 12), rep ('black', 12), rep ('orange', 12)))
legend (59, 5, horiz = FALSE, c ("Mill 1", "Mill 2", "Mill 3", "Mill 4", "Mill 5"),
  inset = c (0,.2), col = c ('gray', 'blue','darkkhaki', 'black', 'orange'), pch = c
  (2, 2, 2, 2, 2), cex = 0.9)
mtext (c (" PC1 values"), side = 2, line = 2, cex = 0.9)
mtext (c ("Mills"), side = 1, line = 0.2, cex = 0.7)
axis (2, seq (-6, 6, 1), cex.axis = 0.9, tck = -0.01, mgp = c (3, .3, 0))
dev.off ()
```

Example 8.2

```
install.packages ("psych")
library (psych)
install.packages ("GPArotation")
library (GPArotation)
# Calculate the variance
round (sapply (Example8_2 [2 : 11], var), digits = 2)
corr.test (Example8_2 [2 : 11])
# calculate Components
Components <- princomp (Example8_2 [2 : 11], cor = TRUE)
Components
Summary (Components) # print variance accounted for
jpeg ("Figure 8.3.jpeg", res = 600, height = 16, width = 16, units = 'cm')
par (mar = c (8, 6, 8, 3))
screeplot (Components, xaxt = "n", yaxt = "n") # scree plot
mtext (c ("Components"), side = 1, line = 2, cex = 0.9)
axis (2, seq (0, 6, 1), cex.axis = 0.9, tck = -0.01, mgp = c (3, .3, 0))
axis (1, seq (1, 10, 1), cex.axis = 0.9, tck = -0.01, mgp = c (3, .3, 0))
dev.off ()
round (unclass (Components $ loadings), digits = 2) #loadings
round (Components $ scores, digits = 3)
#Bar plot for the scores
A = as.data.frame (Components $ scores)
```

#Figure 8.4

```
jpeg ("Figure 8.4.jpeg", res = 600, height = 16, width = 16, units = 'cm')
par (mar = c (7, 6, 5, 2))
barplot (A [,1], xlab = "", ylim = c (-5, 5),ylab =", xaxt = "n", yaxt = "n", col = c
  (rep ('gray', 10), rep('blue', 10)))
legend ("topright", horiz = FALSE, c ("Juru River", "Jejawi River"), col = c ("gray",
  "blue"), pch = c (17, 17), cex = .90)
axis (2, seq (-5, 5, 1), cex.axis = 0.9, tck = -0.01, mgp = c (3, .3, 0))
mtext (c (" PC1 Values"), side = 2, line = 1.5, cex = 0.9)
mtext (c ("Sites"), side = 1, line = 1, cex = 0.9)
dev.off ()
```

#Figure 8.5

```
jpeg ("Figure 8.5.jpeg", res = 600, height = 16, width = 16, units = 'cm')
par(mar = c (7, 6, 5, 2))
barplot (A [,2], xlab = "", ylim = c (-5, 5), ylab =", pch = 19, xaxt = "n", yaxt = "n",
  col = c (rep ('gray', 10), rep ('blue', 10)))
legend ("topright", horiz = FALSE, c ("Juru River", "Jejawi River"), col = c ("gray",
  'blue'), pch = c(17, 17), cex = .90)
axis (2, seq (-5, 5, 1), cex.axis = 0.95, tck = -0.01, mgp = c (3, .3, 0))
mtext (c ("PC2 Values"), side = 2, line = 1.5, cex = 0.9)
mtext (c ("Sites"), side = 1, line = 1.5, cex = 0.9)
dev.off ()
#First and second components
jpeg ("Figure 8.6.jpeg", res = 600, height = 16, width = 16, units = 'cm')
par (mar = c (7, 6, 5, 2))
plot (A [,1], A [,2], xlab = "", ylab =", pch = 8, xaxt = "n", yaxt = "n", col = c (rep
  ('black', 10),rep ('blue', 10)))
abline (h = 0, v = 0)
legend ("topleft", horiz = FALSE, c ("Juru River", "Jejawi River"), col = c("black",
  "blue"), pch = c(8, 8), cex = 0.9)
axis(2, seq(-3, 3, 1), cex.axis = 0.9, tck = -0.01, mgp = c(3, .3, 0))
axis (1, seq (-3, 3, 1), cex.axis = 0.9, tck = -0.01, mgp = c(3, .3, 0))
mtext (c ("PC2 values"), side = 2, line = 2, cex = 0.9)
mtext (c ("PC1 values"), side = 1, line = 2, cex = 0.9)
dev.off ()
```

CHAPTER 9

Example 9.1

```
install.packages ('psych')
library (psych)
install.packages ('GPArotation')
library (GPArotation)
corr.test (Example9_1 [,2 : 19])
variance <- princomp (Example9_1 [,2 : 19],cor = TRUE) # principal components method
Summary (variance)
```

#Figure 9.1

```
jpeg ("Figure 9.1.jpeg", res = 600, height = 16, width = 18, units = 'cm')
par (mar = c (7, 6, 7, 3))
```

```
screeplot (variance, npcs = 18, type = "barplot", xaxt = "n", yaxt ="n")
mtext (c ("Components"), side = 1, line = 2, cex = 0.9)
axis (2, seq (0, 6, 1), cex.axis = 0.9, tck = -0.01, mgp = c (3, .3, 0))
dev.off ()
round (unclass (variance $ loadings),digits = 2)
Hidden <- principal (Example9_1[,2:19], nfactors = 4, rotate = "varimax")
Hidden
Round (Hidden $ scores, digits = 2)
```

#Figure 9.2

```
jpeg ("Figure 9.2.jpeg", res = 600, height = 16, width = 16, units = 'cm')
par (mar = c (7, 6, 7, 3))
barplot (Hidden $ scores [,1], ylim = c (-2, 2), xaxt = "n", yaxt = "n",pch = 17, col =
  c (rep ('gray', 10), rep ('blue', 10)))
legend (16, 2, horiz = FALSE, c ('Juru Estuary', 'Jejawi Estuary'), inset = c (0,.2),
  col = c ('gray', 'blue'), pch = c (17, 17), cex = 0.90)
mtext (c ("The values of F1"), side = 2, line = 1.5, cex = 0.9)
mtext (c ("Sites"), side = 1, line = 1, cex = 0.9)
axis (2, seq (-2, 2, 1), cex.axis = 0.9, tck = -0.03, mgp = c (3, .3, 0))
dev.off ()
```

#Figure 9.3

```
jpeg ("Figure 9.3.jpeg", res = 600, height = 16, width = 16, units = 'cm')
par (mar = c (7, 6, 7, 3))
barplot (Hidden $ scores [,2], ylim = c (-2, 2), xaxt = "n", yaxt = "n", pch = 17, col =
  c (rep ('gray', 10), rep ('blue', 10)))
legend (16, 2, horiz = FALSE, c ('Juru Estuary', 'Jejawi Estuary '),inset = c (0,.2),
  col = c ('gray', 'blue', pch = c(17, 17), cex = 0.90)
mtext (c ("The values of F2"), side = 2, line = 1.5, cex = 0.9)
mtext (c ("Sites"), side = 1, line = 1, cex = 0.9)
axis (2, seq (-2, 2, 1), cex.axis = 0.9, tck = -0.01, mgp = c (3, .3, 0))
dev.off ()
```

#Figure 9.4

```
jpeg ("Figure 9.4.jpeg", res = 600, height = 16, width = 16, units = 'cm')
par (mar = c (7, 6, 7, 3))
barplot (Hidden $ scores [,3], ylim = c (-2, 2) , xaxt = "n", yaxt = "n",pch = 17, col =
  c (rep ('gray', 10), rep ('blue', 10)))
legend (2, 2, horiz = FALSE, c ('Juru Estuary', 'Jejawi Estuary '),inset = c (0,.2),
  col = c ('gray', 'blue'), pch = c(17, 17), cex = 0.90)
mtext (c ("The values of F3"), side = 2, line = 1.5, cex = 0.9)
mtext (c ("Sites"), side = 1, line = 1, cex = 0.9)
axis (2, seq (-2, 2, 1), cex.axis = 0.9, tck = -0.01, mgp = c (3, .3, 0))
```

#Figure 9.5

```
jpeg ("Figure 9.5.jpeg", res = 600, height = 16, width = 16, units = 'cm')
par (mar = c (7, 6, 7, 3))
barplot (Hidden $ scores [,4], ylim = c (-2,2) , xaxt = "n", yaxt = "n",pch = 17, col =
  c (rep ('gray',10), rep('blue', 10)))
```

```
legend (16, 2, horiz = FALSE, c ('Juru Estuary ', 'Jejawi Estuary'),inset = c(0,.2),
  col = c ('gray', 'blue'), pch = c(17, 17), cex = 0.90)
mtext (c ("The values of F4"), side = 2, line = 1.5, cex = 0.9)
mtext (c ("Sites"), side = 1, line = 1, cex = 0.9)
axis (2, seq (-2, 2, 1), cex.axis = 0.9, tck = -0.01, mgp = c(3, .3, 0))
dev.off ()
```

Figure 9.6

```
#Draw the first and second factors
jpeg ("Figure 9.6.jpeg", res = 600, height = 16, width = 16, units = 'cm')
par (mar = c (8, 6, 7, 3))
plot (Hidden $ scores [,1], Hidden $ scores[,2], xaxt ="n", yaxt ="n", pch = 17, ylab = ",
  xlab =", ylim = c (-2.5, 2.5), xlim = c(-2, 2), col = c (rep ('gray',10), rep
  ('blue',10)))
legend (-2, 2.5, horiz = FALSE, c('Juru Estuary', 'Jejawi Estuary'),inset = c (0,.2),
  col = c ('gray', 'blue'), pch = c(17, 17), cex = 0.90)
abline (h = 0,v = 0)
mtext (c ("The values of F2"), side = 2, line = 2, cex = 0.9)
mtext (c ("The values of F1"), side = 1, line = 2, cex = 0.9)
axis (2, seq (-2, 2, 0.5), cex.axis = 0.9, tck = -0.01, mgp = c(3, .3, 0))
axis (1, seq (-2, 2, 0.5), cex.axis = 0.9, tck = -0.01, mgp = c(3, .3, 0))
dev.off ()
```

Example 9.2

```
install.packages ("psych")
library (psych)
install.packages ("GPArotation")
library (GPArotation)
corr.test (Example9_2 [,2 : 9])
variance <- princomp (Example9_2 [,2 : 9],cor = TRUE)
summary (variance)
```

#Figure 9.7

```
jpeg ("Figure 9.7.jpeg", res = 600, height = 16, width = 16, units = 'cm')
Par (mar = c (7, 6, 7, 3))
Screeplot (variance, npcs = 8, type = "barplot", xaxt = "n", yaxt ="n")
mtext (c ("Components"), side = 1, line = 2, cex = 0.90)
axis (2, seq (0, 4, .5), cex.axis = 0.9, tck = -0.01, mgp = c(3, .3, 0))
dev.off ()
round (unclass (variance $ loadings), digits = 2)
Hidden <- principal (Example9_2[,2 : 9], nfactors = 2, rotate ="varimax", covar =
  FALSE)
print (Hidden, digits=2)
round (Hidden $ scores, digits = 2)
```

#Figure 9.8

```
jpeg ("Figure 9.8.jpeg", res = 600, height = 16, width = 16, units = 'cm')
par (mar = c (7, 6, 7, 3))
barplot (Hidden $ scores [,1], ylim = c (-2, 2), xaxt = "n", yaxt = "n", pch = 17, col =
  c (rep ('gray', 10), rep ('black', 10)))
```

```
legend (1, 2, horiz = FALSE, c ("Juru Estuary", "Jejawi Estuary"), inset = c (0,.2),
  col = c ('gray', 'black'), pch = c(17, 17), cex = 0.70)
mtext (c ("The values of F1"), side = 2, line = 1.5, cex = 0.9)
mtext (c("Site"), side = 1, line = 0.2, cex = 0.90)
axis (2, seq(-2, 2, 1), cex.axis = 0.9, tck = -0.01, mgp = c(3, .3, 0))
dev.off ()
```

#Figure 9.9

```
jpeg ("Figure 9.9.jpeg", res = 600, height = 16, width = 16, units = 'cm')
par (mar = c (7, 6, 7, 3))
barplot (Hidden $ scores [,2], ylim = c (-3, 3), xaxt = "n", yaxt = "n", pch = 17, col =
  c (rep ('gray', 11), rep ('black', 11)))
legend (16, 3, horiz = FALSE, c ("Juru Estuary ", "Jejawi Estuary"), inset = c (0,.2),
  col = c ('gray', 'black'), pch = c(17, 17), cex = 0.70)
mtext (c ("The values of F2"), side = 2, line = 1.5, cex = 0.9)
mtext (c ("Site"), side = 1, line = 0.2, cex = 0.9)
axis (2, seq (-3, 3, 1), cex.axis = 0.9, tck = -0.01, mgp = c(3, .3, 0))
dev.off ()
```

#Figure 9.10

```
# Draw the first and second principal component
jpeg ("Figure 9.10.jpeg", res = 600, height = 16, width = 16, units = 'cm')
par (mar = c (8, 6, 7, 3))
plot (Hidden $ scores [, 1],Hidden $ scores [, 2] , ylab = ", xlab =", xaxt = "n", yaxt =
  "n", xlim = c (-3,3) , ylim = c (-3,3), pch = 17, cex = 0.90, col = c (rep ("gray", 10), rep
  ("orange", 10)))
abline (h = 0, v = 0)
legend (-3,3, horiz = FALSE, c ("Juru Estuary", "Jejawi Estuary "),inset = c(0,.2),
  col = c ('gray', 'orange'), pch = c(17, 17), cex = 0.90)
mtext (c ("The values of F2"), side = 2, line = 1.5, cex = 0.90)
mtext (c ("The values of F1"), side = 1, line = 1.2, cex = 0.90)
axis (2, seq (-4, 4, 1), cex.axis = 0.9, tck = -0.01, mgp = c(3, .3, 0))
axis (1, seq (-4, 4, 1), cex.axis = 0.9, tck = -0.01, mgp = c(3, .3, 0))
dev.off ()
```

CHAPTER 10

Example 10.1

```
install.packages ("MASS")
library (MASS)
LDis <- lda (Location ~ Cu + pb + Zn + Cd + Cr + As + Hg, data = Example10_1)
LDis
O.Vs <- predict (DA)
O.Vs        # scores for each sample
B = as.data.frame (O.Vs)
```

#Figure 10.1

```
jpeg ("Figure 10.1.jpeg", res = 600, height = 16, width = 16, units = 'cm')
par(mar = c (7, 6, 7, 3))
```

```
barplot (B $ LD1, xaxt ="n", ylim = c(-5,5), yaxt = "n", pch = 19, col = c (rep
  ('gray', 20),rep ('blue', 20)))
legend (35, 4, horiz = FALSE, c ('Juru River', 'Jejawi River'),inset = c(0,.2), col =
  c ('gray', 'blue'), pch = c(17, 17), cex = 0.90)
mtext (c ("The value of DF"), side = 2, line = 2, cex = 0.9)
mtext (c ("Sampling points"), side = 1, line = 1, cex = 0.9)
axis(2, seq (-5, 5, 1), cex.axis = 0.9, tck = -0.01, mgp = c (3, .3, 0))
dev.off ()
```

Example 10.2

```
Table <- table (Example10_1 $ Location, Scores $ class) #classification Table
Table
prop.table (Table, 1)     # 1 means row percentage, 2 column percentage#correct and
  misclassification percentage
diag (prop.table (Table, 1))     #correct Classification percentage
sum (diag (prop.table (Table)))  # total percent correct
```

Example 10.3

```
Library (MASS)
LDis <- lda (Location ~ EFB+ Potash+ Fibre+ Shell, data = Example10_3)
LDis
O.Vs <- predict (LDis)
O.Vs        # scores for each sample
B = as.data.frame (O.Vs)
Jpeg ("Figure 10.2.jpeg", res = 600, height = 16, width = 16, units = 'cm')
par(mar = c (7, 6, 7, 3))
barplot (B $ x.LD1, xaxt ="n", yaxt ="n", col = c (rep ('gray', 12), rep ('darkgreen',
  12), rep ('orange', 12), rep ('red', 12), rep ('blue', 12)))
legend (62, 20, horiz = FALSE, c("Mill 1", "Mill 2", 'Mill 3', 'Mill 4', 'Mill 5'),
  inset = c (0,.2), col = c('gray', 'darkgreen', 'orange', 'red', 'blue'), pch = c(17,
  17), cex = 0.9)
mtext (c ("The values of DF1"), side = 2, line = 2, cex = 0.9)
mtext (c ("Mills"), side = 1, line = 1, cex = 0.9)
axis (2, seq (-22, 22, 1), cex.axis = 0.9, tck = -0.01, mgp = c(3, .3, 0))
dev.off ()
```

#Figure 10.3

```
jpeg ("Figure 10.3.jpeg", res = 600, height = 16, width = 16, units = 'cm')
par(mar = c (7, 6, 7, 3))
barplot (B $ x.LD2, xaxt = "n", yaxt = "n",pch = 19, col = c (rep ('gray', 12), rep
  ('darkgreen', 12), rep('orange', 12), rep ('red', 12), rep ('blue', 12)))
legend (3, 4, horiz = TRUE, c ("Mill 1", "Mill 2", 'Mill 3', 'Mill 4', 'Mill 5'),
  inset= c (0,.2), col = c ('gray', 'darkgreen', 'orange', 'red', 'blue'), pch = c(17,
  17), cex = 0.9)
mtext (c ("The values of DF2"), side = 2, line = 2, cex = 0.9)
mtext (c ("Mills"), side = 1, line = 0.2, cex = 0.9)
axis (2, seq (-5, 5, 1), cex.axis = 0.9, tck = -0.01, mgp = c(3, .3, 0))
dev.off ()
```

#Figure 10.4

```
#first and second
jpeg ("Figure 10.4.jpeg", res = 600, height = 16, width = 16, units = 'cm')
Par (mar = c (8, 5, 7, 3))
Plot (B $ x.LD1, B $ x.LD2, ylab = ", xlab =", xaxt = "n", yaxt = "n", pch = 17, cex =
  .9, col = c (rep ('gray', 12), rep ('darkgreen', 12), rep ('orange', 12), rep
  ('red', 12), rep ('blue', 12)))
legend (14, 4, horiz = FALSE, c("Mill 1", "Mill 2", 'Mill 3', 'Mill 4', 'Mill 5'),
  inset = c(0,.2), col = c ('gray', 'burlywood', 'orange', 'red', 'blue', 'yellow',
  'black'), pch = c (17, 17, 17, 17, 17), cex = 0.90)
abline (h = 0,v = 0)
mtext (c ("The value of DF2"), side = 2, line = 1.5, cex = 0.9)
mtext (c ("The value of DF1"), side = 1, line = 2, cex = 0.9)
axis (2, seq (-4, 4, 1), cex.axis = 0.9, tck = -0.01, mgp = c(3, .3, 0))
axis (1, seq (-22, 22, 1), cex.axis = 0.9, tck = -0.01, mgp = c(3, .3, 0))
dev.off ()
```

Example 10.4

```
Table <- table (Example10_3 $ Location, Scores $ class) #classification Table
prop.table (Table, 1) # correct and misclassification percentage
diag (prop.table (Table, 1)) #correct Classification percentage
sum (diag (prop.table (Table))) # total percent correct
```

CHAPTER 11

Example 11.1

```
Example11_1[c (-1, -2, -3, -6)]
Euclidean <- round (dist (Example11_1[-1]), digits = 2)
Euclidean
as.matrix (Euclidean)
Cluster <- hclust (Euclidean, method = "single")
Cluster
jpeg ("Figure 11.2.jpeg", res = 600, height = 16, width = 16, units = 'cm')
par (mar=c(10, 6, 9, 3))
plot (Cluster, labels = Example11_1 $ Parameter, font.main = 1, cex.lab = 0.9, cex.
  main=.9, font.sub = 1, cex.sub = .9, cex.axis = 0.9, cex = 0.9) # plot the dendrogram
dev.off ()
```

Example 11.2

```
Euclidean <- round (dist (Example11_2 [2 : 9]), digits = 2)
Euclidean
Cluster <- hclust (Euclidean, "single")
Cluster
Jpeg ("Figure 11.3.jpeg", res = 600, height = 16, width = 16, units = 'cm')
par(mar=c(8, 7, 6, 3))
plot (Cluster, labels = Example11_2 $ Site, font.main = 1, cex.lab = 0.9, cex.main=.9,
  font.sub = 1, cex.sub = .9, cex.axis = 0.9, cex=0.9)
dev.off ()
```

Index

Note: Page numbers followed by *f* indicate figures, *t* indicate tables, and *b* indicate boxes.

Printed in the United States
By Bookmasters